Rapid Prototyping  Manufacturing

Fundamentals of StereoLithography

First Edition

Paul F. Jacobs, Ph.D.
Principal Author

David T. Reid
Publications Administrator

Published by
Society of Manufacturing Engineers
One SME Drive
P.O. Box 930
Dearborn, MI 48121-0930

Published by the Society of Manufacturing Engineers in Cooperation with the Computer and Automated Systems Association of SME.

McGraw-Hill, Inc.

New York San Francisco Washington, D.C. Auckland
Bogotá Caracas Lisbon London Madrid
Mexico City Milan Montreal New Delhi San Juan
Singapore Sydney Tokyo Toronto

Rapid Prototyping & Manufacturing
Fundamentals of StereoLithography

First Edition
Third Printing

Library of Congress Catalog Card Number: 92-81454
International Standard Book Number: 0-07-032433-6
Manufactured in the United States of America

This book is dedicated to

THE UNITED STATES OF AMERICA

where I was born, raised, educated, and have lived my life. My fondest wish is that this book, describing the birth and growth of a new technology, may, in some way, contribute to competitive excellence, which has always been the hallmark of this land.

Paul Francis Jacobs
La Crescenta, California

Contributing Authors

Chapter 1: Introduction to Rapid Prototyping & Manufacturing
- Charles W. Hull and Paul F. Jacobs, Ph.D.

Chapter 2: Basic Polymer Chemistry
- Dr. Max Hunziker and Dr. Richard Leyden

Chapter 3: Lasers for Rapid Prototyping & Manufacturing
- William F. Hug, Ph.D.

Chapter 4: Fundamental Processes
- Paul F. Jacobs, Ph.D.

Chapter 5: Software Architecture
- Grady O. Floyd

Chapter 6: CAD Modeling
- David S. Reynolds

Chapter 7: Introduction to Part Building
- Todd J. Mueller

Chapter 8: Advanced Part Building
- Paul F. Jacobs, Ph.D.

Chapter 9: Postprocessing
- Paul F. Jacobs, Ph.D.

Chapter 10: Diagnostic Testing
- Hop Nguyen, Jan Richter, and Paul F. Jacobs, Ph.D.

Chapter 11: Accuracy
- Jan Richter and Paul F. Jacobs, Ph.D.

Chapter 12: Texas Instruments: An Aerospace Case Study
- Paul Blake and Owen Baumgardner

Chapter 13: Chrysler Corporation: An Automotive Case Study
- Lavern D. Schmidt and William L. Phillips

Chapter 14: AMP Incorporated: A Simultaneous Engineering Case Study
- Thomas A. Kerschensteiner

Chapter 15: DePuy Incorporated: A Medical Case Study
- David G. Trimmer

Chapter 16: Alternate Approaches to RP&M
- Allan Lightman and Adam Cohen

Table of Contents

Foreword

Chrysler first began rapid prototyping in 1989 through the use of Stereo-Lithography. The results were so promising in terms of cost and time savings (not to mention the improved quality of early prototypes) that we haven't looked back. In fact, today Chrysler leads the automotive industry in the application of StereoLithography.

Rapid Prototyping and Manufacturing (RP&M) technology can be used to increase competitiveness in virtually *any* industry, from aerospace to medical. It's well suited to prototype components and products as diverse as rocket thruster gimbals, hip implants, and intake manifolds. I'm certain that we've only just scratched the surface of RP&M's full potential and range of applications.

This book begins with in-depth studies of rapid prototyping theory, techniques, and equipment, and concludes with case studies of pioneering efforts in RP&M. There's something here for everyone:

- Students and professionals will find this to be the only text or reference book that covers the "A to Z" of RP&M. Plus, it was written by the inventors and users of the technology.
- Designers, engineers, model makers, and tool builders, who may be new to RP&M, will find this to be an excellent step-by-step resource guide for using RP&M technology to produce parts at a fraction of the time and cost of conventional methods.
- Advanced users of RP&M will benefit from the trial and error experience of the authors, as well as their "secrets of success" with RP&M.

But, first and foremost, the goal of this book is to encourage the *application* of RP&M. This is a technology that was invented, developed, and produced in the United States to increase industrial competitiveness. RP&M is a vehicle for turning basic research and design work into finished products faster, with higher quality, and at lower cost. However, it only works when applied.

So read this book first. Then use the technology!

Robert A. Lutz
President
Chrysler Corporation

Preface

This book is about a new technology called Rapid Prototyping & Manufacturing, or RP&M. Within the last few years, a diverse array of organizations have documented major productivity gains, substantial cost savings, improved product quality, and the ability to quickly transition from concept to market using RP&M. Although the initial laboratory demonstration occurred as recently as 1984, and the first operational systems did not become commercially available until early 1988, there are already a number of different RP&M approaches being marketed.

Rather than attempt to describe all the various methods in considerable detail, we have chosen to explain one in depth, and the others in a more general manner. The reason for this choice of format was fourfold. First, StereoLithography, conceived, designed, developed, tested, manufactured, and sold by 3D Systems, Inc. of Valencia, California, pioneered this technology. Second, of the total number of RP&M systems currently in operation, about 90 percent were built by 3D Systems. Thus, StereoLithography is presently the worldwide leader in this field by a very considerable margin. Third, of the remaining RP&M systems, roughly half are essentially derivatives of StereoLithography. Comprehensive explanations of each of these systems would therefore be substantially redundant. Fourth and finally, the intent of this text was to provide an introduction to the technology, based upon theory, analysis, experiment, and extensive test data. At this time, the body of scientific analysis and test results available on the subject of StereoLithography exceeds that for all the other approaches combined.

Therefore, the prime focus of this book is on StereoLithography-based RP&M systems. The first six chapters provide a description of the fundamental physical and software processes, materials, and equipment which form the foundation of this new technology. The next five chapters are essentially a step-by-step explanation of how to actually build a prototype as well as a quantitative evaluation of the accuracy levels that can be realistically achieved at this time. This information is followed by four case studies showing practical examples of how RP&M has been utilized in specific applications by leading aerospace, automotive, component, and medical device manufacturing corporations. Finally, the last chapter describes each of the various alternate approaches that are known at this time. The intent is to provide the reader with as much documented information as possible.

Paul F. Jacobs

Acknowledgements

Ideas, sustained effort, and accomplishments are a lot like people; they cannot thrive in a vacuum.

The birth of Rapid Prototyping & Manufacturing technology owes much of its origin and establishment to the creativity, vision, and perseverance of Charles Hull.

Important contributions during the infancy of this new technology were made by Ray Freed, Bob Horrell, Chick Lewis, Dennis Smalley, Harry Tarnoff, Hop Nguyen, Wayne Vinson, Kris Schmidt, Stuart Spence, Paul Marygold, Adam Cohen, and Mike Lockard.

Growth accelerated through the efforts of Joe Allison, Bart Williams, Henry Schultz, Diana Kalisz, Frank Little, Jeff Thayer, Yehoram Uziel, Grady Floyd, Ray Bradford, Ben Modrek, Dave Snead, Ron Raviv, Tom Almquist, Kurt Kremidas, Bert Evans, and Thomas Pang.

Significant technical maturity has been achieved with the work of Rich Leyden, Max Hunziker, Adrian Schulthess, Paul Bernhard, Manfred Hofmann, Bettina Steinmann, Berndt Klingert, Jan Richter, Justin Bronk, Chris Manners, Jon Tindel, Ara Bernardi, Myron Bezdicek, Ed Murphy, Carl Deckard, Joseph Beaman, Bill Cromwell, Frost Prioleau, Dave Flynn, Al Dewitt, Dave Tait, Ed Gargiulo, Richard Chartoff, Dick Aubin, Mike Mc Evoy, Michael Feygin, and many, many others.

Furthermore, the principal author would like to express his true appreciation to all the chapter authors who have put forth such a considerable effort in a relatively short time. The enthusiasm associated with writing the first text on such a new and exciting subject was evident. It is not necessary to list each author here, as their names, titles, and affiliations appear at the start of each chapter.

The real adulthood of Rapid Prototyping & Manufacturing will ultimately be defined by the extent of its practitioners and the breadth of its applications and benefits. It is not possible, within the limitations of brevity, to detail the contributions of many individuals at over 300 corporations, service bureaus, government agencies, and universities currently involved in RP&M. The references at the conclusion of each chapter are an attempt in this direction. Nonetheless, the efforts of some may not have been directly cited. The authors assure that any such omissions were not intentional.

Also, the principal author would like to express his thanks to Robert King, David Reid, and all the people at SME who not only made this book possible, but were so capable and cooperative throughout the past months. Finally, I would like to express appreciation to my wife, Starla, and daughter Heather, for their sustaining positive spirit and encouragement.

chapter 1

Introduction to Rapid Prototyping & Manufacturing

To everything there is a season, and a time to every purpose under heaven.

Ecclesiastes 3:1

1.1 Perspective

People are generally fascinated by innovation, especially if the end result can provide real benefits. This book discusses the birth of a new technology capable of directly generating physical objects from graphical computer data. The technology is called *Rapid Prototyping & Manufacturing*. The initial term—established by 3D Systems (Valencia, CA), the company that pioneered this technology—was "StereoLithography," or three-dimensional printing.

Industry is already using this technology in numerous engineering prototype and manufacturing applications. The growing field has been widely referred to in technical and industrial publications as "rapid prototyping." Since the technology is advancing to encompass many applications of manufacturing beyond only that of prototyping, it was felt that *Rapid Prototyping & Manufacturing* would be properly descriptive. For brevity, the initials RP&M are used. These terms should be interpreted as synonymous.

The book also discusses the infancy and maturing of RP&M technology. As a specific example of the accomplishments possible using these methods, *Figure*

By ***Charles W. Hull****, President, and* ***Paul F. Jacobs****, Ph.D., Director of Research and Development, 3D Systems, Inc., Valencia CA.*

1-1 is a photograph of a prototype turbine blade built on a StereoLithography apparatus, or SLA-250, manufactured by 3D Systems. *Figure 1-2* shows the same prototype turbine blade after fine sanding and polishing. The elapsed time from the initial concept development of the turbine blade to *Figure 1-2* was less than two weeks!

Old axioms notwithstanding, the world does not always beat a rapid path to the door of the person who invents a better mousetrap. Despite impressions to the contrary, a review of relevant historical cases (References 1 through 3) shows that some truly great inventions have not been promptly utilized by the general public. The following are five specific examples:

John Harrison. The chronometer was invented in Yorkshire, England in 1735 as part of John Harrison's attempt to win a prize of 20,000 pounds sterling proposed by the British Admiralty. The prize was to be awarded to anyone who could determine longitude to within half a degree. Furthermore, the winning entry had to function on an actual ship of the line throughout the duration of a voyage from England to the West Indies and back. The correct evaluation of longitude is directly related to the accurate measurement of time. Thus, the task required developing a clock with an error of less than two minutes per year, while maintaining proper function on a pitching, rolling platform with frequent significant temperature and humidity variations!

Despite having achieved an incredibly small error of only 0.2 minutes (12 seconds) on the first trial, the Admiralty haggled, changed the rules in the middle of the game, and required numerous trials with multiple chronometers. Finally, the prize was awarded to Harrison some 40 years after his invention.

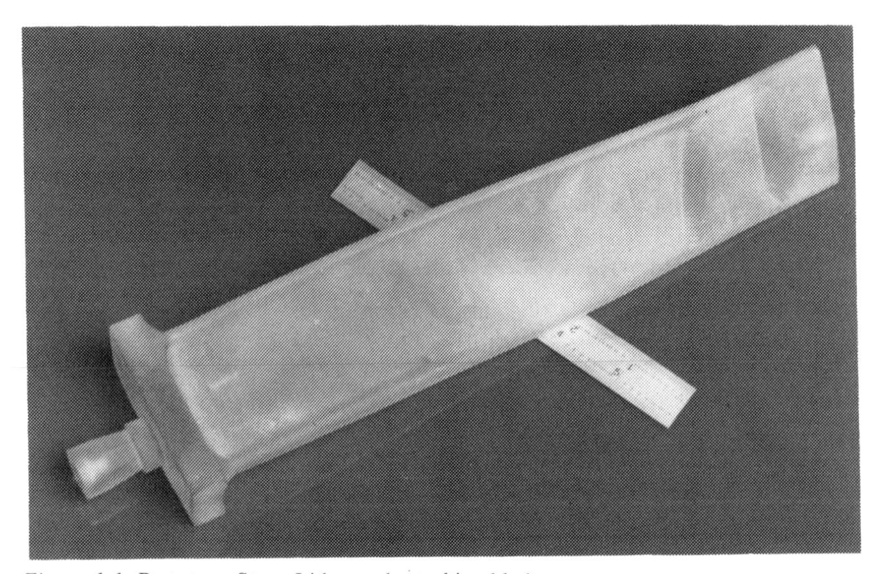

Figure 1-1. Prototype StereoLithography turbine blade.

Thomas Alva Edison. Thomas Edison invented the first practical incandescent lamp in Menlo Park, New Jersey in 1879. This lamp successfully burned without interruption for over 40 hours. However, to utilize the new light source, a complete system of electrical distribution was necessary. The first such electrical plant was completed in New York City, but widespread use of the incandescent lamp as a means of practical illumination did not occur until after the turn of the century, or about 25 years after the actual invention.

Gottleib Daimler and Karl Benz. Both men are generally credited with virtually simultaneous invention of the internal combustion engine used to power an operational automobile. This was accomplished by Daimler in Stuttgart, Germany and by Benz in Mannheim, Germany in 1885. Daimler altered a wooden, four-wheel horse carriage and fitted it with a 1.5 horsepower, four-cycle gasoline engine. The vehicle was then successfully operated on local roads near Stuttgart. Benz utilized a specially constructed steel, three-wheel carriage adapted with a 0.75 horsepower, two-cycle gasoline engine, and ran it on the main thoroughfare of Mannheim.

By 1901, Benz & Cie. had sold about 2,700 small, open Tonneau vehicles. That same year, Daimler Motoren Gesellschaft introduced the first Mercedes, named after the daughter of one of their major investors. However, it was not until Henry Ford introduced the Model T in 1908 that mass production enabled widespread use of the automobile, almost 23 years after the original invention.

Chester Carlson. The first working xerographic dry copier was invented by Chester Carlson in Astoria, New York in 1938. He then spent almost 16 years attempting to gain financial backing. His demonstrations of the new technique

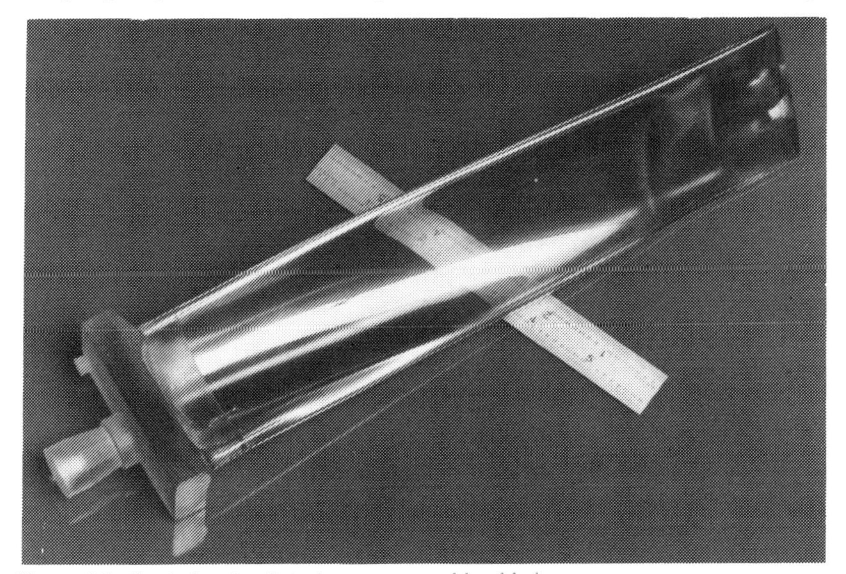

Figure 1-2. Sanded and polished prototype turbine blade.

were met with varying levels of indifference by numerous large corporations apparently convinced that a significant copier market would never exist.

Finally, joining forces with the tiny Haloid Company, the first commercial plain paper copier, the Xerox 914, was introduced in 1959, some 21 years after the initial invention.

John Von Neumann. A Hungarian mathematician and professor at Princeton University, John Von Neumann is generally considered to be the originator and team leader of a group that pioneered the design and development of the first high speed electronic computer. The Mathematical Analyzer Numerical Integrator And Computer (MANIAC) was first successfully tested at the Institute for Advanced Studies, Princeton, New Jersey in 1947.

However, it was not until 1960 that IBM produced the first commercial mainframe computers, and it was not until 1980 that the first personal computers became widely available to the general public. This involved a gap of about 33 years after the original demonstration.

This is quite a remarkable list of inventions. It is hard to imagine our lives today without clocks, lightbulbs, automobiles, copiers, or computers. Yet, it is evident that none of these inventions broke any speed records in achieving their current significance. It is also clear that for these inventions, the average time until major impact was about 25 years.

Thus, it is important when studying RP&M and its remarkable potential to remember that in 1992 only eight years have elapsed since the first primitive demonstration of StereoLithography (SL). The fact that over 300 SL systems now operate in 17 countries on five continents is especially significant in relation to these examples. To the authors,and others working in the field, progress and acceptance may seem to move slowly. Based upon historical perspective, however, the contrary is true. RP&M is just beginning to see widespread use, and this is occurring much more rapidly than many other well-known innovations.

1.2 Practical Applications

As a technology, RP&M is still quite young. During the first three years after the commercial introduction of SL, the practical applications of rapid prototyping were limited. The "common denominator" of these early applications was that they involved the prototyping step, but they did not yet lead to the fabrication of functional parts.

Visualization

No matter how experienced one may be at reading blueprints and computer-aided design (CAD) images of a complex object, it is still very difficult to visualize exactly what the actual part will look like. Blind holes, complex interior passageways, compound curved surfaces, etc. often lead to interpretation difficulties. The saying, "a picture is worth a thousand words" has been updated by various SL users to suggest that "one real prototype is worth a thousand pictures."

As an example, a 2.4 liter automotive engine block was recently modeled using CAD by Chrysler Corporation. Initially, the CAD views all appeared correct. However, when the first prototype was built on an SLA-500, it was discovered that a flange intended to be positioned only in the space between cylinders 2 and 3 had also been inadvertently positioned between cylinders 1 and 2. This was an error that had escaped detection despite the review of numerous engineers and designers.

Thus, the reduction of errors through improved part visualization can be substantial. Simply stated, there is still no better way to be certain that a complex part contains exactly those features intended for it than to hold it in your hand, turn it around a few times, and look at it from all sides.

Verification

Manufacturing engineers, designers, and managers have heard much about the need for improving the quality of products. A possible cause for poor quality is the unfortunate tendency to compromise a design when faced with an urgent timetable to "get it out the door." Under the conventional scenario, generating a fully comprehensive series of prototypes to prove the validity of a design often involves spending more time and money than is available. Thus, when "minor" errors are detected, there is a temptation to accept the design "as is."

The project team members are faced with a dilemma. Should they try "doing it over again until we get it right" or fail to meet the project's financial or schedule requirements? Under these kinds of demands, things that are "not quite right" are often allowed to proceed down the path to production. Ultimately, quality, product reputation, and market share suffer.

In contrast, an RP&M prototype can be generated so quickly that it becomes simple to verify that the design does indeed contain the features desired and, conversely, does not contain any features that are not desired. The example of the 2.4 liter engine block previously cited illustrates this point. Through inspection of the SL prototype, a design error in the engine block was detected very early in the design cycle. The CAD model was corrected and the part was rebuilt, all within one week. When completed, the geometry of the redesigned part was verified by visual inspection (*Figure 1-3*).

Obviously, verification of other characteristics such as strength, operational temperature limits, fatigue, corrosion resistance, etc., will have to wait for the results of tests on a fully functional prototype. However, designers and engineers will be able to hold in their hands, at a much earlier date, a part which has been geometrically verified. They can then spend additional time functionally testing other important characteristics which may result in further improvements in quality and, ultimately, in the reputation of the product.

The prescription for improved part quality can be summarized as follows:

1. Design the prototype on CAD.
2. Build the prototype with RP&M.
3. Inspect the RP&M part for errors.
4. Correct the errors in CAD.
5. Verify the corrected RP&M part.
6. When correct, build a functional test model.
7. Perform functional testing.
8. When satisfactory, proceed to manufacture.

Iteration

One can sense from this eight step procedure that the use of RP&M technology can be a crucial component of part quality improvement. However, what may be more important is the ability to do rapid design iteration.

Previously, the development of a prototype was so expensive and time-consuming that often few resources and little time was left for design iteration. However, with RP&M technology, it is now possible to go through multiple design iterations within a week or two. Numerous examples of accomplishing two or even three design iterations in a single week have been *documented* by various SL users. This is an incredible improvement over traditional methods. For example, in the design of an inlet manifold for an automobile engine, as shown in *Figure 1-4*, it is now possible to build a nearly transparent RP&M prototype within a few days.

One now can perform fluid dynamic flow tests on this prototype within the physical limits of the cured photopolymer material. If a problem is detected with

Figure 1-3. Prototype engine block.

Figure 1-4. Prototype RP&M inlet manifold.

the flow pattern, the geometry can be modified in the CAD model, the RP&M part rebuilt, and additional flow testing accomplished within a few days. This process, in combination with engineering analysis, can often lead to a substantially improved design solution after three or four iterations. Formerly, this kind of multiple prototype iteration of a design would be time-consuming and expensive, and only in rare cases would this approach be followed.

With RP&M, the previous prescription for improved part quality can be amended to account for the extraordinary iteration capability of this new technology. Specifically:

1. Design the prototype on CAD.
2. Build the prototype with RP&M.
3. Inspect the RP&M part for errors.
4. Correct the errors in CAD.
5. Verify the corrected RP&M part.
6. Iterate, using RP&M, to improve the design.
7. When acceptable, build a functional test model.
8. Perform functional testing.
9. When satisfactory, proceed to manufacture.

Optimization

The sequence just described provides a practical method for improving the quality of products through RP&M technology. However, design optimization can improve the situation further. As with iteration, optimization can be a very expensive and time-consuming process. When the design team finally does achieve an inlet manifold without any obvious flow problems, what is the next step? Probably, the design for functional testing is approved. If it works, the task is considered complete, and the team will likely celebrate.

But, what about the possibility of an even better design? Building perhaps dozens of prototypes is so expensive and time-consuming that, with the possible exception of the hull design for an America's Cup racer or a critical spacecraft component, the search will likely go no further. The chance that the designers and engineers will have hit upon a true optimum on the first attempt is very small.

One of the great advantages of RP&M technology is that it allows the design team to play the "what if" game without spending a fortune, an eternity, or both. Having achieved an "acceptable" design through rapid prototype iteration, one has the opportunity to attempt to optimize the design. Returning to our example, one might try modifications to the manifold geometry to see which, if any, results in the minimum pressure drop or some other figure of merit. Since prototypes can be generated rapidly, a range of test models that might have previously taken a year can now be completed in weeks..

At this point, one is no longer just iterating, since an "acceptable" design is in hand. One is now truly optimizing to achieve a better design. If none of the variations prove to be any better, then the original design must have been a good one. Further, the sensitivity of the design to a number of variations has been explored. If the performance did not change significantly from one design to the next, this implies that small production tolerances will not be fatal. This alone is useful information, which might justify relaxed specifications and thereby result in a legitimate product cost reduction.

However, if one of the variations is better, then the team has advanced the quality of the product, assuming that the design passes the functional tests. Finally, if the design is a great deal better, a breakthrough may have been achieved, which is really cause to celebrate.

An appreciation for the potential benefits of the RP&M procedure should now be evident, and the prescription for improving part quality can be further modified as follows:

1. Design the prototype on CAD.
2. Build the prototype with RP&M.
3. Inspect the RP&M part for errors.
4. Correct the errors in CAD.
5. Verify the corrected RP&M part.
6. Iterate, using RP&M, to improve the design.
7. Optimize, testing multiple RP&M design variations.
8. When optimized, build a functional test model.
9. Perform functional testing.
10. When satisfactory, proceed to manufacture.

Fabrication

Visualization, verification, iteration, and optimization are available to SL users without a subsequent fabrication process. Since SL has come into widespread use, considerable work by users has expanded the role of rapid prototyping to the point where the fabrication and ultimate manufacture of "real parts" is becoming common. Fabrication is done either by using a material with appropriate properties in the RP&M process itself, or by using the prototype as a pattern or mold for a following process.

Once an optimized prototype has been developed by the methods described, it is important to fabricate a functional test model (FTM). Because this model has not yet been fabricated or tested, it is not known if it will pass functional test requirements. It may be wasteful to spend the time and expense of generating final tooling. Again, RP&M can help significantly.

At present, there are a number of techniques that have been successfully used to go from an RP&M prototype to an FTM part in a relatively rapid and cost-effective manner. Some of these techniques are as follows:

- Silicon Room Temperature Vulcanizing (RTV) molding.
- Vacuum casting.
- Form block casting.
- Spray metal molding (Tafa process).
- Resin transfer molding (epoxy tool molding).
- Sand casting of aluminum and ferrous metals.
- Investment "flask" casting.
- Investment "shell" casting.
- Abrading die EDM tools (Hauserman process).

Some of the details of these techniques are described in References 4 through 8, and are presented in Chapters 12 through 15. These chapters include specific RP&M application case studies. The technique that will be most cost-effective depends upon the size and geometry of the prototype, the type of material needed for the FTM, and the required accuracy. It is particularly important to realize that each of the nine methods listed have already been successfully employed by various users of RP&M technology. These methods now enable one to proceed directly from prototypes to functional test models.

Let us return to the prototype turbine blade shown in *Figures 1-1* and *1-2*. Since the resulting RP&M prototype was geometrically correct and no further iteration was required, an FTM was pursued using a combination of the resin transfer molding technique and the investment "shell" casting method.

As shown in *Figure 1-5*, a mold box was made, a female mold was created in Ciba-Geigy REN mold epoxy, and multiple wax patterns were then created using the mold box. The wax patterns were then used for the generation of functional test models in aluminum, using investment shell casting. A final aluminum turbine blade FTM is shown in *Figure 1-6*.

We can now list the current prescription for improved part quality, based upon the recent advances achieved in fabricating functional test parts through the use of RP&M.

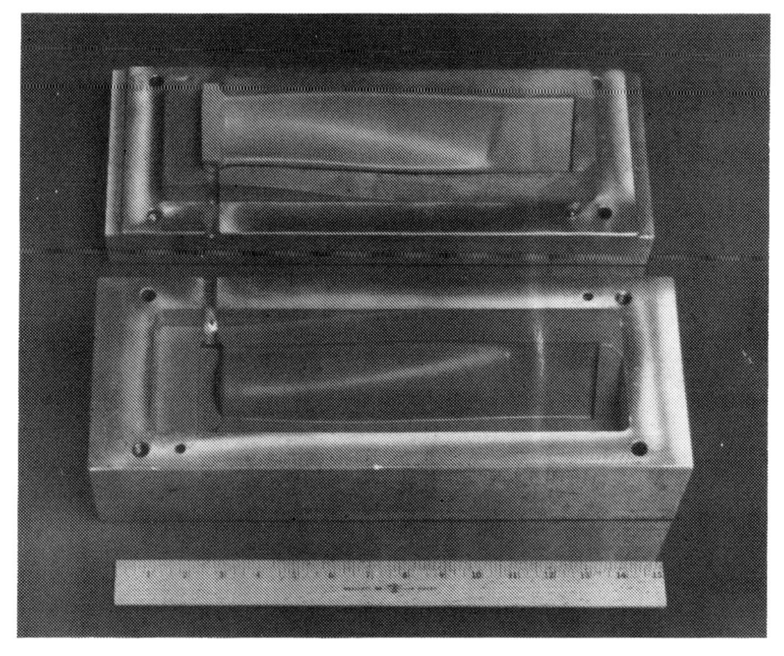

Figure 1-5. Turbine blade mold box and wax pattern.

Specifically

1. Design the prototype on CAD.
2. Build the prototype with RP&M.
3. Inspect the RP&M part for errors.
4. Correct the errors in CAD.
5. Verify the corrected RP&M part.
6. Iterate, using RP&M, to improve the design.
7. Optimize, testing multiple RP&M design variations.
8. Fabricate an FTM utilizing the RP&M prototypes.
9. Perform functional testing on the FTM.
10. When satisfactory, proceed to manufacture.

Note that in this final version, six of the ten steps directly involve RP&M technology. The overall savings in both time and money through use of this prescription can be spectacular. Specific examples will be given in Chapters 12 through 15.

Remember, the British Admiralty once ruled the waves. They offered a prize for the accurate determination of longitude because accurate longitude is the key to accurate navigation, which is vital if one is attempting to rendezvous multiple sailing vessels in preparation for a naval encounter at sea. If the ships do not assemble, the situation could change from numerical superiority to localized

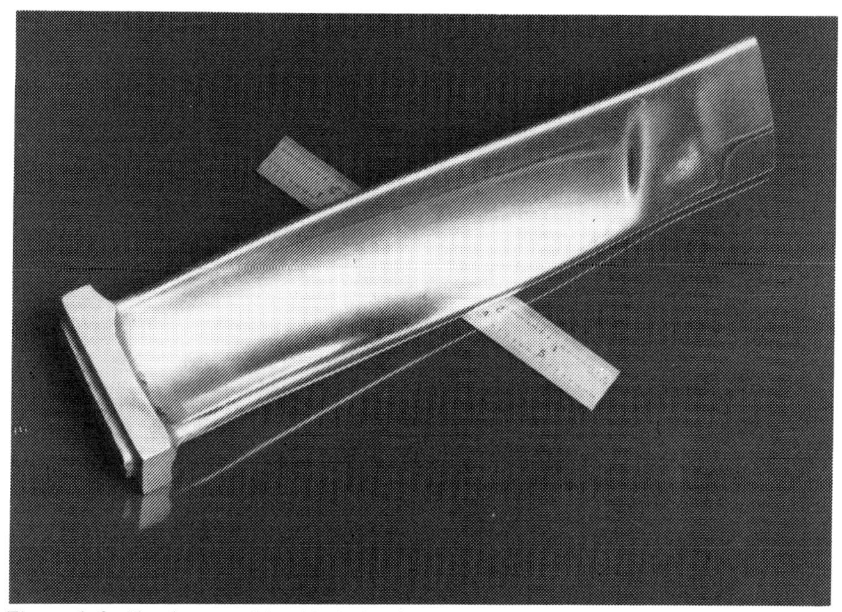

Figure 1-6. Aluminum turbine blade functional test model (FTM).

jeopardy. Courage and good seamanship aside, it never hurts to have more immediate firepower than the enemy.

Fully aware of this, the Admiralty offered a prize which was a small fortune at the time. Yet, it took 40 years to recognize excellence when Harrison placed his chronometer before them. A similar theme exists in manufacturing today. It is essential to develop high quality products quickly so that a company can bring maximum firepower to the competitive marketplace. And today, just as in the past, when new timekeeping technology was required to assemble the fleet, RP&M technology is required to assure that high quality products can be developed quickly.

As author Cyril Charney explains:

"How to do more" was emphasized in the '60s. "How to do it cheaper" became important in the '70s. "How to do it better" was certainly the theme of the '80s. But "how to do it quicker" will be the key in the 1990s. This is not to say that doing more, doing it for less per-unit cost, and doing it better do not apply anymore. They still do. But meeting the increasing time demands of customers will become paramount.[9]

1.3 Basic Operation

About 90% of the RP&M systems sold through the end of 1991 are SLA units designed, developed, and tested by 3D Systems. Furthermore, an additional 5% (approximately) are systems sold by Sony, Mitsubishi, Mitsui, or EOS GmbH.

They are functionally very similar to SLA units. Thus, about 95% of all RP&M systems sold through 1991 operate in the same (or in nearly the same) manner as StereoLithographic systems.

The remaining 5% (approximate) are quite different. The DTM approach involves sintering or fusing particulates with a high-power infrared carbon dioxide laser. The Stratysys approach involves extruding heated plastic filament material which solidifies upon cooling. The Helisys technique involves cutting nonphotoreactive materials, such as paper sheets, with a high-power carbon dioxide laser. The Cubital approach involves flash exposure with continuum (nonlaser) ultraviolet radiation, and the Perception Systems method involves ballistic particle fusion. Each of these alternative approaches are described in Chapter 16.

This section is devoted to the operation of SL and SL-like systems. Many of the operational steps are covered in great detail in later chapters of this book.

CAD Model

The first step in the RP&M process is virtually identical for all of the various systems, and involves the generation of a three-dimensional computer-aided design model of the object. A detailed discussion of CAD modeling procedures is given in Chapter 6. The important points are as follows:

1. The entire RP&M process begins with a CAD Model.
2. An experienced CAD engineer/designer is an indispensable component of success.
3. A good, preferably solid modeling CAD system is also a key component of success.

Translator

Usually, the CAD file must go through a CAD to RP&M translator. This step assures that the CAD data is input to the SL machine in the ‘‘tessellated’’ STL format, which has become the standard of the RP&M field. Here, the boundary surfaces of the object are represented as numerous tiny triangles. Tessellation is discussed further in Chapters 5 and 6.

Supports

The next step involves generating supports in a separate CAD file. CAD designers/engineers may accomplish this task directly, or with special software such as ‘‘Bridgeworks,’’ as discussed in Chapters 5, 6, and 7.

Supports are used for the following three reasons:

1. To ensure that the recoater blade will not strike the platform upon which the part is being built.
2. To ensure that any small distortions of the platform will not lead to problems during part building.

3. To provide a simple means of removing the part from the platform upon its completion.

Slice

At this point, both the part and the supports must be "sliced"—the part is mathematically sectioned by the computer into a series of parallel horizontal planes like the floors of a tall building. Also, during this step, the layer thickness, the intended building style, the cure depths, the desired hatch spacing, the linewidth compensation value, and the shrinkage compensation factor(s) are selected. These items are discussed in detail in Chapters 5, 6, 7, 8, and 10.

Merge

In this step, the supports, the part, as well as any additional supports and parts (in the case where multiple parts are being built concurrently on the same platform) have their computer representations merged. Details are presented in Chapters 5 and 7.

Prepare

In the prepare step, certain operational parameters are selected, such as the number of recoater blade sweeps per layer, the sweep period, and the desired "z-wait." *Z-wait* is the amount of time, in seconds, that the system is instructed to pause after recoating. The purpose of this intentional pause is to allow any resin surface nonuniformities to undergo fluid dynamic relaxation. The output of this step is the selection of the relevant parameters. Default parametric values may be used.

Build

As the name implies, the build step is where resin polymerization begins and a physical, three-dimensional object is created. The build operation consists of the following steps:

Leveling. If liquid resin did not shrink upon polymerization, this step would only be necessary at the start of build, to insure that the resin was at the proper z level for optimum laser focus. However, typical SL resins undergo about 5% to 7% total volumetric shrinkage. Of this amount, roughly 50% to 70% occurs in the vat as the result of laser-induced polymerization, and the remaining polymer volumetric shrinkage occurs during the postcuring step (see Chapter 9).

Consequently, a level compensation module is built into the SLA system. Upon completion of laser drawing on each layer, a sensor checks the resin level. In the event that the sensor detects a resin level that is not within the tolerance band, a plunger is activated by means of a computer-controlled precision stepper motor. The motion of the plunger then corrects the resin level through simple fluid displacement. When the resin level lies within the tolerance limits, the leveling operation is completed, and the system proceeds to the next step.

Deep Dip. Under computer control, the z-stage motor moves the platform down (about 0.3″ for the SLA-250 and about 0.7″ for the SLA-500). This is done to insure that even those parts with large flat areas, comparable in size to the entire platform, can be properly recoated. When the platform supporting such a part is lowered, a substantial "depression" is generated on the resin surface. The time required to "close" this surface depression has been determined from both viscous fluid dynamic analysis and experimental test results. It has been found to be proportional to the so-called "critical circle radius" for that layer cross section (see Chapter 5 for a definition and graphic description of the critical circle concept) to the resin viscosity, and inversely with the square of the depth of the depression. Since the critical circle radius for that layer and the resin viscosity are both fixed, an effective means of reducing the time for resin closure is to "deep dip."

Elevate. Under the influence of gravity, the resin fills the depression created during the previous step. The z-stage, again under computer control, now elevates the uppermost part layer above the free resin surface. This is done so that during the next step, only the excess of resin beyond the desired layer thickness need be moved. Otherwise, additional resin would be disturbed, further extending the required z-wait. When the elevation step is completed, the uppermost surface of the previously cured layer is positioned one layer thickness below the bottom edge of the recoater blade.

Sweep. At this point, the recoater blade traverses the vat from front to back, or vice versa, and "sweeps" the excess resin from the part. As soon as the recoater blade has completed its motion, the system is ready for the next step. This is the approach currently in use on the SLA-250 and SLA-500. The major advantages are as follows:

- Relatively rapid operation. For the majority of part geometries, the optimum sweep period is about 5 seconds. Only for parts involving so-called "trapped volumes" does the total time for recoater blade sweeping increase beyond this value. Trapped volumes occur in parts whose topology is such that pockets of liquid resin, within the part interior, are unable to communicate with the rest of the liquid resin in the vat. A simple example of a part containing a trapped volume is a coffee mug built in the normal right-side-up geometry. While building a layer of the mug, the resin inside cannot equalize any level differences between itself and the resin in the vat. This inability is the source of difficulty in recoating such parts.
- Reasonably good thin layer uniformity. Measured variations in layer thickness are generally less than one mil peak to peak.

The disadvantages are as follows:

- The potential for increased layer nonuniformity when recoating trapped volumes.

- Finite surface tension effects cause resin to adhere to the trailing edge of the blade during sweeping. Increased viscous drag forces encountered when traversing the leading edge of a solid layer of the part cause some of this resin to be separated from the blade. The result is a small but definite surface nonuniformity or ''bulge'' just downstream of the leading edge of the part. This phenomena is referred to as *leading edge bulge.* The amplitude of the leading edge bulge is a function of the resin viscosity and surface tension and will undergo exponential decay with time.

Move to Build Position. Upon completion of sweeping, the z-stage, still under computer control, moves downward. When it comes to rest, the top of the layer of recoated resin is at the free surface level of the resin in the vat. At this position, the top of the previous layer is now at a depth equal to one layer thickness.

Z-Wait. Once the platform has moved to the build position, in principle, the resin on top of the previous layer should blend seamlessly with the free surface of the resin in the vat. Unfortunately, due to finite surface tension effects, it is common to notice a small but distinct ''crease'' around the perimeter of the part, at the solid to liquid interface. This effect is best observed by viewing the resin surface through the SLA window at a very low angle when the recoater blade is ''parked'' at the rear of the vat and does not obscure the sightline. Careful observation of the reflection of a normally straight edge off the resin surface will clearly show these surface perturbations.

Continued observation reveals that the amplitude of the crease will tend to decay with a finite relaxation time constant that again is dependent upon the resin viscosity and surface tension. The z-wait interval is intended to give the fluid surface adequate time to essentially eliminate these nonuniformities. Since the approach to a planar resin surface shows a classical exponential time decay, one must trade surface nonuniformity against build time. Small nonuniformities tend to be more problematic with thinner layers. Therefore, in general, the thinner the layer thickness used to build parts, the greater the required z-wait. However, lest the reader gain the incorrect impression that these steps take a long time, it is important to point out that utilizing current software, the average time for an SLA-250 to compute and adjust the resin level per layer is less than one second.

Furthermore, the ''deep dip'' interval, including the pause at the bottom, is about 11 seconds. The elevation step takes another six seconds. The sweep period is most often five seconds, except when dealing with trapped volumes, when this time interval may increase to 15 to 25 seconds, depending upon the number of sweeps required. The return of the platform to the final position for laser drawing takes another two seconds. Finally, the z-wait interval is the most variable. Commonly employed z-wait values lie in the range from 15 to 30 seconds.

Based upon a 20 second z-wait, the resulting total recoating ''overhead'' time is about 45 seconds per layer. While this represents close to a 50% decrease from the values of a few years ago, the task of further reducing this overhead time per layer continues. These items also are discussed in Chapter 7.

Laser Drawing. Having established a quasi planar photopolymer resin surface, the system now proceeds to laser drawing. The first step involves drawing the part borders for the given layer cross-section. Having preselected the desired cure depth for borders during the slice operation, the computer automatically calculates the correct laser scan speed for that cure depth and resin. Once the borders have been drawn, which normally takes only a few seconds, the system then proceeds to ''hatch,'' or fill-in those areas that will eventually become solidified. The great majority of the laser drawing time for most parts is spent in hatching. Hatching is a critical step in the eventual accuracy of the RP&M part.

Finally, on virtually all up-facing and down-facing surfaces, so-called ''skin fills'' also are drawn. Skin fills involve drawing a series of very closely spaced parallel vectors, such that the finite cured linewidths actually contact one another laterally to form a continuous ''skin.'' With some of the original part building styles, this was essential to avoid the unintentional draining of resin from interior portions of the part. However, resin drainage is no longer a problem with the newer advanced part building methods, and, consequently, skin fills are no longer critical. Nonetheless, they may still improve the strength of down-facing surfaces and the quality of up-facing surfaces.

Part Completion and Draining

Once the laser drawing sequence is completed for a given layer, the previously described steps will be repeated for subsequent layers of the part. Experience reported by a wide range of users indicates that once a part is started on an SLA, over 90% of the time the system will complete the build cycle. The other 10% most commonly involves users intentionally stopping the build cycle, power failures, file errors, malfunctions, etc. This is a remarkable ratio for the successful creation of complex three-dimensional objects. When the last layer is completed, the computer will, upon request, activate the z-stage. This elevates both the platform and the attached part above the free surface of the resin in the vat.

It is wise to let excess resin drain back into the vat since the resin is valuable. Also, if the part happens to contain trapped volumes, it is then appropriate to tip the platform on its support arms to facilitate drainage of the uncured resin back into the vat. Once the vast majority of excess liquid resin has been recovered, this step is complete.

Removal, Cleaning, and Rinsing

The platform, with the part still attached, can now be removed from the SLA. To avoid excessive contact with the resin, use rubber gloves. Also, to avoid dripping resin, shallow-rimmed stainless steel trays (from which resin may be cleaned) or cellulose padding (either disposed of as hazardous waste or cured in sunlight) are recommended.

Next, excess liquid resin should be wiped from both the part and the platform with a paper towel or Q tips. Wiping away excess resin from the part and platform extends the life of the cleaning solvent.

Finally, the part and the platform are placed in a solvent cleaning apparatus. The conclusion of this step should be a part that has been both cleaned of excess resin and rinsed of cleaning solvent without suffering any ill effects upon part accuracy. Part cleaning and rinsing are discussed in detail in Chapter 9.

At this point, the part and the platform are best dried with a stream of low-pressure compressed air. This helps speed the drying process, especially for complex parts with numerous recesses.

The last portion of this step is the removal of the part from the platform. Since RP&M is about improving productivity, do be careful. The best methods for part removal vary with the characteristics of the resin. Generally, a variety of flat bladed knives work well. An X-acto knife or even a fine sharp scissor also may prove handy for very delicate parts.

Postcure

To this point, the object has only been partially polymerized, and is in the ‘‘green state.’’ Much of the strength of an SL part is achieved with exposures well beyond those delivered by the laser. Thus, green SL parts are postcured to essentially complete the polymerization process and to improve the final mechanical strength of the prototype.

Postcuring is accomplished using broadband or ‘‘continuum’’ ultra violet (UV) radiation in a specially designed postcuring apparatus (PCA). Details of the postcuring step are presented in Chapter 9. The intent is to optimize the output wavelengths of the PCA to achieve the most uniform postcuring possible, with minimal temperature rise of the polymer and maximum part accuracy consistent with reasonable duration.

Data in Chapters 10 and 11 shows that provided one uses the latest in advanced part building techniques, the PCA is very effective at minimizing postcure distortion. Most parts can be completely postcured within an hour or two, although very large parts may require up to 10 hours.

Part Finishing

Depending upon the application intended, various levels of part finishing may be appropriate. For visualization and concept modeling, simply removing the supports is adequate. For iteration and optimization, more extensive finishing such as hand sanding, mild glass bead blasting, or some combination of the two is probably appropriate. Those parts intended for one of the various RP&M

applications, leading to the generation of functional test models, will almost certainly benefit from more extensive touch work such as polishing, painting, or spray metal coating.

Furthermore, some of the more recent urethane acrylate resins (see Chapter 2) will also enable multiple machining operations such as drilling, boring, tapping, and milling. The details of the finishing steps will depend upon the intended application of the RP&M part, and the characteristics of the resin used.

1.4 RP&M Benefits

As noted, many significant inventions did not begin to yield major benefits for 20 to 25 years. The publication of this book, in mid-1992, marks a period of only a little more than four years since the installation of the first beta test units. Chapters 12, 13, 14, and 15 present case studies of benefits already realized by four different corporations representing aerospace, automotive, component, and medical applications of RP&M technology.

The following are some examples from these chapters:

Defense Electronics Group of Texas Instruments, Inc.

- ''The first real benefit [of solid modeling CAD and RP&M], and perhaps the most obvious one, is the enhanced visualization capability. Engineers, designers, technicians, and managers now discuss scale plastic prototype parts, not two-dimensional drawings.''
- ''Texas Instruments has experienced typical cycle time reductions of two weeks for simple parts and 20 weeks for complex machined parts in the fabrication of prototypes. Also, customer concept communication models have been fabricated for half the cost and in half the time of conventional models.''
- ''A cost and cycle time graph [was prepared] comparing the three different manufacturing methods of SL pattern casting (RP&M), traditional wax tool/pattern casting, and machining from stock material. The SL pattern casting clearly demonstrated a significantly lower cost and shorter cycle time.'' On a single project, documented savings of over $300,000 were realized.

Chrysler Corporation, Jeep and Truck Engineering Division

- ''SL'' allowed us to quickly produce prototype parts, within 24 hours of receipt of CAD data, directly from 3D CAD models into a finished part, bypassing the traditional methods where print errors, interpretations, machining and inadequate cut sections delay the building of a correctly toleranced prototype.''

- "The StereoLithography equipment was installed in early January 1990, and we have been using it nonstop, 24 hours a day, seven days a week, for about two years, building well over 1,500 parts in 500 different geometries."
- "Without an SL part, a prototype shift knob would have been manufactured before the desired changes [in size and shape] could have been discovered. A second prototype would have been required. Based upon outsourcing's estimates, the SL model saved over $40,000 dollars and 18 weeks of design time."

AMP Incorporated, Automotive/Consumer Business Group

- "A major U.S. company initiated a design competition for a next generation interconnection system. This procurement solicitation involved five competitors. Each company was given four hours to make a presentation of its proposed solution to the customer's need. SL samples of proposed product were shown by only one company.... As a result, AMP Inc. was awarded all the business.... Based upon this experience, our organization will not issue a major customer quotation package without a StereoLithography model."
- "A 16 position, two-part connector with 9 circuits erroneously lining up with 7 circuits [was discovered]. The cost and time saved were estimated to be $80,000 and four months."
- "While it is sometimes difficult to admit ones mistakes, one engineer, upon finding an error, said, '[The SL system] is the best thing we've ever purchased.'"

DePuy, Inc.

- "The goal of rapid prototyping is to bridge the [communication] gap by providing actual full-size physical models that each party can touch, analyze, and use for further development. This capability has revolutionized the design review meeting. The constructive dialogue among the members, as they pass the models to one another, suddenly allows everyone to express concerns and suggestions in a manner based upon a common level of visual understanding."
- "We no longer have the luxury of waiting over two years to get new ideas to the marketplace. If the market preferences change, or competitive producers get established, the product has a very small chance of providing the expected return on investment. The shoulder [prosthesis] project was the first large-scale product to fully utilize all stages of RP&M. The return from launching a single product several months early pays for the entire technology investment."

"Rapid prototypes have allowed a paradigm shift in Concurrent Process and Product Development. Each person's job is no longer just another step in the chain of events. Every individual can see the goal, the product, and can relate to the customer. They are a vital part of a team that is alive and functioning. The job is more challenging and rewarding, as they are empowered to make decisions and see the direct results of their collective efforts. If we are to continue as a society that manufactures things, then RP&M is essential to our survival."

It has been truly rewarding to be associated with a new technology that has provided such a diverse array of benefits to an extraordinarily broad range of customers within only four years. With further advances in resin properties, system accuracy, and overall productivity, we are confident that RP&M will play an increasing role as a practical tool for enhanced manufacturing competitiveness.

1.5 Chronology

When the time for an idea has arrived, it is common for numerous implementations to be developed at about the same time and completely independent of each other. Rapid prototyping is no exception.

In the late 1970s and early 1980s, A. Hebert of 3M in Minneapolis, H. Kodama of the Nagoya Prefecture Research Institute in Japan, and C. Hull of UVP (Ultra Violet Products, Inc.) in California worked independently on rapid prototyping concepts based on selectively curing a surface layer of photopolymer and building three-dimensional objects with successive layers.

Both Herbert and Kodama had difficulty maintaining ongoing support from their research organizations, and they each stopped their work before proceeding to a commercial or product phase. UVP continued to support Hull, and he worked through numerous problems of implementing photopolymer part building until he developed a complete system that could automatically build detailed parts. Hull coined the term StereoLithography, or three-dimensional printing. This system was patented[10] in 1986, at which time Hull and R. Freed, jointly with the stockholders of UVP, founded 3D Systems, Inc. to develop commercial applications in three-dimensional printing.

Venture capitalists financed the start up of 3D Systems, to combine computers, lasers, and photopolymer chemistry to provide better ways to design products. During a period of rushed innovation from late 1986 through late 1987, many of today's rapid prototyping concepts were developed. The roles of hatching, up-facing and down-facing skins, and near flat skins were defined.

Also, many of the modes of part distortion were identified, and early methods to control these distortions were developed. The requirements of postprocessing were established. A precision imaging system was designed, including geometric calibration and drift correction. Computer graphic slicing software was written, and user interfaces to guide the process were designed, completed, and installed.

The SLA-1, the first commercial rapid prototyping product, was designed at 3D Systems throughout this period. It was publicly introduced at the AUTO-FACT Show in Detroit in November, 1987. The SLA-1 beta program, in early 1988, included AMP, General Motors (CPC and Fisher Guide), Baxter Health Care, Eastman Kodak, and Pratt & Whitney (commercial and military). The year 1988 saw swift expansion and significant growth for rapid prototyping. Many more large companies joined the ranks of early SL users, contributing to the accumulating knowledge of rapid prototyping applications. A StereoLithography Users Group was formed by the SL users themselves so that these companies could formally share information and provide a uniform voice to 3D Systems about future product direction.

In addition, several "service bureau" companies were founded and began providing engineering and rapid prototyping services to the general manufacturing community. 3D Systems went into partnership with Ciba-Geigy, Ltd. of Switzerland and began a program of advanced photopolymer resin development. SL became available in Europe and Japan, reflecting the international nature of today's industrial environment.

The SLA-250, similar to the SLA-1 but with an upgraded resin recoating system, was announced in 1989. *Figure 1-7* shows the SLA-250 system. In 1992, the SLA-250 is, by far, the most widely used RP&M system in the world.

The SLA-500, a larger and faster machine, became available in 1990. *Figure*

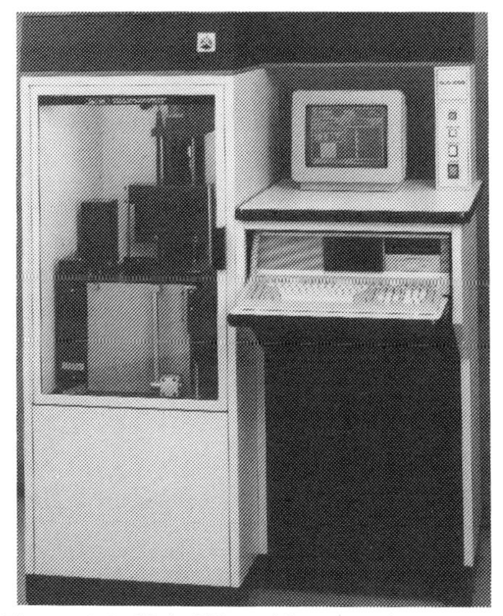

Figure 1-7. SLA-250 rapid prototyping system.

1-8 shows the system. With working volume about eight times greater than the SLA-250, the SLA-500 has the ability to produce significantly larger prototypes.

During this period, the range of applications of SL technology expanded to include medical imaging, prototype investment casting patterns, communication tools for concurrent engineering, and many classes of soft tooling patterns.

At about that time, the first of the competitive systems reached the market. Details of the various alternative approaches to RP&M are presented in Chapter 16.

SL involves a unique combination of polymer chemistry, laser physics, optics, material science, viscous fluid dynamics, computer science, and electrical and mechanical engineering. During 1990, the science teams devoted much of the year to the study of the basic interactions and relationships of the technology. The results of this program formed the foundation for steady improvements in part accuracy and resolution, as well as greatly expanding our knowledge of many of the fundamental processes of SL. A very significant outgrowth of this effort was the development of the advanced part building method known as WEAVE™, which quickly led to significant improvements in part accuracy.

At the same time that improved system accuracy and improved material properties were becoming available, easier to use and more efficient system software, more comprehensive solid modeling CAD systems and better and easier to use chemical postprocessing systems were evolving. This pattern of constantly improving capability, available to users of early equipment as well as brand new systems, has done much to expand the overall industrial acceptance of RP&M.

Figure 1-8. SLA-500 rapid prototyping system.

REFERENCES

1. Crawford, W.P., *Mariners Celestial Navigation*, Miller Freeman Publications, San Francisco, CA, 1972, Chapter 11, pp. 167-168.
2. Bridgwater, W. and Kurtz, S., *The Columbia Encyclopedia*, Third Edition, Columbia University Press, New York, NY, 1968.
3. Nitske, W.R., *Mercedes-Benz : A History*, Motorbooks Int., Osceola, WI, 1978, pp. 8-30.
4. Mueller, T., *Applications of StereoLithography in Injection Molding*, Second International Conference on Rapid Prototyping, University of Dayton, Dayton, OH, Conference Proceedings, June 23-26, 1991, pp. 327-333.
5. Phillips, W.L., *StereoLithography and Conventional Tooling Applications*, Second International Conference on Rapid Prototyping, University of Dayton, Dayton, OH, Conference Proceedings, June 23-26, 1991, pp. 163-164.
6. Prioleau, F.R., *Applications of StereoLithography in Investment Casting*, Second International Conference on Rapid Prototyping, University of Dayton, Dayton, OH, Conference Proceedings, June 23-26, 1991, pp. 149-151.
7. Trimmer, D., *The Exploitation of Rapid Prototyping*, Second International Conference on Rapid Prototyping, University of Dayton, Dayton, OH, Conference Proceedings, June 23-26, 1991, pp. 169-171.
8. Cromwell, W.E., *Prototype Casting Fabrication by StereoLithography*, Second International Conference on Rapid Prototyping, University of Dayton, Dayton, OH, Conference Proceedings, June 23-26, 1991, pp. 103-148.
9. Charney, C., *Time To Market: Reducing Product Lead Time*, Society of Manufacturing Engineers, Dearborn, MI, 1990, p. 1.
10. Hull, C., *Apparatus for Production of Three-Dimensional Objects by StereoLithography*, U.S. Patent 4,575,330, March 11, 1986.

chapter 2

Basic Polymer Chemistry

Every passing hour brings the Solar System forty-three thousand miles closer to Globular Cluster M13 in Hercules—and still there are some misfits who insist that there is no such thing as progress.

—Kurt Vonnegut, Jr.
The Sirens of Titan

2.1 Introduction to Radiation Curable Polymers

Radiation curable polymer technology became a significant industry about 20 years ago. This occurred when government mandates were issued to reduce volatile organic components in coating formulations. In the early 1970s, systems were developed for coated printed stock. In the mid-1980s, UV-curable inks became widely used. Currently, increased usage is occurring in such areas as clear coating for furniture.

StereoLithography (SL) resins represent a small but growing segment of this market. Overall, the radiation-curables industry, both electron beam and ultraviolet (UV) has grown to a market of about $250 million in U.S. sales in 1989 and $500 million in worldwide sales.[1] These sales continue to grow at a 15% to 20% annual rate.

The advantages of liquid photocurable resins extend beyond their composition involving fully 100% reactive components. These systems are also energy efficient, requiring 50 to 100 times less energy than that consumed in thermally cured coatings. A variety of properties can be achieved for different applications using the wide range of existing monomers and oligomers.

By ***Dr. Max Hunziker****, Chief of Photopolymer Resin Development, Ciba-Geigy, Ltd., Marly, Switzerland, and* ***Dr. Richard Leyden****, Chemical Development Manager, 3D Systems, Inc., Valencia, CA.*

At first glance, rapid prototyping models may seem unrelated to paints, adhesives, and inks. However, resin shares many common features with these materials. First, the resin is always applied in thin layers. Second, spreading properties and viscosity are important issues. Furthermore, increased speed, improved toughness, and reduced shrinkage are objectives common to all these applications. SL has benefited from the monomers and oligomers developed for these earlier applications. On the other hand, the precisely defined exposure conditions that arise from the use of laser sources both demand and allow a deeper understanding of the basic photopolymerization processes. Ironically, other applications may eventually benefit from the development efforts directed toward RP&M materials.

2.2 Photopolymerization

In this section, a brief overview is given of the basic processes involved in photopolymerization. Definitions of terms used throughout this chapter are provided to benefit the nonchemist. This is not intended to be a tutorial on chemistry. The general aspects of photoinitiated radical polymerization have been reviewed thoroughly, and readers may gain a deeper understanding by utilizing References 2, 3, and 4.

Polymerization is the process of linking small molecules (monomers) into larger molecules (polymers) comprised of many monomer units. This process is illustrated in Scheme 2-1 for vinyl-type monomers.

Vinyl monomers are broadly defined as monomers containing a *carbon-carbon double bond.* The vinyl group may be attached to other molecular structures, represented by "R" in Scheme 2-1. The "R" group may contain one or more other vinyl groups, in which case the monomer is called multifunctional (difunctional, trifunctional, etc.). Polymerization of multifunctional monomers results in a *cross-linked polymer*, as shown in Scheme 2-2.

Chemical reaction of the vinyl group allows each carbon atom in the carbon-carbon double bond to form a new bond, typically with a carbon atom from another monomer molecule. In going from loose Van der Waals interactions with neighboring monomers to a network of covalent bonds, many bulk properties change. As shear strength increases, the system changes from liquid to solid. The average distance between groups decreases, resulting in an increase in density (shrinkage). For multifunctional systems, these changes will tend to occur at a lower degree of reaction.

$$\begin{array}{ccc} H_2C = \underset{\underset{\textstyle R}{|}}{CH} & \rightarrow & -CH_2\text{–}\underset{\underset{\textstyle R}{|}}{CH}\text{–}CH_2\text{–}\underset{\underset{\textstyle R}{|}}{CH}\text{–}[CH_2\text{–}\underset{\underset{\textstyle R}{|}}{CH}]_n - \end{array}$$

Scheme 2-1. Vinyl-type monomer.

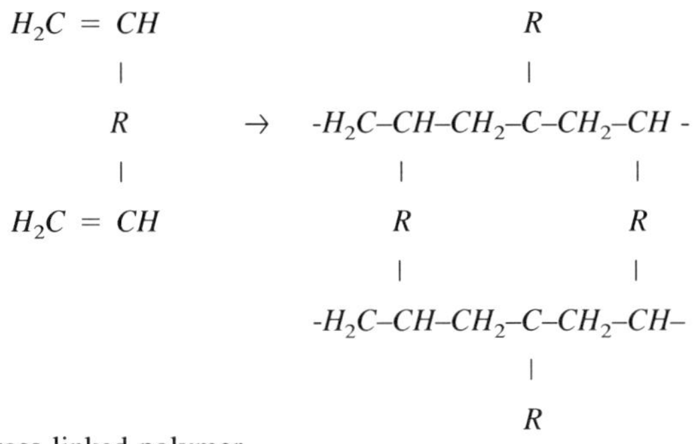

$$
\begin{array}{lcl}
H_2C = CH & & \qquad\qquad\qquad R \\
\quad\;\;\; | & & \qquad\qquad\qquad | \\
\quad\;\;\; R & \rightarrow & \text{-}H_2C\text{–}CH\text{–}CH_2\text{–}C\text{–}CH_2\text{–}CH\text{ -} \\
\quad\;\;\; | & & \qquad\quad | \qquad\qquad\qquad\;\; | \\
H_2C = CH & & \qquad\quad R \qquad\qquad\qquad R \\
 & & \qquad\quad | \qquad\qquad\qquad\;\; | \\
 & & \text{-}H_2C\text{–}CH\text{–}CH_2\text{–}C\text{–}CH_2\text{–}CH\text{–} \\
 & & \qquad\qquad\qquad | \\
 & & \qquad\qquad\qquad R
\end{array}
$$

Scheme 2-2. Cross-linked polymer.

Acrylate monomers are a subset of the vinyl family with high reactivity and versatility due to the carboxylic acid group (-COOH) attached to the carbon-carbon double bond. Attachment of various chemical segments (through acrylate ester functionality) is a relatively easy transformation. This has lead to the great variety of acrylate functionalized monomers.

Polymerization of an acrylate monomer is an energetically favorable or "exothermic" reaction. The heat of reaction is near 85 kJ/mole. Despite the large potential driving force, acrylate formulations can be stabilized to remain unreacted indefinitely at ambient temperature. A catalyst is required for the polymerization to proceed at a meaningful rate.

For acrylate resin systems, the usual catalyst is a free radical. The radicals may be generated either thermally or photochemically. The source of the photochemically generated radical is a photoinitiator, which reacts with an actinic photon, as shown in Scheme 2-3. This produces radicals (indicated by a large dot) that catalyze the polymerization process. In a sense, an actinic photon is the ultimate catalyst for transformation of the liquid monomer to solid polymer.

Using photons as a catalyst is remarkably leveraged from an energy standpoint. On the average, for every two actinic photons (roughly 1.2×10^{-18} Joule of energy), one radical will be produced. *Each radical will result in the polymerization of more than 1000 acrylate monomers*. For comparison, the heat

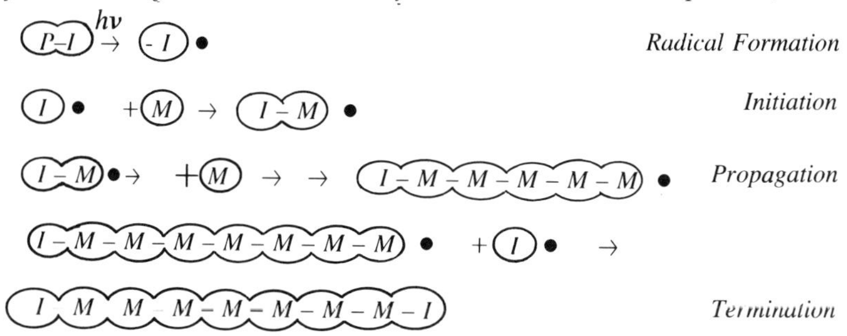

Scheme 2-3. Photoinitiator reacts with an actinic photon.

released by these 1000 acrylic monomers is over 100 times greater (130 x 10^{-18} Joules). This efficiency of photopolymerization allows productive use of relatively low-power ultraviolet lasers in rapid prototyping. RP&M approaches involving powder sintering, for example, require approximately 1000 times more laser power.

Radical Polymerization

Let us briefly discuss the three essential steps in polymerization for SL: initiation, propagation, and termination.

Initiation. By a photochemical process, free radicals ($R\bullet$) are formed from the photoinitiator (I) upon absorption of photons of appropriate frequency. These radicals react with a monomer (M) to initiate a chain reaction:

$$\begin{aligned} I + h\nu &\rightarrow 2R\bullet \\ R\bullet + M &\rightarrow (R - M)\bullet \end{aligned} \qquad (2\text{-}1)$$

where h = Planck's constant, ν is the frequency of the photon, and $h\nu$ is the photon energy. For laser induced photopolymerization, an initiator must be chosen that absorbs efficiently at the frequency of the incident monochromatic photons, and then generates the initiating species with a high quantum yield.

Propagation. The addition of a monomer (M) to a macroradical, ($M_n\bullet$), results in a larger macroradical ($M_{n+1}\bullet$),

$$(M_n\bullet) + M \xrightarrow{k_p} (M_{n+1}\bullet) \qquad (2\text{-}2)$$

where k_p is the reaction rate for polymerization.

In the highly functionalized, solvent-free systems used in SL, polymerization takes place in microregions near the site of radical initiation. The growing macroradical becomes a cross-linked gel at a relatively low degree of conversion. The rate of polymerization steadily decreases as the concentration of unreacted monomer decreases and the viscosity of the gel increases (the Tromsdorff effect). The mobility of free monomer to the site of the macroradical is thus restricted at higher conversion.

In principle, the polymerization rate could be controlled by suppressing the onset of viscous gel formation through the selection of monomers with differential reactivity. A higher degree of conversion (improved green strength) might be achieved by choosing small monomers that could more easily diffuse through the gel to react at the final stages of reaction. Unfortunately, the differential reactivity of various acrylate types is small.

Termination. Principal reactions leading to chain termination are recombination, disproportionation, and occlusion, as shown in equation 2-3:

$$\rightarrow M_{m+n} \qquad \text{(recombination)} \qquad (2\text{-}3)$$

$$(M_m\bullet) + (M_n\bullet) \rightarrow M_m + M_n \qquad \text{(disproportionation)}$$

$$\rightarrow \{M_m\bullet\} + \{M_n\bullet\} \qquad \text{(occlusion)}$$

Recombination involves two radicals joining to form a nonreactive molecule. Disproportionation involves the transfer of a β-hydrogen from one radical to the other, producing two polymer molecules, one with a saturated end, the other with an unsaturated end. For laser-induced polymerization, the reaction continues for about one second in the dark, although initiating radicals are no longer produced.[5]

Electron paramagnetic resonance (EPR) measurements showed that trapped radicals are still present in a crosslinked polymer long after irradiation. Occlusion or "frozen mobility" of the macroradical within the polymer network thus represents a major termination pathway for SL resins.

The lifetime of these radicals can be as long as several months. A slow polymerization may, therefore, take place in the dark since unreacted monomers and dangling double bonds are present in cross-linked polymers. The long-term stability of a part depends on the extent to which these reactions can continue despite an extended postcure.

2.3 Laser Exposure

In a typical coating or resist layer, the thickness is normally tuned to absorb only part of the incident radiation,[6] according to the Beer-Lambert exponential law of absorption. A sufficient amount of exposure to induce cure is delivered to the bottom of the layer, which then adheres to the substrate. A higher exposure will gradually increase the degree of cure, but the cure depth is limited by the thickness of the layer.

In the SL process, the starting material is a liquid in a deep container which can be considered to be of infinite thickness. The first support layer is drawn on the platform (a metallic punched plate) and subsequent layers are stacked on previous cross sections. Here, the layer thickness to bc polymerized is given by the amount of liquid which has been recoated onto the part, and any excess laser radiation that penetrates this layer only acts to slightly increase the curing of the previous layers. On layers that start an overhang or bridge (which are larger at some places than the underlying structure), the laser beam shines right down into the resin pool, attenuated essentially by the initiator absorption. According to the Beer-Lambert law of absorption, the laser exposure $E(\text{mJ/cm}^2)$ will decrease exponentially with depth z in the following manner:

$$E(z) = E_o \exp(-z/D_p) \qquad (2\text{-}4)$$

where D_p is the resin "penetration depth" at the laser wavelength and E_o is the laser exposure at the resin surface ($z = 0$). In practice, the polymerization does not proceed beyond a limited depth where the laser exposure is below a threshold value; this is primarily due to oxygen inhibition, which imposes a minimal threshold to start the polymerization. The exposure level where the gel point is reached is slightly higher still (where actual solid material is formed and becomes attached to the previous layer). The exposure threshold for the formation of gel is known as the "critical exposure," E_c.

Exposure is defined as radiant energy per unit area; it is proportional to the laser power (which is normally fixed), and inversely proportional to the product of the laser beam width at the liquid surface and the laser scan speed. For a more detailed estimation of the cured "profile" and its local cure degree, other factors like power distribution within the beam profile should also be considered. This is discussed in Chapter 4.

Depending on photopolymer composition, and especially the photoinitiator characteristics, deviations from the straightforward exponential law may occur, particularly when the initiator shows "photo bleaching," or when the whole formulation is optically inhomogeneous and scatters the laser beam. In the first case, the polymerization depth will be larger than expected. In the second case, it is evident that the tightly focused laser beam will disperse its irradiance peak, and penetration becomes limited by lateral and backward radiation scattering[7]. The exact formalism is beyond the scope of this discussion (see Reference 8). Other effects, which are considered in more detail for photopolymer exposure, are wavelength and exposure time (see Chapter 4). In an optimum case, one would like to have a uniform cure from the surface down to the specified layer thickness, while below that depth no photoconversion would occur. Using the presently known high-efficiency photoinitiators, this is not possible due to the absorption profile shown in *Figure 2-1*. Nevertheless, the penetration of a single wavelength laser beam is still the most reasonable approximation, and the importance and advantages of monochromatic light for a more uniform cure of relatively thick layers has been recognized.[9]

Lasers can deliver high irradiance values, compared to conventional incandescent or arc lamps, especially when they are focused. In blanket exposures or mask patterning of photopolymer layers, irradiation typically lasts several seconds, and the polymerization rate at the beginning of the reaction is generally limited by the photon flux. We estimate for a typical case in SL, the laser irradiance and the resulting primary photoproducts, using SLA-250 specifications: laser power (P_L) = 15 mW, spot size ($2W_o$) = 0.25 mm (multimode helium cadmium profile, where W_o is the $1/e^2$ (0.135) half width of the laser spot on the resin surface), and a variable speed scanning system. From this, we derive an average irradiance density of:

$$H_{av} = P_L/(\pi W_o^2) = 30.56 \text{ Watts/cm}^2 \tag{2-5}$$

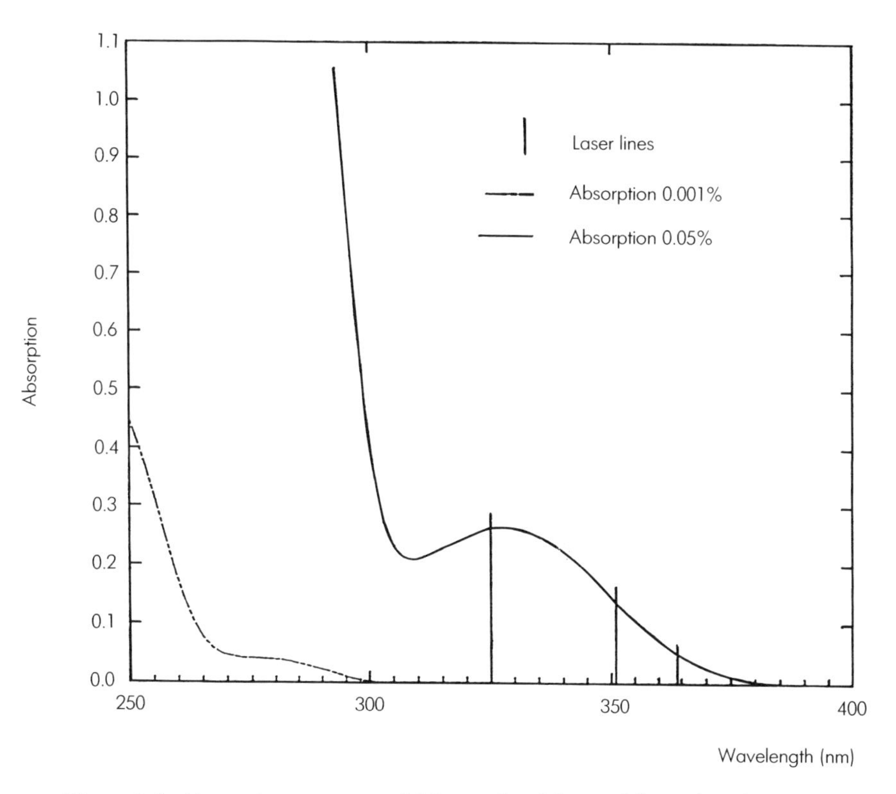

Figure 2-1. Absorption spectrum of 1-benzoylcyclohexanol in methanol.

At a typical scan speed V_s = 200 mm/second, the exposure time for a spot along the center of a line is approximately given by:

$$t_e \approx 2W_o/V_s = 1.25 \text{ msec} \quad (2\text{-}6)$$

In this period, the irradiated area will receive an average laser exposure of:

$$E_{av} = H_{av} * t_e = 38.2 \text{ mJ/cm}^2 \quad (2\text{-}7)$$

at the specified wavelength of 325 nm. We can now calculate the photon flux N_{ph}, using the Planck equation for the photon energy

$$E_{ph} = \text{hc}/\lambda = 6.1 \times 10^{-19} \text{ J/photon} \quad (2\text{-}8)$$

where h is Planck's constant (6.62×10^{-34} Joule-sec), c is speed of light (3.00×10^{10} cm/sec) and $\lambda = 3.25 \times 10^{-5}$ cm.

$$N_{ph} = E_{av}/E_{ph} = 6.3 \times 10^{16} \text{ photons/cm}^2 \quad (2\text{-}9)$$

In chemical terms, using a typical penetration depth, $D_p = 0.17$ mm, for a volume element near the surface, this corresponds to a photon concentration of:

$$I_a = N_{ph}/(N_A * D_p) = 0.006 \text{ Einsteins/liter} \quad (2\text{-}10)$$

where N_A = Avogadro's number = 6.02×10^{23} atoms per mole and an Einstein is defined as 6.02×10^{23} photons. This is a substantial concentration if we take into account the short interaction time and a photochemical efficiency of about 0.5 for a good photoinitiator.

Also, short exposures with high photon flux require some special attention. In a series of investigations,[6,8] Hoyle et al have laser polymerized quite simple acrylate formulations using a pulsed actinic laser at variable repetition rates and found a dependence of the polymer molecular weight distribution on both photon fluence and pulse frequency. When the time between pulses is shorter than the "natural" termination of all growing polymer chains, subsequent pulses will provide more radicals, which can terminate old growing chains or start new ones. Thus, at high laser pulse repetition rates, shorter chains will result on average.

In a complex mixture including multifunctional components, a more complicated pattern resulting in different mechanical properties has to be expected. Pulse repetition rate and resulting polymer compositions have also been used to derive absolute reaction rates.[10,11,12] So it can be expected that the structure of the polymer network being formed will depend on the specific laser power density (absorbed radiant power per unit volume). This in turn can influence the properties of finished parts (mainly mechanical, but also diffusion of solvents and monomers, swelling, refractive index, etc.). Investigations of model calculations for such laser photopolymerization processes, including their temporal influence on chemical and thermal profiles, are in progress.[13] To date, however, no clear correlation with the actual process has been shown.

2.4 Concepts and Methods for the Evaluation of Resin Properties

Passing a laser beam of maximum irradiance H_o and dimensionless distribution $I(x,y)$ with a velocity V_s over the resin surface generates a polymer string of width L_w and cure depth C_d. Integration of the intensity distribution in the direction of the laser movement (x) leads to the total exposure $E(y)$ experienced by a point at a distance y from the centerline of the scan, and at $z = 0$, on the resin surface:

$$E(y) = (H_o/V_s) \int I(x, y)\, dx \quad (2\text{-}11)$$

Since $E(y)$ is subject to variations in the beam profile, the difficulty of achieving a well-defined exposure may be overcome by scanning a small area by a series of overlapping strings separated by a hatch separation increment hs, with hs being smaller than the laser beam radius. In this situation the average exposure at the resin surface is given by (see Chapter 10):

$$E_{av} = P_L/V_s\, h_s \tag{2-12}$$

By varying the laser scan velocity, we obtain the dependence of the cure depth upon the incident exposure at the resin surface, which represents a useful definition of photosensitivity. The cure depth, C_d, of the polymerized scanned area can be measured with a micrometer (see Chapter 10).

The cure depth, C_d, depends upon the absorption characteristics of the photopolymer. It follows from equation 2-4 that the incident exposure at depth z is:

$$E(z) = E_{max} * \exp(-z/D_p) \tag{2-13}$$

For the photopolymer to gel, $E(z)$ must reach at least the critical threshold exposure E_c, when $z = C_d$:

$$E(C_d) = E_{max} * \exp(-C_d/D_p) = E_c \tag{2-14}$$

from which we obtain the dependence of C_d upon the maximum exposure at the resin surface:

$$C_d = D_p \ln(E_{max}/E_c) \tag{2-15}$$

A plot of C_d versus $\ln(E_{max})$ should thus yield a straight line with a slope equal to D_p and an exposure intercept equal to E_c, provided that both D_p and E_c are constants of the resin. This is equivalent to assuming reciprocity (twice the power for half the time will produce the same results). As a representative example, *Figure 2-2* shows the experimental "working curves" for Cibatool SL resin XB 5081-1 in air and in an argon atmosphere. The results imply that the threshold exposure E_c is mainly determined by the concentration of dissolved oxygen. It is worth noting that reciprocity will only extend over a finite range of parameters. If the scan velocity gets so low that the exposure duration becomes comparable to the characteristic time for oxygen to diffuse back into the scanned region, one would expect reciprocity failure. Fortunately, such limitations do not pose practical problems with SL systems.

The photosensitivity, as given by equation 2-15, may be modified by a change in initiator concentration $[I]$ since both D_p and E_c depend upon $[I]$: D_p is

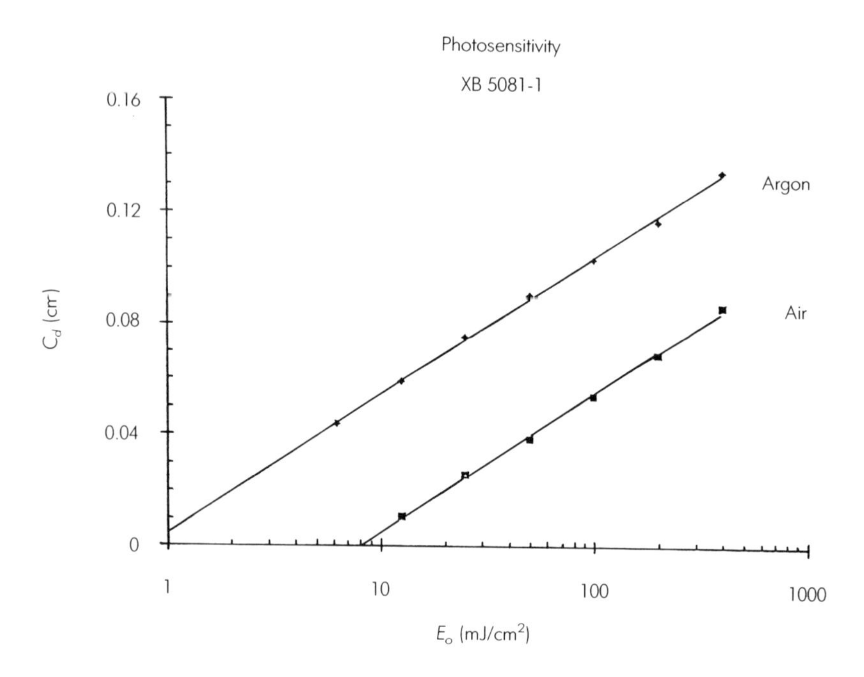

Figure 2-2. Cure depth versus incident HeCd laser exposure for Cibatool SL XB 5081-1.

inversely proportional ($D_p = 1/(2.3 * \epsilon [I]$ where ϵ is the molar extinction coefficient of the initiator). It can be argued that E_c should behave inversely with concentration $[I]$ (E_c,2 $=$ E_c,1 $*$ (I_1/I_2)) *since it corresponds to the amount of actinic laser energy per unit area required to generate the critical number of radicals for O_2 consumption and hence for the material to gel*. This means that, given a resin with two initiator concentrations I_1 and I_2, the photosensitivity curves will be invariant at an exposure E^* and at a cure depth Cd* given by

$$E^* = E_{c,1} \{I_1/I_2\}^{[I_1/(I_1-I_2)]} \quad (2\text{-}16a)$$

$$C_d^* = D_{p,1} \{I_1/(I_1-I_2)\} \ln (I_1/I_2) \quad (2\text{-}16b)$$

In practice, this means that below the invariant point a high initiator concentration will increase the photosensitivity whereas above the crossing point the behavior is reversed. These considerations are likely to become more important in the future as the technology moves toward thinner layers for better spatial resolution.

As discussed in section 2.3, there is some published work indicating that the rate of photopolymerization is dependent on actinic intensity. For reaction in dilute solutions, the rate of polymerization is proportional to intensity to the half power. Doubling the intensity would only increase the rate of polymerization by a factor of 1.4. For multifunctional acrylates, typical of those used in rapid prototyping, Decker has found that the polymerization rate follows close to a first order relationship with intensity.[14] In these highly functionalized systems, first order termination by radical occlusion predominates over radical recombination.

For systems like the SLA 250, which are operated over a relatively narrow power range, deviations from reciprocity can scarcely be measured. For the higher power systems such as the SLA 500, laser power can potentially be varied by more than one order of magnitude. Even small deviations from first order dependence could affect part accuracy.

Measurement of SLA-500 working curves over a range of laser power levels shows (*Figure 2-3*) small but measurable effects. The critical exposure E_c, increases slightly at higher laser power. The penetration depth, D_p, on the other hand, remains nearly unchanged.

The tensile modulus of laser cured pull bars was also measured as a function of laser power. In the series shown in *Figure 2-4*, raster scanned single layer pull bars were drawn with the laser power varied between 5 and 80 mW. The total exposure was kept constant in all cases by increasing the drawing speed in proportion to the laser power. The maximum tensile modulus occurred at intermediate power (20 mW). Lower green strength at lower power levels can be explained by a slight increase in oxygen rediffusion between successive laser scans. At yet higher power levels, there is a slight decrease in green strength with laser power. This may be due to a slight reciprocity failure, as previously mentioned.

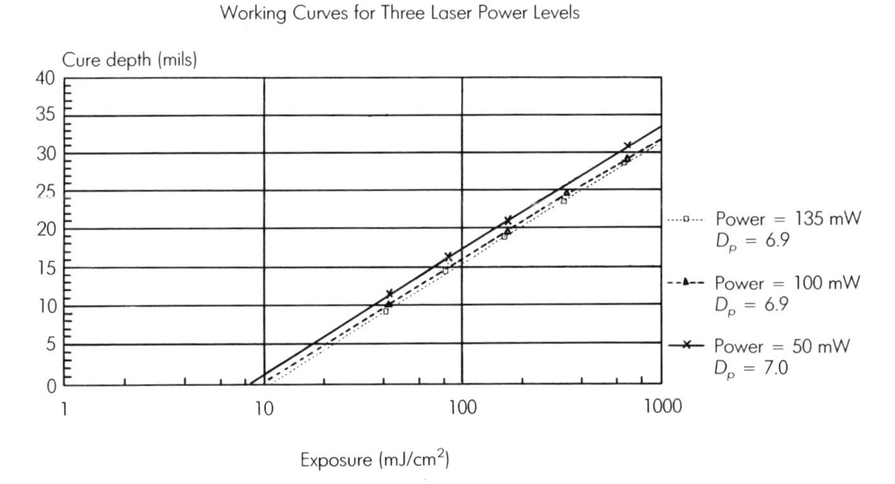

Figure 2-3. Working curves for Cibatool XB 5131 at three argon-ion laser power levels.

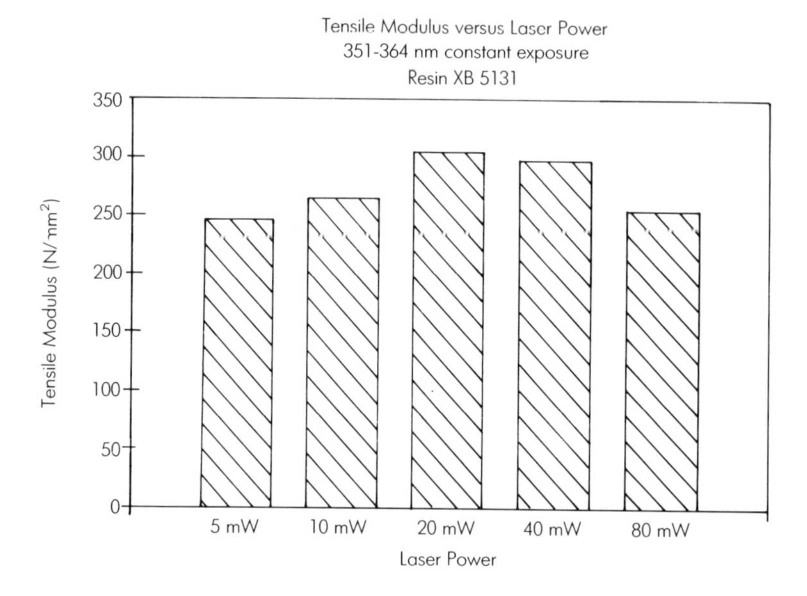

Figure 2-4. Tensile modulus, measured green, for Cibatool XB 5131 at five argon-ion laser power levels.

Green Strength

An important consequence of the limitation of cure depth by absorption (typical cure depths are 0.1 mm to 0.5 mm) is the inhomogeneity in chemical and physical properties within the laser-cured part. Examples are mechanical properties, conversion of monomers, residual reaction enthalpy, extent of shrinkage, rate of polymerization, diffusion of components, etc. In turn, these inhomogeneities are significant in part building.

"Green strength" is a general term that comprises all the mechanical properties of the laser-cured part including modulus, strain, strength, hardness, and layer-to-layer adhesion. It must be sufficiently large that a part will not distort before or during the final postcure.

At the gel point E_c, the material has no mechanical strength. Therefore, the crucial parameter responsible for gaining strength is the energy E_x in *excess* over E_c, which is being absorbed in the upper part of the polymer string:

$$E_x = (1/C_d)\int_{o}^{C_d} (E(z)-E_c)\, dz$$

$$= E_c \{(D_p/C_d)\,(\exp(C_d/D_p)-1)\ -1\} \qquad (2\text{-}17)$$

This function is shown in *Figure 2-5*.

It has the following important properties:

1. The excess curing energy, E_x, is directly proportional to E_c. If E_c is lowered for the benefit of shorter building times, while D_p is kept constant, we will be faced with a decrease in green strength, which may be unacceptable.
2. Decreasing the initiator concentration and hence increasing the penetration depth, D_p, leads to a decrease in green strength for a given cure depth. Conversely, *decreasing* D_p increases green strength.
3. E_x/E_c increases rapidly only at $C_d/D_p > 2$. This is confirmed by the fact that at lower ratios of C_d/D_p it is very difficult if not impossible to determine mechanical parameters of the green material (see Chapter 4).

A quantitative relationship between the mechanical parameters (modulus, strength, strain) and exposure would, in principle, allow for a prediction of the integrated green strength of the part. This would then allow one to optimize for sufficient strength while achieving short building time, although the calculation would be complex considering the geometry of the part, overcure, areas of multiple exposure, and varying cure depths within a part. The concept of the photomodulus describes the dependence of the resin's Young's modulus upon exposure and is useful to characterize green strength. It also provides a link between the properties of the green and the postcured material.

Measurement of green strength can be taken in a number of ways, depending on the intended purpose. For careful modeling studies, to derive the photomodulus coefficient, sample specimens should be virtually uniformly exposed, as described in Chapter 4. However, for a simple comparison between several resin formulations, direct measurement of an inhomogeneous specimen is more convenient. Test results on the tensile properties of a single laser-drawn string give reasonably reproducible data for a given cure depth as long as the laser profile remains constant.

Unfortunately, He-Cd laser profiles can vary significantly as the laser ages. The effects of beam profile can be minimized by *overlapping* scan lines. A useful single layer pull bar can be formed with as few as 15 overlapping scan lines. The measured properties of this type of specimen strongly depend on cure depth (see *Figure 2-5*). To properly compare two resins, the exposure should be adjusted so that both samples have equal cure depth. Green strength measurements of this type are hence reported with their corresponding cure depth. For cure depths (0.3mm to 0.5 mm) Cibatool XB 5081 exhibits a Young's modulus of 40 to 200 N/mm^2 or only a few percent of the value for the fully cured material (3500 N/mm^2). The green strength of greatest interest to the users of SL is that of multilayer parts, which tend to have higher values than single layer parts because of overlapping exposures both vertically and horizontally. The mechanical properties of green multilayer parts depend on both the resin properties and the drawing style.

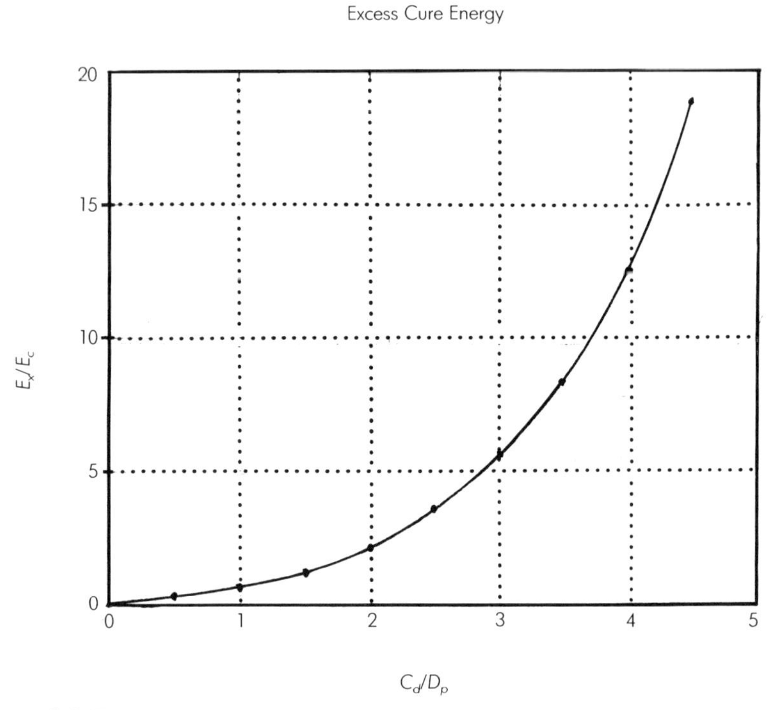

Figure 2-5. Exposure ratio (E_x /E_c) as a function of cure depth to penetration depth ratio (C_d/D_p).

Flexural modulus is also a meaningful measure of the structural strength of multilayer parts. The forces acting on actual green parts are almost always flexural rather than tensile. A small laser-formed test strip, 150 mm long, 2.5 mm wide, and 10 mm high (40 layers) was developed to measure properties of multilayer parts with a minimum volume of resin. In all cases, specimens built with WEAVE™ have much higher flexural modulus than crossed hatched samples.

Curl Distortion

"Curl" is a type of distortion that can occur in all RP&M processes that build parts in successive layers where the solidifying material undergoes shrinkage. *Figure 2-6* shows the sequence of steps leading to curl distortion. Experiments have shown that the first layer drawn, forming an unsupported cantilever, will not initially show any upward deformation. In fact, for a single layer, the shrinkage forces will tend to deform the layer slightly downward, particularly at high exposures. The single unsupported layer is free to shrink without inducing distortion stress. However, the second and successive layers are each bonded to the layer below them. If there is shrinkage after these upper layers become bonded to the layer below, a bending moment is introduced, which can cause upward displacement of the unsupported ends of the layers. The first few

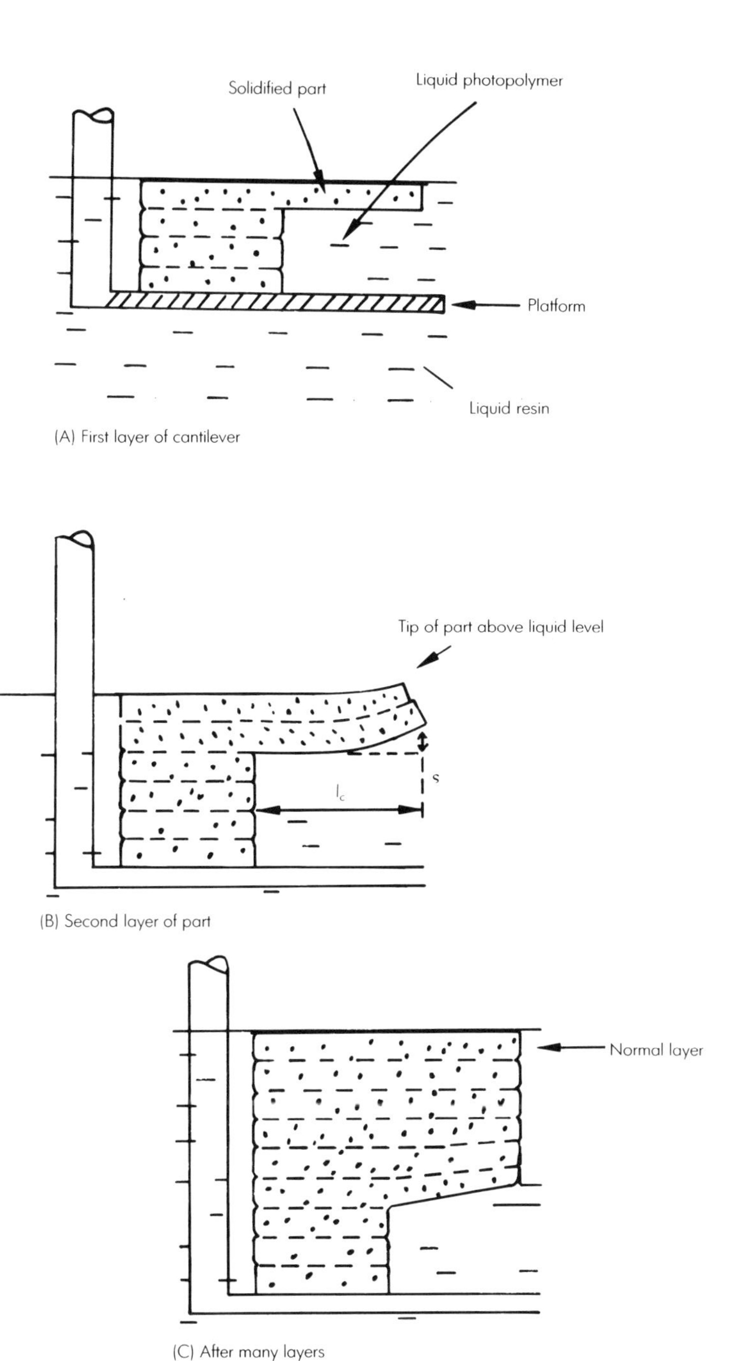

Figure 2-6. Schematic side view of part building leading to curl: (A) first layer of cantilever, (B) second layer of cantilever and, (C) after many layers.

unsupported layers will curl above the liquid surface level. As additional layers of liquid are applied, a self-correcting effect occurs. Less liquid resin is applied to upward deflected regions, and the thicker section below is better able to resist the distortion forces.

Curl distortion can manifest itself in a variety of situations. *Figure 2-7* shows a number of curl-related distortions. The curl forces are generated throughout the part regardless of whether or not a true cantilever is present. If curl induced stresses are large enough, distortions can occur within any part. In extreme cases, these stresses can break the supports holding the part to the machine platform, or cause delamination within the part. Residual stresses generated during part building may be responsible for ''creep'' distortions, which occur long after the part is built. This phenomena is discussed in Chapter 10.

Curl will occur in a part depending on a wide range of parameters including the curl factor of the resin (unrestrained curl); the design, number, and strength of the supports; the part-building style used to form the layers; and the laser profile and power. The diagnostic test used to separate these factors is discussed in Chapter 10.

In developing resins for RP&M, the resin specific curl properties should be separate from other process parameters. A test that attempts to focus on only resin specific curl properties is called the ''Twin Cantilever'' diagnostic test. Here, the curl factor is defined as the vertical distortion distance, δ, divided by the length of the cantilever, L, from the closest support. For convenience, this number may be expressed as a percentage.

$$C_f(\%) = (\delta/L) * 100 \tag{2-18}$$

There are very few published works that attempt to explain the origins of curl distortion. One of the more extensive attempts is by Marutani.[15] His proposed mechanism of curl is illustrated in *Figure 2-8*. The solidification process is shown as continuing after the passage of the laser beam (the postirradiation shrinkage area is shown cross-hatched). The shrinkage continues after the layer has become attached to the layer below. This shrinkage generates stress perpendicular to the beam path, which presumably causes the attached layer to curl.

The quantitative relationship between curl distortion and shrinkage has been derived by Hull.[16] This model assumes that a line drawn over a previously drawn line adheres to that line when it reaches gel. Additional shrinkage past gel is transmitted to the line below, causing it to curl upwards since the top surface is now shorter than the bottom surface, as shown in *Figure 2-9*. This behavior is very much like a classic bimetallic strip, where the top metal of the strip shrinks more than the bottom metal upon cooling, resulting in upward curling of the entire strip. The curled layers can be modeled as an arc of radius R; where the layer thickness, a; and the linear shrinkage, s, are related as follows:

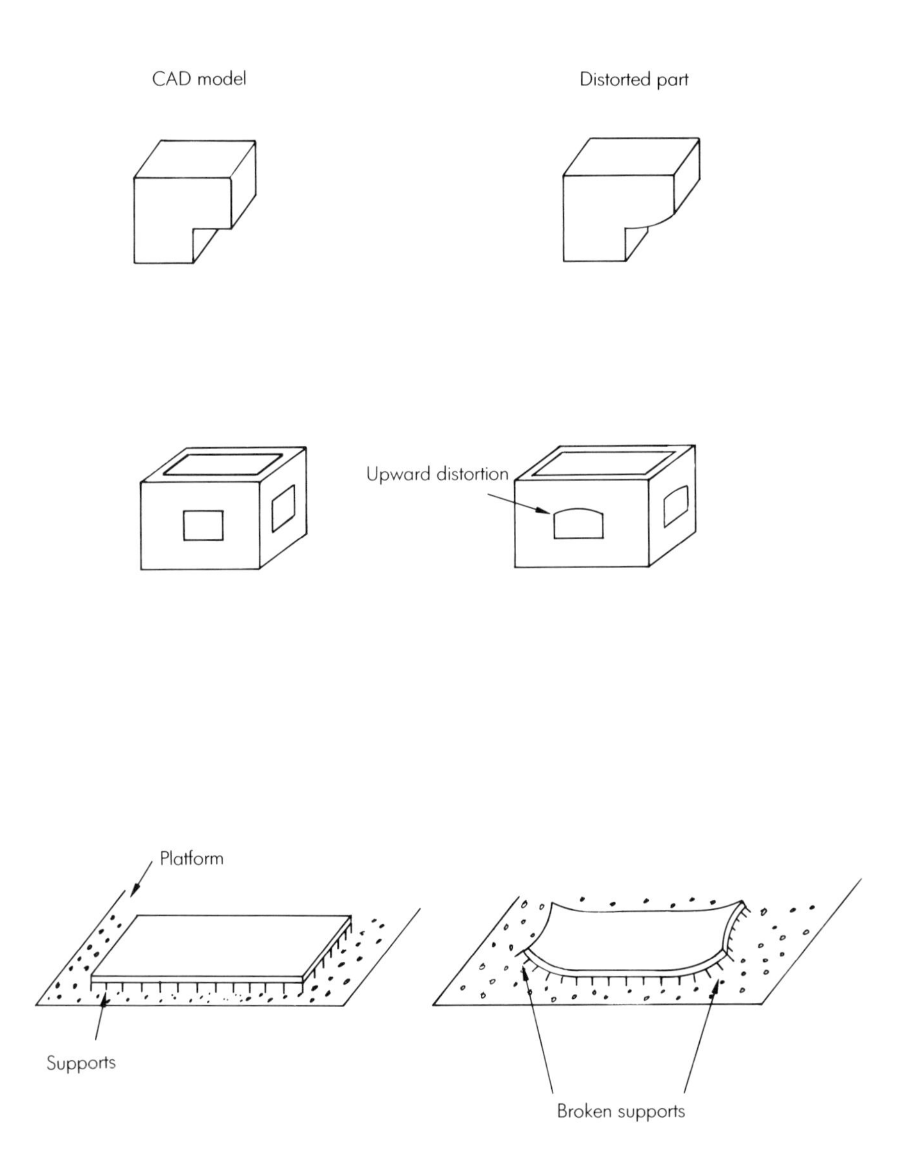

Figure 2-7. Various manifestations of curl; comparison of CAD models with distorted parts.

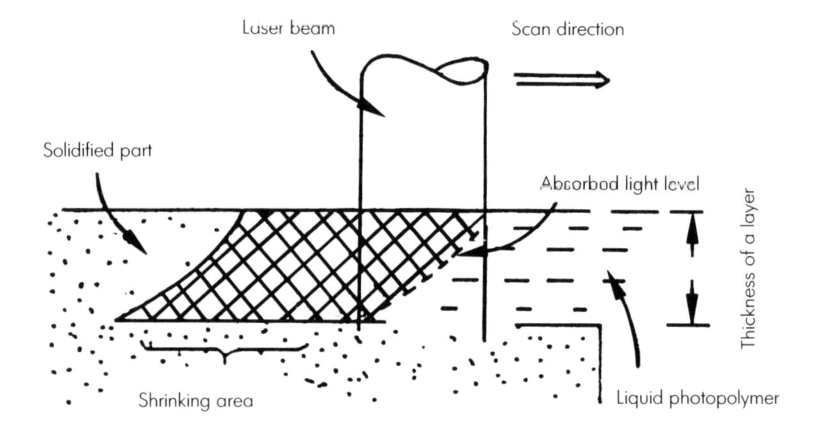

Figure 2-8. Enlarged schematic side view of moving laser beam incident on photopolymer.

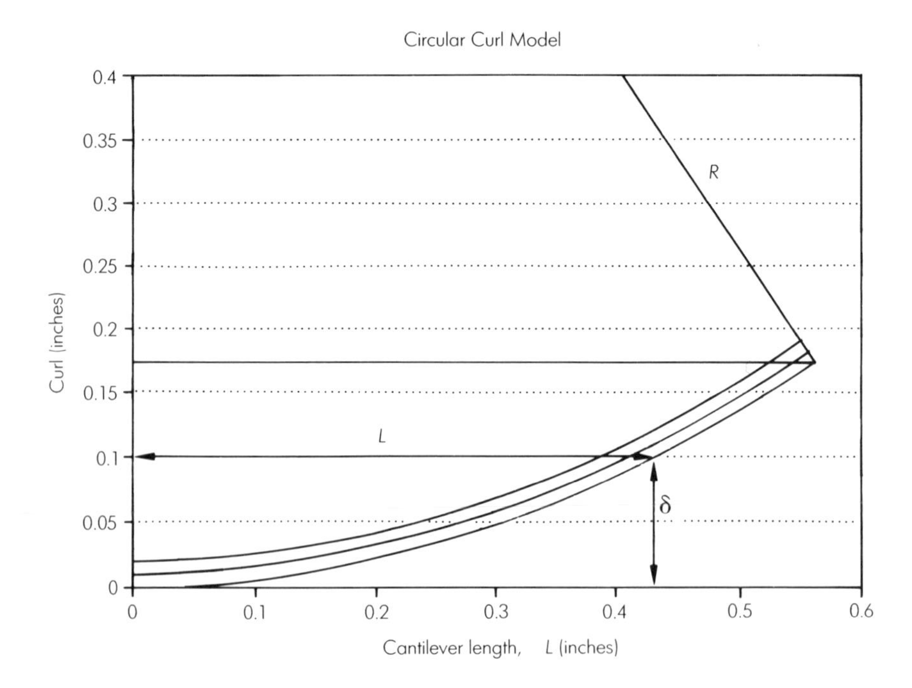

Figure 2-9. Circular curl model where distortion is represented by an arc with radius R and distortion distance δ.

$$R = a/s \tag{2-19}$$

For example, if $a = 0.01''$ and $s = 0.01$ (1% linear, 3% volume shrinkage during laser exposure), an arc of radius 1″ will be formed. The distortion distance, δ, is given by:

$$\delta = R(1-\cos(L/R)) \tag{2-20}$$

where L is the length of the unsupported cantilever.

Note, for $L/R << 1$ we may write:

$$\cos(L/R) = 1 - 1/2(L/R)^2 + \ldots\ldots$$

Neglecting higher order terms leads to the result:

$$\delta = R\,[1/2\; L^2/R^2] = 1/2\; L^2/R = 1/2 * s * L^2/a \tag{2-21}$$

Thus, this model predicts that, for $L/R << 1$ (cantilever lengths much smaller than the radius of curvature), the curl distortion should scale as the square of the unsupported length, L, directly with the linear shrinkage, s, and inversely with the layer thickness, a. As we shall see in Chapter 10, the correlation between curl and the square of the unsupported length is very good. The inverse dependence upon layer thickness is generally correct, but exceptions have been found. The direct proportion with linear shrinkage is *not* in agreement with total shrinkage from liquid to solid. Unfortunately, the portion of the shrinkage relevant for this model is experimentally difficult to measure.

For $L = 6$ mm, $a = 0.25$ mm, and $s = 0.01$, the model predicts a value of $\delta = 0.72$ mm or a curl factor of 12%. These values are close to those commonly encountered. Total shrinkage from liquid to solid ranges from 6% to 8% by volume, with 50% to 70% occurring during laser curing. A curl factor of 12% at 6 mm is reasonable for commercial resins.

Marutani makes a number of predictions of how curl may be reduced. These predictions form a useful device for presenting experimental results. His predictions are as follows:

1. Use high exposure and slow scan speed such that polymerization is essentially complete under the laser spot.
2. Use a resin with a faster rate of polymerization.
3. Decrease laser power to decrease scan speed for a given exposure.
4. Use a low-shrinkage resin.
5. Increase layer thickness to increase the strength.

Despite the intuitive appeal of the Marutani model, experimental results show only weak or even negative correlation. Variations on a simple three-component resin formulation can be used to illustrate the subtle interactions between formulation and curl. Table 2-1 shows the compositions of three formulations used to test the predictions of Marutani.

Variation of Curl with Exposure. In *Figure 2-10*, the curl factors for these three formulations are plotted versus overcure. The cantilever test part is built using 10 mil layers. Cure depth is expressed in terms of "overcure." If a part could be built with zero overcure, the layers would be just touching. In this case, there would be almost no adhesion between the layers, the stresses due to shrinkage of the upper layer would not be transmitted to layers below, and very little curl would result. However, useful parts require adhesion and, hence, overcure to avoid layer delamination.

In contradiction to the prediction of the Marutani model, increasing the exposure or overcure typically results in large increases in curl. As discussed in

Table 2-1
Percent Composition by Weight

Formulation	N 3700	SR 349	SR 348	bis-GMA	D 1173
GA-EA	48	48			4
GA-EM	48		48		4
GM-EA		48		48	4

N 3700 = glycidyl-bisphenol-A-diacrylate, Sartomer
bis-GMA = glycidyl-bisphenol-A-dimethacrylate, Freeman
SR 349 = ethoxylated-bisphenol-A-diacrylate, Sartomer
SR 348 = ethoxylated-bisphenol-A-dimethacrylate, Sartomer
D 1173 = 2-hydroxy-2-methyl-1-phenylpropan-1-one, Ciba-Geigy

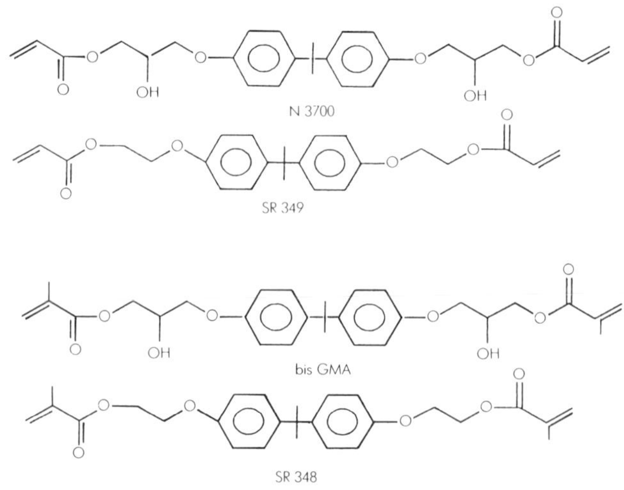

N 3700

SR 349

bis GMA

SR 348

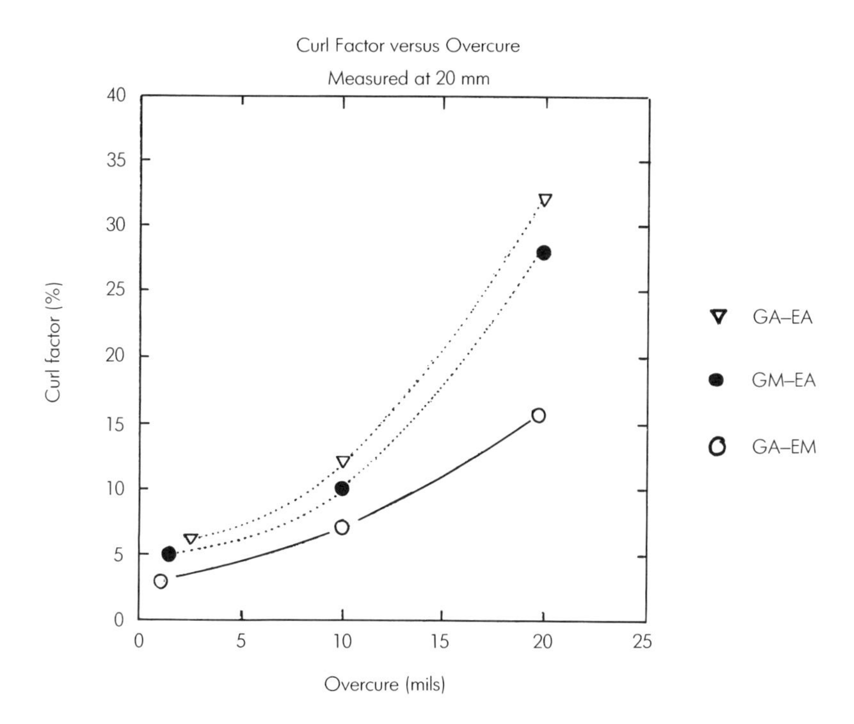

Figure 2-10. Curl factors versus overcure for the three model resin systems.

Chapter 4, the laser drawing speed is exponentially slower at higher cure depths, and more of the resin will certainly be cured while the laser spot is still exposing a given area. To explain the steep increase in curl with exposure using the model in *Figure 2-8* would require that the rate of dark reaction decreases much faster with exposure than drawing speed decreases with exposure. In this case, the length of the cross-hatched region in *Figure 2-8* would increase.

Polymerization rate is known to decrease dramatically at the gel point (the Tromsdorff effect), so this explanation is plausible. Since the Tromsdorff effect is general for all resins, one must conclude that curl will always increase with increased exposure despite the increased shrinkage which may occur directly under the laser spot. Thus the simple predictions based on Marutani's model are incorrect, but the general form, lagging shrinkage after attachment, may still apply.

Variation of Curl with Polymerization Rate. The three formulations shown in Table 2-1 are very similar, differing only by partial replacement of the acrylate reactive groups by methacrylates. Each has only two reactive groups and a photoinitiator. The reactive groups are a glycidyl acrylate or glycidyl methacrylate (GA or GM) and an ethoxy acrylate or ethoxy methacrylate (EA or EM). The overall shrinkage of these three combinations varies only slightly. The polymerization rate of simple acrylate monomers is typically 3 to 10 times faster than the analogous methacrylate monomer.[17] *Figure 2-10* shows that in both cases the methacrylate formulations, GM-EA and GA-EM, display lower curl than the faster acrylate formulation.

Again, this contradicts the simple prediction. A probable explanation is that *cure shrinkage significantly lags the chemical reaction*. Kloosterboer introduced the concept of excess free volume[18] showing that more excess free volume (shrinkage lag) is generated at high exposure rates. It is plausible that greater excess free volume could occur with higher reaction rates driven by kinetic factors rather than intensity. The shrinkage lag often has a greater impact on curl than total shrinkage from liquid to solid. Further data would be required to show if shrinkage lag and, hence, curl generally increases with faster polymerization rates. However, these examples are sufficient to show that the predicted decrease in curl with faster reaction rate is *not* generally true.

Variation of Curl with Exposure Rate. According to the Marutani model, curl should increase when building at higher laser power with the drawing speed adjusted to give constant exposure. This has not been observed on the SLA-250. Experiments have been done using Cibatool XB 5081-1 while varying the laser power by a factor of 4.5 (2 to 9 mW) with the curl factor changing by less than the error in the measurement. In this experiment, the power was adjusted using neutral density filters so the beam profile did not significantly change. Other work done on the SLA-500 extended the range of laser power up to several hundred milliwatts with similar results. A very small increase was observed but could be due to unavoidable secondary effects, such as changes in working curve or beam profile. Certainly, no first order effects were seen.

A possible explanation for this negative result is that the reaction rate for the dominant processes leading to curl are slow compared to the drawing speed, even at the lowest power. The key process effecting curl is, almost certainly, the shrinkage after the layer being drawn becomes attached to the layer below. This post-gel-point polymerization often continues for a number of seconds after exposure.

Variations of Curl with Total Shrinkage. Examples can be found of formulations with low curl and low total cure shrinkage as well as formulations with high curl and relatively high shrinkage. It is tempting to conclude that curl distortion is directly dependent on linear shrinkage, as predicted by the Hull model. However, it is probably impossible to find pairs of formulations that differ only in cure shrinkage, while leaving other potential variables unchanged.

Formulations GA-EA and GA-EM have nearly identical *total* shrinkage from liquid to solid, about 6% by volume, and yet they have different curl behavior. Since laser polymerization may only result in 60% conversion, differing shrinkages at intermediate exposures could, in principle, account for the variation in behaviors. To test this possibility, thin, uniformly exposed films of the model resin were prepared, and their densities were determined by neutral buoyancy measurements.

Figure 2-11 shows that shrinkage at each exposure level tracks closely for the two formulations. Only at the lowest measured exposure (about 25 mJ/cm^2) is there a significant difference. When the curves are normalized for cure depth (there is a slight difference in critical exposure between the resins), the difference is even smaller.

Other Possible Sources of Curl Distortion. Thermal effects have also been proposed as a possible driving force for curl. In this model, the heat of reaction raises the temperature of the polymer gel at the time of attachment to the layers below. Eventual shrinkage due to cooling will induce forces similar to cure shrinkage.

Measured heats of reaction by DSC are 31 cal/gm for GA-EM and 49 cal/gm for GA-EA. The relative difference between these formulations is in accord with lower heat of reaction for methacrylates although both values are only about half the calculated energies using handbook values. Calculated adiabatic temperature rises at 50% conversion are 42°C and 66°C.

Measurements of the actual temperature rise using both microthermocouples placed in the beam path and thermal imaging equipment show less than a 25°C increase for cure depths smaller than 30 mils. The coefficient of thermal expansion can be estimated at .01% per °C for partially cured resin, leading to overall thermal shrinkage of 0.25% (CF(6mm) = 1% for a = 0.75 mm, from

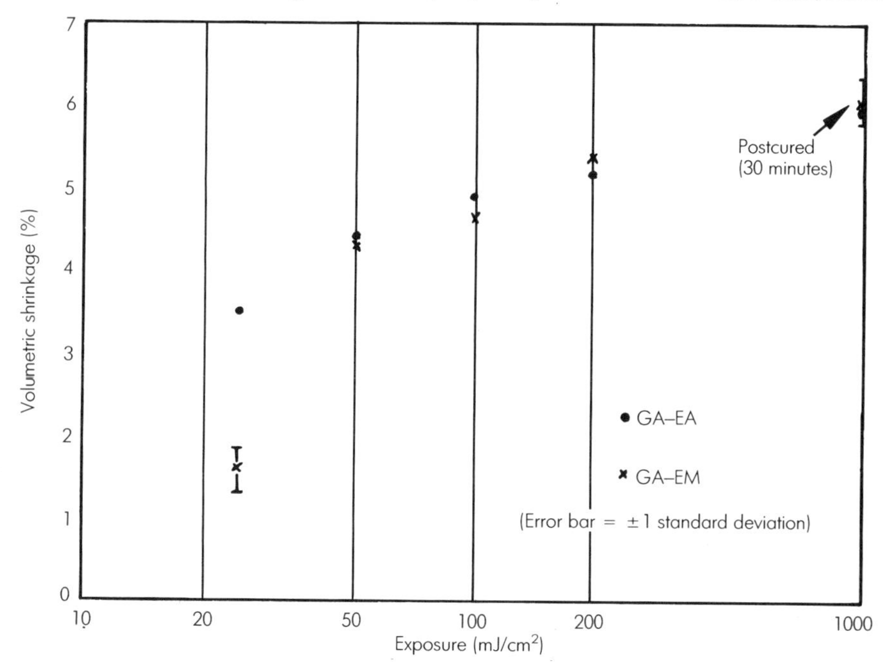

Figure 2-11. Volumetric shrinkage of uniformly exposed thin films versus exposure for two model resins.

equation 2-21). Thus, thermal effects, while potentially worth investigating, cannot account for more than a small portion of the curl effects observed.

If nothing else, this discussion shows the *strong dependence of curl distortion on relatively subtle changes in formulation.* A complete understanding of curl distortion appears to require an accurate picture of the temporal dynamics of cure shrinkage occurring on the same time scale as the laser scanning process. Unfortunately, these measurements are experimentally complex and only recently have test methods been developed to address this time domain.[19]

Degree of Monomer Conversion

The problem of cure characterization is omnipresent in radiation technology.[20,21,22] The fully cured SL resins may be characterized the same way as other crosslinking systems (by thermogravimetric, and dynamic mechanical methods). However, the unique situation involving partially laser cured (green) material encountered in SL obviates the successful application of some techniques due to difficulties in sample preparation and handling.

Differential scanning calorimetry (DSC) is used to characterize the residual reaction enthalpy in green parts. Simply stated, a thin layer of photopolymer is exposed to a known dose of focused laser radiation under the same conditions as in StereoLithography. The material is then fully cured in a modified DSC apparatus equipped with a suitable actinic source (preferably the same laser radiation), and the residual reaction enthalpy is determined. This leads to an accurate relationship between the input laser exposure and the residual enthalpy. *Figure 2-12* shows the experimental relationship for the commercial resin Cibatool XB 5081-1.

Direct information about double bond conversion processes can, in principle, be obtained from infrared spectroscopy by measuring the intensity of the signals associated with the vibrations of the (meth)acrylate groups.[14] While this method has proven successful in simple systems, its effectiveness in multicomponent systems depends upon the quality of the signal assignments and baseline correction.

Extraction experiments, although somewhat tedious, provide the most direct access to the extent of reaction of individual components. We have performed such analyses on partially cured layers of photopolymers and analyzed the individual components by high-pressure liquid chromatography (HPLC), interpreting the results with a simple statistical model. Using the conversion from DSC at a given exposure, this model predicts the amount of extractable material. The fraction f_i of unreacted monomer i with functionality n_i is given by:

$$f_i = (1-\phi)^{n_i} \qquad (2\text{-}22)$$

where ϕ is the DSC conversion fraction. The total amount of extractable material, f_w, is thus given by the weighted sum of all these contributions:

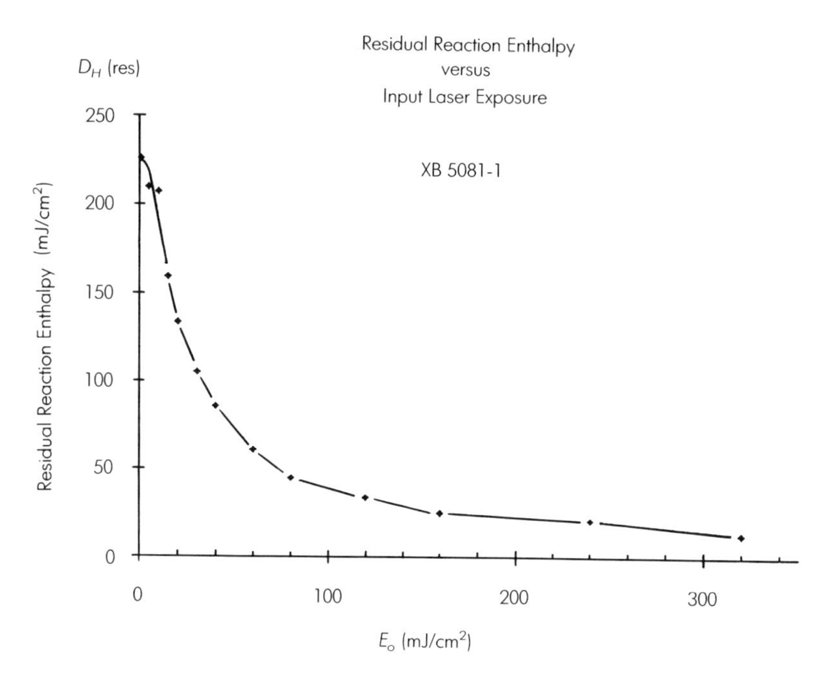

Figure 2-12. Residual reaction enthalpy versus incident HeCd laser exposure for Cibatool XB 5081-1, measured by Photo-DSC.

$$f_w = \sum_{i=1}^{N} f_i \, w_i \qquad (2\text{-}23)$$

where w_i represents the percentage by weight of component i, and N is the total number of components in the system. *Figure 2-13* shows that the amount of extractable material predicted by equation 2-23, using DSC data (the solid line), compares reasonably well with the experimental values and the analysis of the individual components, revealing no drastic deviation from their "statistical" reactivity.

These results show that a consistent picture of the (partially) cured polymer can be obtained by a combination of various analytical methods focusing on reaction enthalpy, shrinkage, chemical composition, and mechanical strength. As the "green state" is particularly relevant to SL, special efforts should be made to elaborate methods that characterize these soft and residually photosensitive materials.

Extent of Shrinkage

Polymerization leads to an increase in density of the material since double bonds and van der Waals distances are converted to single bonds. The volume reductions associated with the conversion of 1 Mol of double bonds in a series

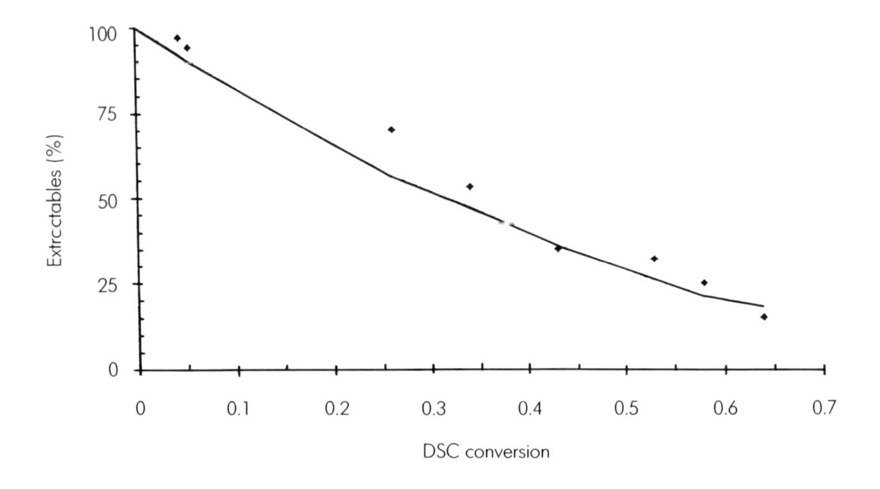

Figure 2-13. Percentage of extracted material from Cibatool XB 5081-1 as a function of conversion. Solid line is predicted from Photo-DSC.

of monofunctional (meth)acrylates cluster around 22 cm^3 as calculated from the densities of the monomers and the polymers. If polymerization is coupled with other chemical reactions (ring opening), the usual increase in density may be compensated for or at least reduced. Shrinkage can lead to serious problems such as distortions, cracks, or long-term instability in parts. Moreover, the inhomogeneous cure within a polymer string will be reflected in inhomogeneous shrinkage and may manifest itself as curl. As noted earlier, curl does not appear to be a simple function of gross shrinkage, but it will tend to be greater for resins with more gross shrinkage. Methods to quantify the extent of shrinkage as a function of energy are therefore needed, and some have been described previously.[23]

The density change of a material can be determined by the change in buoyancy during polymerization of a deformable sample of known weight immersed in a liquid of known density. A practical approach involves the exposure of a thin layer of the photopolymer between two thin UV transparent glass plates or mylar sheets separated by a precision spacer (typically 50 μm). The cell is suspended horizontally by very thin threads on a light-weight frame, which is fixed to an analytical balance. A beaker is placed on a separate support under the cell and is filled with the test liquid until the cell is barely immersed. After weight stabilization, the cell (approximately 20 mm x 20 mm) is scanned with the focused laser beam, and from the change in weight, the absolute shrinkage for a given incident energy is obtained quantitatively. Shrinkage data as a function of exposure for Cibatool XB 5081-1 is given in *Figure 2-14*.

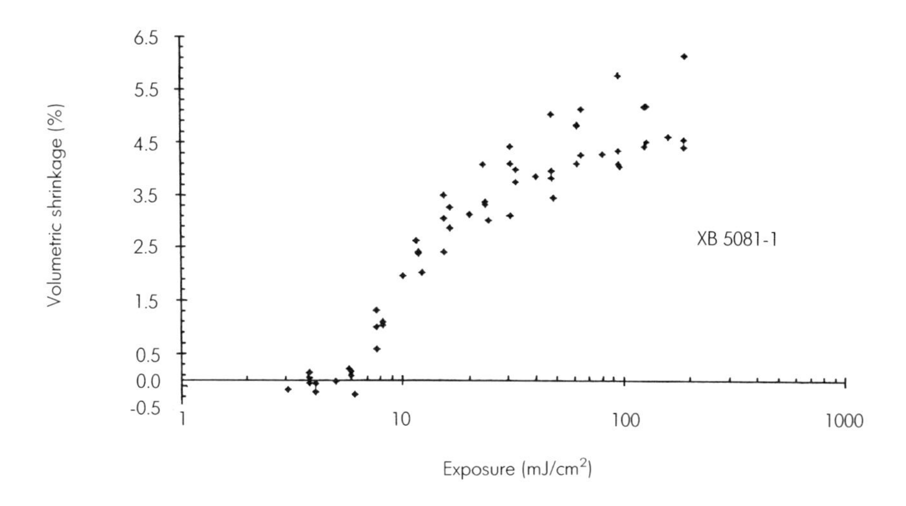

Figure 2-14. Volumetric shrinkage of Cibatool XB 5081-1 versus HeCd laser exposure, as measured by buoyancy method.

Above a threshold energy, which appears to be slightly below the value of E_c in the photosensitivity curve (see *Figure 2-2*), the density change sets in very sharply, implying that there may be some shrinkage below the gel point exposure. It can be concluded from a comparison with green strength measurements that *the shrinkage is virtually complete when the material just begins to gain significant green strength.*

A second experimental method uses a similar deformable cell which is placed on a stable support, but here the cover slide has a reflective coating. A HeNe laser beam is split into two beams, one of which is reflected at the metal coating. When the cell is irradiated from the bottom with UV laser radiation, shrinkage produces a vertical movement of the reflecting glass plate. Interference of the two HeNe beams, essentially forming a Michelson interferometer, produces a moving fringe pattern, which can be captured by measuring the intensity variation at a suitably placed photocell. The extent of shrinkage in the vertical direction is given by the number of fringes that have moved past the photocell.

This highly sensitive and elegant method enables the measurement of extremely small displacements. Further, the interferometric technique offers the possibility of measuring cure shrinkage on the same time scale as the laser exposure. This type of information is critical to understand curl distortion. At this stage, little is known about the behavior of SL resins at high power densities, but it would be clearly desirable to extend such measurements to this range. The interferometric setup, in conjunction with higher output lasers coupled with a fast data acquisition system, may allow actual resin shrinkage to be tracked on the millisecond time scale.

Kinetic Aspects

Thus far, we have concentrated on the static properties of SL resins; that is, the properties that are determined after polymerization has ended. There are several arguments on why kinetic aspects of the SL process are of interest:

1. The lifetime of a radical (or cation) should be short enough that diffusion processes over a distance comparable to the laser beam width can be neglected. Otherwise, the linewidths could grow beyond the region exposed and accuracy would be lost.
2. The residence time of the laser, which ranges from about 30 microseconds to 10 milliseconds, is in the same time scale as the chemical reaction.
3. Characteristics of the ''dark reaction.'' The longer the dark reaction, the lower the resolution.

A promising new tool to follow the decrease in double bond concentration by real time infrared spectroscopy (RTIR) has been described.[14] Studies at power densities up to 0.2 W/cm^2 were performed. This method offers good access to the kinetics of a photopolymerization reaction. Differences in reactivity of monomers, simple mixtures, the efficiencies of initiators, and the influence of oxygen can be studied using this method.

Faster methods are based on refractive index changes occurring during polymerization. The principles of generating holographic diffraction gratings are well known.[24,25] Recently, the method has been used to unravel mechanistic questions in polymerization reactions.[26] Results from Ciba-Geigy's studies on SL resins have been reported (see Reference 19).

The experimental setup consists of two interfering Argon UV (photo reactive) laser beams of equal intensity to generate the holographic diffraction grating, and an analyzing HeNe (nonreactive) beam, which is diffracted by the grating. The apparatus includes an appropriate rapid detection and data acquisition system to follow the time evolution of the diffracted intensity. The sample consists of a drop of resin between two glass plates separated by precision spacers (typically 40 μm). The intensity ratio, I_d/I_o of the diffracted HeNe beam (the diffraction efficiency) is given by:

$$I_d/I_o = \sin^2 (K \Delta n) \qquad (2\text{-}24)$$

$$K = \pi d_o/(\lambda_o \cos q)$$

where Δn, d_o, λ_o and q represent the amplitude of the refractive index variation in the resin, the sample thickness, the wavelength of observation, and the Bragg angle, respectively.

A typical hologram growth curve for Cibatool resin, XB 5081-1 is given in *Figure 2-15*. After an inhibition period, t_i, during which oxygen is consumed, the efficiency rapidly increases before reaching a maximum. The slope of this rise is a good measure of the cure rate of a resin since it is related to the change in refractive index, which changes as the density of the resin changes.

Access to information regarding the conversion of monomers is clearly less direct than in RTIR spectroscopy, but the simplicity of the experiment, especially the sample preparation, makes it a useful tool in kinetic studies, particularly where comparative information is sufficient.

Another potentially rapid technique to follow polymerization uses a suitable fluorescent probe dissolved in the polymerizing material.[27,28] The fluorescent behavior changes markedly as a function of viscosity. Application of this technique to SL resins would require a judicious choice of the fluorescence probe, since the absorption characteristics of the resin should not be altered by its presence.

2.5 Cationic Photopolymerization

Whereas radical photopolymerization is a well-established technology and a subject of intense scientific and industrial activities, cationic cure remains largely a subject of research. Few industrial applications currently exist. The development of cationic photoinitiators began only about 15 years ago. These initiators are capable of generating Lewis acids and Broenstedt acids efficiently.

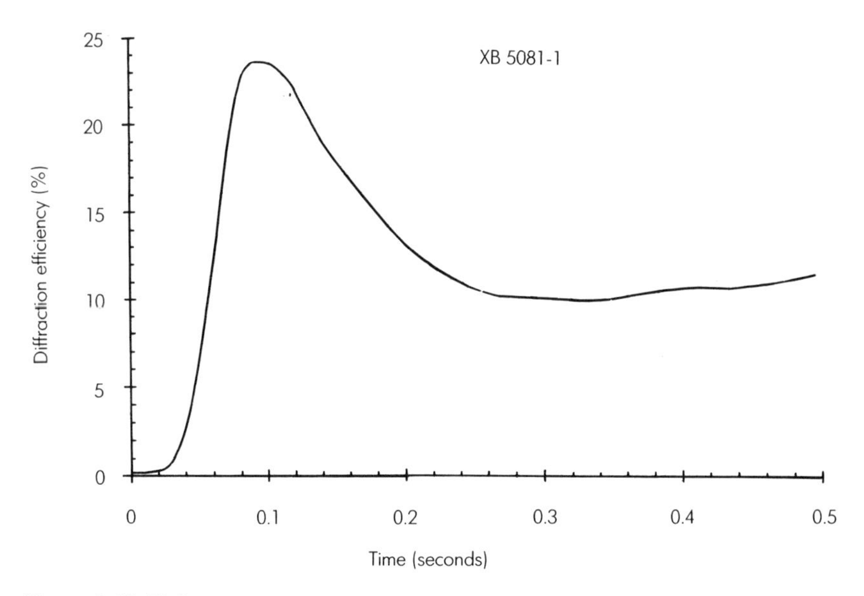

Figure 2-15. Hologram growth curve for Cibatool XB 5081-1. Incident power density 1.5 W/cm^2 at 351 nm.

In principle, cationic resins hold great promise due to their attractive physical properties, which are well known from epoxy compounds. The lower volumetric shrinkage (2% to 3%) of epoxy compounds is a potentially major advantage over acrylic formulations (5% to 7%).

Cationic Mechanism

Mechanistic studies have been undertaken with iodonium salts as cationic initiators.[29,30] For most cationic initiators, a scheme similar to that in equation 2-25 is proposed. The major products formed are an aryliodide and a diaryl species, the proton being abstracted typically from the solvent.

$$
\begin{aligned}
&(Ar_2\text{–}I)^+X^- &&\rightarrow [Ar_2\text{–}I^+\ X^-]^* \\
&[Ar_2\text{–}I^+X^-]^* &&\rightarrow (Ar\text{–}I)^+X^- + Ar\bullet \\
&(Ar\text{–}I)^+X^- + RH &&\rightarrow (Ar\text{–}I\text{–}H)^+X^- + R\bullet \\
&(Ar\text{–}I\text{–}H)^+X^- &&\rightarrow (Ar\text{–}I) + H^+\ X^- \qquad (2\text{-}25) \\
&2\ Ar\bullet &&\rightarrow (Ar\text{–}Ar) \\
&Ar\bullet + RH &&\rightarrow (Ar\text{–}H) + R\bullet
\end{aligned}
$$

The character of the anion (X-) plays no major role in the formation of the initiating species, since the quantum yields of a series of iodonium- and sulfonium salts are very similar (approximately 0.2 at 365 nm). Those photoinitiators with very low nucleophilic anions like ClO4-, BF4-, AsF6-, SbF6-, or CF3SO3- are suitable. Anions like Cl-, Br-, or I- would immediately quench the formed cation and inhibit polymerization.

Termination of the cationic polymerization process occurs primarily as a result of impurities in the formulation. Bases of all kinds, *including water* and other hydroxyl-containing compounds, are known to terminate or inhibit cationic polymerization. Other functional groups with easily abstractable protons, like alcohols, are known to act as transfer agents. Unlike radical polymerizations, *oxygen is not an inhibitor* for cationic polymerizations.

Hybrid systems take advantage of the best properties of both acrylate/methacrylate and epoxy systems. Photoinitiator systems have been proposed for concurrent radical and cationic polymerization.[31]

Cationic Initiators

Comprehensive lists of photoinitiators usable in cationic polymerizations have been published in several articles.[32,33,34] The only three classes of cationic photoinitiators commercially available are:

1. **Diaryliodonium salts.** These salts were introduced by 3M[35] and General Electric.[36] Due to their high quantum yield and initiation efficiency, they are used with most cationically curable monomers. Iodonium salts show only very weak absorbance above 350 nm. But, they can be efficiently sensitized into the longer wavelength range by special dyes like acridine orange, benzoflavin (see Reference 29), and anthracene and its derivatives, particularly 9,10-diethoxyanthracene.[37]
2. **Triarylsulfonium salts.** Several sulfonium salts are commercially available from companies like General Electric, Union Carbide, and Degussa. A mechanism analogous to that for the iodonium salts has been proposed for the photolysis of sulfonium salts. It has been possible to isolate the radical cation as a crystalline diarylsulfinium salt.[38] Most properties of these two classes of photoinitiators, such as quantum efficiency, reactivity against monomers, and sensitization are comparable.
3. **Cyclopentadiene-Fe-Arene complexes.** This new class of initiators has been introduced recently by Ciba-Geigy.[39] A commercially available example is Irgacure 261.

Cationic Monomers and Oligomers

Many monomers could, in principle, be polymerized cationically including acrylates, but two classes are actually used (epoxies and vinylethers). Other cyclic ethers like tetrahydrofurane or lactones cannot be polymerized at room temperature. Epoxies and vinylethers are discussed in the following.

Epoxies. This is the most versatile class of cationically polymerizable monomers. The availability of different multifunctional oligomers and low-viscosity monofunctional and multifunctional monomers is important for commercial formulations. These oligomers and monomers provide finetuning of the mechanical end properties. Oligomers like bisphenol-A-diglycidylethers, advanced compounds of the same structure, or epoxynovolacs are available from companies like Dow Chemical, Shell, Union Carbide, and Ciba-Geigy. These suppliers also have a wide selection of monofunctional and multifunctional monomers.

Vinylethers. Until recently, only monofunctional and multifunctional monomers were commercially available from the companies BASF and GAF. Since oligomeric compounds were not available, vinylethers were mostly used as reactive diluents in epoxy formulations where they increase the cure rates.[40,41] Recently, Allied Signal has introduced a complete line of reactive diluents and oligomeric urethane-vinylethers under the trade name Vectomer.[42] A Stereo-Lithographic resin based on vinyl ether monomers, Exactomer 2201, was introduced by Allied Signal in March 1992. This product is noted for its low curl and low viscosity. A potential disadvantage is the high critical exposure, about 25 to 50 mJ/cm^2, which requires that either the laser drawing speed be about four to eight times slower than typical acrylate resins, or correspondingly higher laser power is needed to maintain the same drawing speed.

A bisphenol-A based crystalline vinylether is also available from Nisso-Maruzen. Propenylethers show cure rates similar to vinylethers.[43] Vinylethers and iodonium and sulphonium salts or Fe-arene complexes show cure rates similar to those of acrylate polymerizations.

2.6 Conclusions and Outlook

By a very wide margin, the most commonly used materials in modern RP&M technologies are photopolymer systems. Unfortunately, all of the more common resins used in these systems experience shrinkage upon changing from liquid to solid. This results in stress and subsequently in strain deformations. To control this problem, materials with reduced shrinkage are currently being developed. However, part accuracy is also a function of how the laser energy is introduced into the photopolymer. New part building styles employing recently developed software algorithms (see Chapter 8) allow for more stress relaxation in each individual layer, and also cure the photopolymer to a higher average degree. This, in turn, reduces postcure shrinkage and the resulting postcure distortions (see Chapter 10). The mechanical properties of most current materials are well suited for concept modeling and for limited functional testing as well as for some casting applications, but further improvements are needed for many additional applications.

Future polymer developments are likely to include materials with improved impact resistance and higher elongation at break, resulting in less brittleness. Other features may be greater flexibility, improved photospeed, and increased green strength. Especially important is better overall part accuracy. Expanding the number of applications for RP&M may also require the development of resins with electrical conductivity, high temperature resistance, biocompatibility, or solvent resistance, to name some possibilities. In the past, rapid prototyping has gained significantly from developments made on photocurable materials for other applications. In the future, the situation may reverse. The diverse and demanding requirements of resins for RP&M will provide impetus for expanded use of photocurable resins in many other fields.

ACKNOWLEDGMENTS

The authors gratefully acknowledge the assistance of Paul Bernhard, Bernd Klingert, Adrian Schulthess, Manfred Hofmann, and Bettina Steinmann, all from Ciba-Geigy, Ltd., for their significant contributions to this chapter. In addition, we wish to acknowledge the kind permission of Chuck Hull, Hop Nguyen, Thomas Pang, and Kris Schmidt to use their previously unpublished results.

REFERENCES

1. Relsch, M.S., *C & E News*, October 14, 1991, p. 29.
2. Reiser, A., *Photosensitive Polymers*, John Wiley, New York, 1989, p. 102.
3. Mood, G. and Solomon, D.H., (1990) *Aust. J. Chem.* 43, 215.
4. Pappas, S.P., ed., *UV Curing, Science and Technology*, Vol. I, Technol. Marketing Corp., Stamford, CT, 1978, p. 2 ff.
5. Decker, C. and Moussa, K., (1990) *Makromol. Chem.* 191, 963.
6. Hoyle, C.E., Trapp, M., and Chang, C.H., (1987) *Polym. Mater. Sci. Eng.* 57, 579.
7. DuPont (1990) Eur. Pat. Appl. 0 393 674 A1; 0 393 675 A1; 0 393 676 A1 (DuPont).
8. Hoyle, C.E., Chawla, C.P., Chatterton, P., Trapp, M., Chang, C.H., and Griffin, A.C., (1988) *Polymer Preprints* 29 (1), 516.
9. Decker, C., (1987) *J. Coatings Technology* 59 (751), pp. 97-106.
10. Olaj, O.F., Bitai, I., and Gleixner, G., (1985) *Makromol. Chemie* 186, 2569.
11. Olaj, O.F., Bitai, I., and Hinkelmann, F., (1987) *Makromol. Chemie* 188, 1689.
12. Pascal, P., Napper, D.H., Gilbert, R. G., Piton, M.C., and Winnic, M.A., (1990) Macromolecules 23 (24), 5161-5163.
13. Flach, L. and Chartoff, R.P., (1990) *Proc. RadTech '90*, Chicago, March 25-28.
14. Decker, C. and Moussa, K., (1988) *Makromol. Chem.* 189, 2381; (1988) ibid 191, 963 (1990).
15. (a) Marutani, Y. and Nakai, T., (1989) *Laser Research* 17, 410. (b) Nakai, T. and Marutani, Y., (1988) *Trans. Inst. Electron. Inf. Commun.* J71-D, 416.
16. Hull, C., 3D Systems Memo, Feb. 25, 1990.
17. Korus, R. and O'Driscoll, K., *Polymer Handbook*, John Wiley & Sons, 1975, pp II-45.
18. Kloosterboer, J.G. and Lijten, G.F.C.M., *Cross-linked Polymers*, ACS Symp. Ser. 367, 1987.
19. Bernhard, P. and Hunziker, M., (1990) *Proc. 1st Nat. Conf. Rapid Prototyping*, Dayton, Ohio, June 4-5, pp. 79-85.
20. Provder, Th. (1989) *J. Coatings Techn.* 61, 33.
21. Williams, D.R.G., Allen, P.E.M., and Simon, G.P., (1989) *Materials Forum* 13, 108.
22. Kloosterboer, J.G. and Lippits, G.J.M., (1986) *J. Imag. Science* 30, 177.
23. Kloosterboer, J.G., van der Hei, G.M.M., Gossink, R.G., and Dortant, G.C.M., (1984) *Polymer Commun.* 25, 322.
24. Tomlinson, W.J. and Chandross, E.A., (1980) *Adv. Photochem.* 12, 201.

25. Braeuchle, C. and Burland, D.M., (1983) *Angew. Chem. Int. Ed. Engl.* 22, 582.
26. Carre, C., Lougnot, D.J., and Fouassier, J.P., (1989) *Macromolecules* 22, 791.
27. Valdes-Aguilera, O., Pathak, C.P., and Neckers, D.C., (1990) *Macromolecules* 23, 689, and references therein.
28. Paczkowski, J. and Neckers, D.C., (1991) *Macromolecules*, 24, 3013.
29. Crivello, J.V., Lam, J.H.W., (1978) *J. Polym. Sci*, 16, 2441-51.
30. Pappas, S.P., (1982) *Proc. Soc. Photogr. Lic & Eng.*, 22nd Fall Symp., Washington, DC, Nov. 15-18, 1982, p. 46.
31. Crivello, J.V., (1984) *Adv. Polym. Sci.*, Vol. 62, Springer Verlag, Berlin, Heidelberg, 1984, p. 3.
32. Crivello, J.V.; Lam, J.H.W., and Volante, C.N., (1977) *J. Radiat. Curing* 4, 2.
33. Green, G.E., Stark, P.P., and Zahir, S. A., (1981-82) *J. Macro. Sci.-Revs. Macro. Chem.* C21, 187.
34. Crivello, J.V., ed., *UV Curing, Science and Technology.*
35. Smith, G.H., (1975) Belg. Patent 828,841.
36. Crivello, J.V., (1976) USP 3,981,897.
37. Schulthess, A., unpublished results, Ciba-Geigy AG, Fribourg, Switzerland.
38. Mani, J.R. and Shine, H.J., (1975) *J. Org. Chem.* 40, 2756.
39. Meier, K. and Zweifel, H., (1986) *J. Imaging Sci.* 30, 174.
40. Dougherty, J.A., Vara, F.J., and Anderson, L. R., (1987) *Polym. Paint. Colour* 177, 593.
41. Lapin, S.C. and Olivares, M., (1989) US 4,845,265.
42. Brautigam, R.J., Lapin, S.C., and Snyder, J.R., (1990) *Radtech '90 Conf. Papers*, Radtech Intl., Chicago.
43. Crivello, J.V., Lee, J.L., and Conlon, D.A., (1988) *Makromol. Chem. Macromol. Symp.* 13/14, 145.

chapter 3

Lasers for Rapid Prototyping & Manufacturing

And God said, Let there be light: and there was light.
And God saw the light, that the light was good.

GENESIS 1:3

3.1 Introduction

A variety of technologies are presently competing in the rapid prototyping systems market. Some of these new technologies utilize lasers to expose photopolymers, sinter particulates, or cut nonphotosensitive materials. All rapid prototyping systems form solid parts as the result of sequential stacking of thin two-dimensional image laminae. Most of these technologies involve lasers and beam steering devices as well as computers, image generation, and manipulation software. They owe much to the laser printer and color separation systems of the early 1980s, when many of the present two-dimensional, digital image generation concepts were developed.

The focus of this chapter is the development of lasers specifically for SL systems. At present, about 90% of all Rapid Prototyping & Manufacturing (RP&M) systems worldwide are SLA units designed, developed, tested, and sold by 3D Systems, Inc. Roughly, another 5% are systems derived from SL Technology (such as those produced by Sony, Mitsubishi, and EOS), which also use helium-cadmium or argon ion lasers. The only RP&M processes that do not

*By **William F. Hug, Ph.D.**, CEO & President, Omnichrome Corp., Chino, CA.*

use either of these lasers employ carbon dioxide lasers (DTM and Helisys) or no lasers at all (Cubital, Stratasys, Perception Systems, and Light Sculpting). These systems cumulatively account for only about 5% of total RP&M system sales at this time. Accordingly, the major emphasis of this chapter will be directed toward the lasers used by the majority of current operational systems.

Lasers were developed in the late 1950s. However, it is only since about 1973 that commercial products emerged using these sources. Since that time, lasers of adequate reliability and diversity have evolved to the point where they are acceptable for commercial applications. This evolution continues as these sources become more reliable and as additional requirements arise demanding the special properties provided by lasers. Since about 1986, a growing number of applications, together with the introduction of SL, provide the economic drivers for development of commercially practical lasers in the ultraviolet. These are necessary for exposing the photopolymer resin media which form the basis of SL. Laser action has been demonstrated in a large fraction of the elements of the periodic table.[1] Very few elements, however, have received much attention beyond initial demonstration of oscillation. In part, this is the result of poor performance or theoretical prospect discouraging further development of lasers based on many of these elements. Lasers such as krypton, neon, frequency tripled neodymium-YAG, and direct frequency doubled diode lasers are being evaluated. However, helium-cadmium and argon lasers are currently the lasers of choice for SL.

3.2 The StereoLithographic Exposure Process

The key to the universality of the SL system design concept is the ability to rapidly direct focused radiation of appropriate power and wavelength onto the surface of a liquid photopolymer resin, forming patterns of solidified photopolymer. *Figure 3-1* is a block diagram showing the elements of a direct laser writing StereoLithographic system. The beam of a continuous wave laser emitting radiation of power P_L at a suitable wavelength, λ, is sent through a beam-expanding telescope to fill the optical aperture of a pair of cross-axis, galvanometer driven, beam scanning mirrors. The beam comes to a focus at a distance L, on the surface of a liquid photopolymer where the $1/e^2$ beam radius is W_o. To obtain high speed, the galvanometer driven scanning mirrors must be of small inertia and small size. The source of radiation must have very high radiance to provide a tightly focused spot on the surface of the photopolymer located at a substantial distance. This explains the need for a laser.

Time required to generate a part depends on the spectral responsivity of the photopolymer, speed of the galvanometer mirrors, power and wavelength of the laser, and the recoating time of each lamina, which depends on the viscosity and surface tension of the photopolymer and the characteristics of the

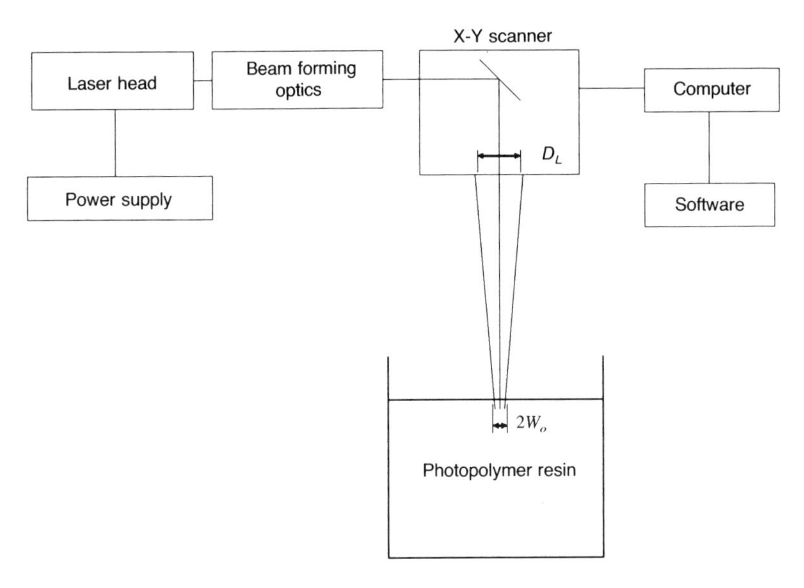

Figure 3-1. Elements of a stereolithographic system.

recoating system. The minimum feature size of a part is determined by the diameter of the focused laser spot on the photopolymer surface. The resolution of a part depends on the angular accuracy of the galvanometers, the ability to control the intensity distribution, and the laser power.

The solidification of the liquid resin depends on the optical energy per unit area or "exposure" deposited during motion of the focused spot on the surface of the photopolymer. Moving with a scan velocity V_s on the surface of the liquid photopolymer, the laser beam with power P_L deposits a peak exposure, E_{max}, which is given by (see derivation in Chapter 4):

$$E_{max} = \sqrt{\frac{2}{\pi}} P_L / W_o V_s \tag{3-1}$$

Figure 3-2 shows the cure depth, C_d, versus the maximum exposure, E_{max}, for Ciba-Geigy photopolymer XB 5081-1 being exposed at a wavelength of 325 nm, using a helium-cadmium laser. Several features of this figure are important. First, there is a threshold exposure, E_c, at which the photopolymer transitions from the liquid to the gel phase. Second, the 1/e penetration depth of the laser beam into the liquid resin surface, D_p, is independent of exposure. Third, the depth of photopolymer cured by the laser exposure can be derived from the Beer-Lambert law (see Chapter 4):

$$C_d = D_p \ln (E_{max}/E_c) \tag{3-2}$$

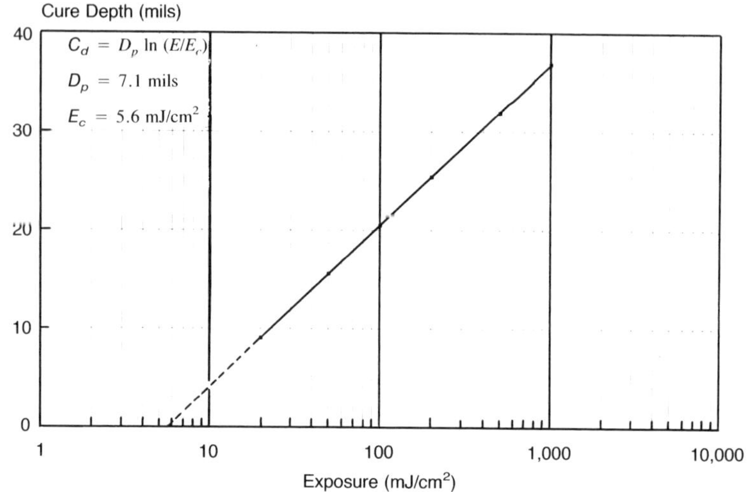

Figure 3-2. Cure depth versus exposure.

This equation describes the "working curve" for a given photopolymer and laser combination since D_p and E_c are constants for a given resin and output wavelength of the laser.

Figure 3-3 shows the specific energy, E/C_d, (the amount of laser energy required to solidify a unit volume of resin) as a function of exposure (mJ/cm^2) on the surface of the photopolymer. The specific energy is a measure of the effectiveness of the laser beam at solidifying resin. The most important feature of this figure is that there exists a minimum specific energy, indicating an exposure value for which part generation efficiency is optimized. This optimum value occurs, for this resin, at an exposure of about 15 mJ/cm^2 when $C_d = D_p$. For XB 5081-1, D_p equals 7.1 mils or 180 microns.

To maintain accuracy and consistency during StereoLithographic part formation, the cure depth and the cured linewidth must be controlled with great accuracy, as discussed in Chapters 8 and 10. Accurate exposure and focused spot size are essential. The ultimate system resolution is determined by control of the edge of a focused spot. These parameters are coupled in the basic scanning equation 3-1, which establishes the relationship between scan velocity, laser power, and laser beam focusing capability. Pulsed lasers can achieve adequate control of energy deposition if the pulse frequency is high enough to provide a large number of pulses per resolution element. At typical SL scan speeds, this implies pulse repetition rates in excess of 25 kHz and limits the appropriate type of laser to continuous wave (CW) lasers at the present time.

Figure 3-4 shows the wavelength dependence of the specific energy for two different ultraviolet photopolymer resins. The wavelength dependence is primarily determined by the choice of photoinitiator. A number of these materials are

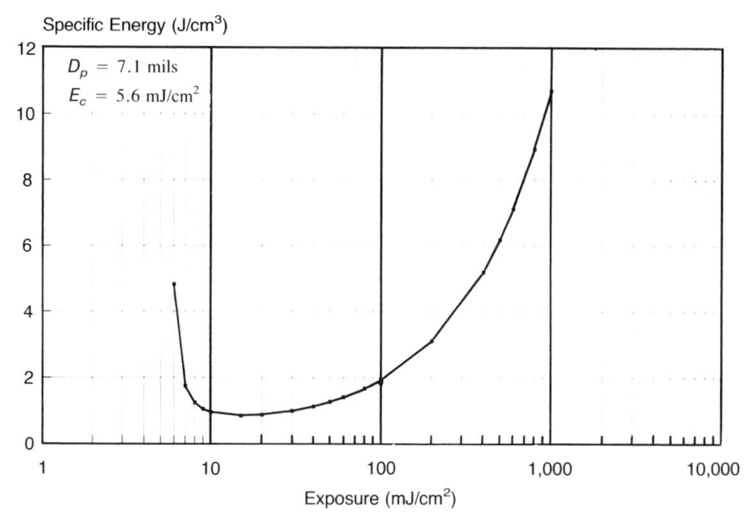

Figure 3-3. Specific energy versus exposure.

presently used for the manufacture of StereoLithographic resins. While it is possible to adjust their concentrations to achieve response at other wavelengths, the most effective photoinitiators presently have superior response in the ultraviolet.

It currently appears that ultraviolet sensitive photopolymers will dominate this technology for the next five years. This is predicated on the issues that arise with visible photopolymers such as:

- The need for safelight handling of visible photopolymers;
- The decreased stability and increased cost of visible photopolymers;
- The greater difficulty of achieving good batch to batch consistency of D_p and E_c, which leads to problems with advanced part building techniques.

As we learn about the exposure process, it is becoming increasingly clear that more raw power from lasers is not the best answer to faster, higher quality part building. The presumed advantage of greater laser power from visible lasers compared to UV lasers is not as important as originally thought for four reasons:

1. The trend toward thinner layers for greater part resolution has significantly reduced laser power needs.
2. New part building techniques utilize reduced cure depths.
3. The exposure varies exponentially with cure depth. Therefore, the laser power requirement decreases exponentially with reduced cure depth.
4. Once it is recognized that lower exposure values are often adequate, higher laser power can only be utilized if much higher scan speed galvanometers become available, or if future photopolymer sensitivity decreases.

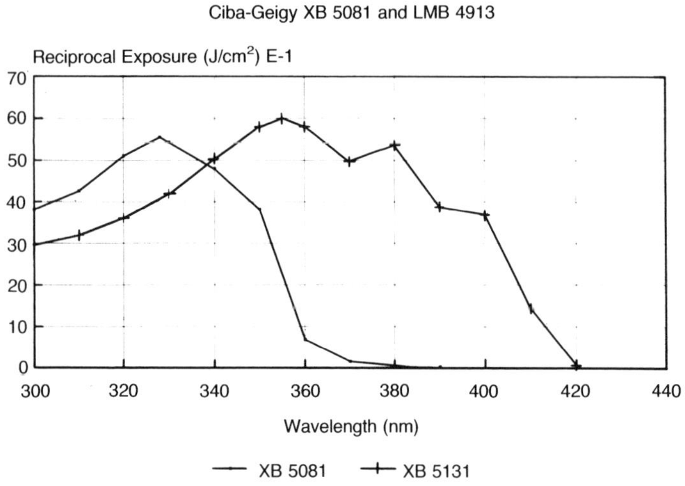

Figure 3-4. Reciprocal exposure versus wavelength.

3.3 Background of Lasers for RP&M

Lasers come in a variety of forms including solid, liquid, and gas. Each basic type of laser is usually identified by the material within it, which provides the active light amplifying medium. In the case of argon lasers, the amplifying material is argon gas. In helium-cadmium lasers, the amplifying material is a mixture of helium gas and cadmium vapor. Each of these lasers is available in a variety of configurations to satisfy the needs of a specific application.

The performance characteristics of a laser include: output power, wavelength, beam diameter, beam divergence, mode shape, optical noise, oscillation bandwidth, input power, ruggedness, reliability, and lifetime to some performance specification. Important financial characteristics include acquisition and life cycle cost. All of these characteristics are interrelated in a complex manner with the design and operating variables of a laser. These include the choice of laser medium, geometric variables, and operating parameters such as gas and/or metal vapor pressures, laser tube currents, and voltages.

Gas lasers are presently the primary sources of ultraviolet laser radiation. A gas laser system generally consists of three parts or subsystems:

1. A laser head consisting of a gas discharge tube providing optical gain for the desired emission wavelengths, and a resonator structure which supports and maintains alignment of the mirrors at each end of the laser tube.
2. A heat exchanger for dissipating the waste heat generated by the laser tube.
3. A power supply for igniting, exciting, and regulating the electrical current as well as the gas and metal vapor pressures within the laser tube.

The interaction between the various parts of the system establishes the ultimate performance of a laser. The most important element in the system is the gas discharge tube or plasma tube. This tube determines the essential performance characteristics of a laser and is housed in the laser head along with a resonator structure.

The most common type of tube is a positive column gas discharge laser tube generally consisting of a long narrow cylindrical capillary with the anode near one end and the cathode near the opposite end. Typical gas discharge laser tubes have a reservoir of gas and/or metal and means for sensing and controlling the pressure within the tube. The capillary is filled with a mixture of gases and/or metal vapors whose atoms are excited to a variety of energy states by electrons flowing from the cathode to the anode. To produce optical gain and laser action, the excitation cross sections and radiative lifetimes of energy states within the excited gas and/or metal vapor atoms inside the capillary must be such that the steady state population density of the upper laser energy states is much larger than that of the lower laser energy states. This population inversion produces optical gain and laser action, typically on radiative transitions between several pairs of upper and lower states. The laser output wavelengths depend on the choice of gas, the operation conditions, and the use of wavelength-selective laser mirrors.

The optical radiation generated within the gas discharge created in the capillary is directed repeatedly through the discharge region by laser mirrors at each end. The gas discharge constitutes the laser gain medium, which amplifies the radiation to produce the desired laser output. The laser resonator, including the laser mirrors, provides the feedback to produce laser oscillation at the desired wavelength or wavelengths. The radii of curvature and spacing between the laser mirrors as well as the diameter, length, and position of the discharge capillary determine the intensity distribution or transverse mode structure and focusing capabilities of the laser beam.

The output power, P_L, of a laser is given approximately by the relation (see Reference 2):

$$P_L = A_m \, p_s \, T \, [(2k_g \, L_b/d(a + T))-1] \tag{3-3}$$

where A_m is the average laser beam cross sectional area within the active portion of the gas discharge capillary, p_s is the gain saturation parameter shown in Table 3-1, T is the transmission of the output laser mirror at the wavelength of interest, k_g is the gain coefficient given in Table 3-1, L_b is the length of the active portion of the gas discharge capillary, d is the diameter of the gas discharge capillary, and a is the total scatter and absorption losses of laser mirrors, brewster angle windows, and limiting optical apertures within the laser cavity.

Table 3-1
Optimum Laser Parameters[3]

Optimum Values	Helium-Cadmium (at 325nm)	Argon (at 351/364nm)
Pressure Parameter (pd)	9.0 Torr-mm	1.2 Torr-mm
Tube Current Parameter (I/d)	.05 A/mm	30 A/mm
Tube Voltage Parameter (Vd/L_b)	7.0 volts	2.0 volts
Input Power Parameter (W/L_b)	3.5 W/cm	550 W/cm
Typical Output Power (W/L_b)	.10 W/m	5.0 W/m
Gain Parameter (k_g)	1.4×10^{-4}	5×10^{-4}
Gain Saturation Parameter (p_s)	700 mW/mm^2	1500 mW/mm^2

The values of the above parameters are identified only for optimum output conditions of each laser. Because of the low heat dissipation of helium-cadmium lasers, these lasers are typically operated at or very near optimum conditions. However, due to excessive heat dissipation and its deleterious effect on lifetime, argon lasers are typically operated at less than optimum values.

Helium-Cadmium Lasers

Helium-cadmium lasers were originally demonstrated in the late 1960s at visible and ultraviolet wavelengths. The visible and ultraviolet laser lines of cadmium are excited by a process called Penning ionization. This process involves electron excitation of metastable states of the neutral helium atoms. Through collisions with neutral cadmium atoms, the excited helium atoms ionize and excite the upper state of the cadmium ion from which the 441.6 nm visible and 325.0 nm ultraviolet laser lines emanate. This helium to cadmium collisional excitation process is relatively efficient and results in lasers that operate at low power densities, low discharge currents, and low temperatures compared to argon lasers. There is, however, the added complexity of regulating the flow and distribution of cadmium vapor within the laser tube.

The excitation mechanism in a helium-cadmium laser is a positive column glow discharge. Other tube configurations such as the hollow cathode discharge have been developed, producing a variety of laser emission lines. These have not been developed adequately for commercial application at the present time, however.

A typical positive column helium-cadmium laser tube, as illustrated in *Figure 3-5*, has a capillary diameter of about 2 mm (0.08″) and operates at an optimum laser tube current of about .090 amperes and an input power per unit of capillary length of about 3.5 W/cm. The helium pressure in the tube is typically about 4 mm of mercury, or Torr, and the cadmium vapor pressure in the capillary is about 5 milli-Torr. This compares to a typical ultraviolet argon laser with a capillary diameter about 3 mm (0.12″), which operates at an optimum laser tube current of about 90 amperes, an input power density about 550 W/cm, and an argon pressure less than 1 Torr. The output power of a typical 60 cm long helium-cadmium laser is about 50 mW at 325 nm for an input power to the laser discharge of 250 watts. The 442 nm visible output is between three and four times greater for a given size laser. Optimum operating conditions of helium and cadmium pressure and tube current are nearly the same for visible and ultraviolet output.

The output power and optical noise of helium-cadmium lasers are sensitive to the operating conditions of the laser including helium pressure, cadmium pressure, and laser tube discharge current. Each of these parameters must be independently regulated to maintain optimum performance and long lifetime. Discharge current is typically regulated by the power supply to remain constant. The helium pressure is regulated using an independent feedback control subsystem consisting of a pressure sensor inside the laser tube and a controllable source of helium. Cadmium vapor pressure is regulated using the laser tube discharge voltage as a measure of the cadmium pressure.

Helium pressure regulation is necessary since this gas is typically lost during operation of a laser due to inevitable permeation through the glass walls of the laser tube. As shown in *Figure 3-5*, helium is stored at high pressure (about 2,280 Torr or 3 atmospheres) in a sidearm reservoir. This reservoir is separated from the main body of the laser tube by a temperature-dependent permeable glass membrane. When the pressure sensor indicates helium pressure below a preset regulation value, a heater is turned on which raises the temperature of the permeable membrane located between the helium reservoir and main tube body.

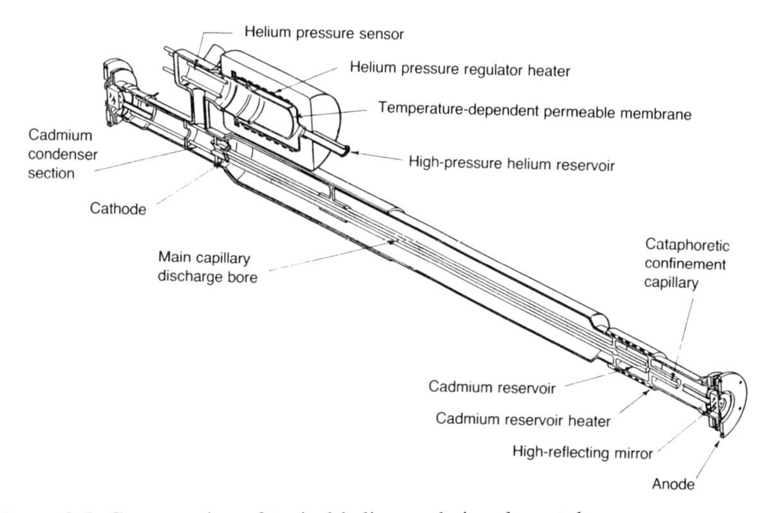

Figure 3-5. Cross-section of typical helium-cadmium laser tube.

This allows helium to flow at a very slow rate from the reservoir into the main tube body until the pressure regains its regulation value.

Cadmium pressure sensing and regulation is even more complex. Cadmium is a solid metal at the normal operating temperatures of these lasers. The metal is stored in a reservoir near the anode end of the laser tube. When a helium-cadmium laser is turned on, a discharge is ignited through the main capillary, initially filled with helium gas only. A heater surrounding the cadmium reservoir is turned on, heating the solid metal stored within. As the cadmium sublimes and its vapor pressure increases, the metal atoms diffuse into the discharge region. The consumption rate of cadmium is about one milligram per hour, and these lasers are normally supplied with a quantity sufficient for over 10,000 hours of operation. As the vapor pressure in the discharge capillary increases, the tube voltage decreases. This voltage is a very sensitive measure of the cadmium pressure in the capillary, with approximately a two volt change corresponding to a 0.05 °C (0.09°F) change in cadmium temperature. A closed loop regulation circuit which utilizes this inverse tube voltage sensitivity is the basis of the cadmium vapor pressure control system.

Figure 3-6 shows the output power and optical noise of a typical helium-cadmium laser as a function of tube voltage. Optimum output power occurs when the tube voltage is approximately 75% of the pure helium tube voltage which occurs on startup of a laser. Normally, it takes about three to five minutes for the tube temperature and cadmium vapor pressure to come to equilibrium during the warm-up of a laser. The optical noise indicated in *Figure 3-6* normally occurs at relatively high frequencies (250 kHz) and is of little consequence in SL.

Helium-cadmium lasers predominantly use cadmium of natural isotopic abundance. The use of single isotope cadmium, such as Cd(114), can signifi-

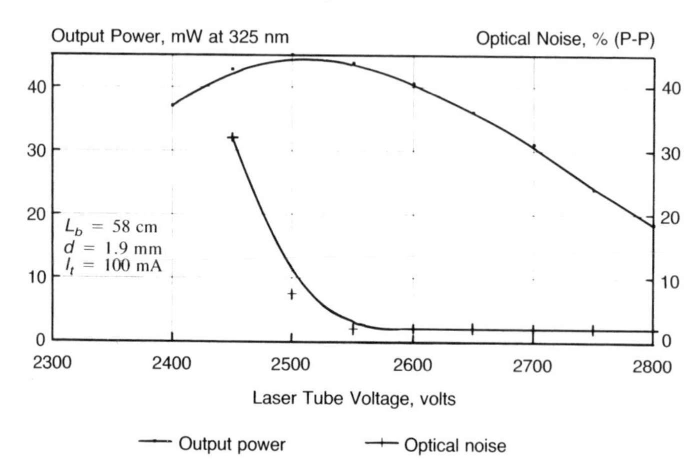

Figure 3-6. UV HeCd output power and optical noise versus tube voltage.

cantly increase the output power of these lasers. Theoretical estimates of the potential increase in P_L are between two and four times. This would enable output levels over 200 mW from existing laser configurations and sizes. Unfortunately, the cost of single isotope cadmium is prohibitive, being nearly $4000 per gram, or about $40,000 per laser tube. Alternate methods of isotopic refinement are being evaluated with the prospect of future cost reductions.

Argon Lasers

Argon lasers were originally demonstrated in 1964 for visible emission lines and in 1965 for ultraviolet emission lines. The argon laser is normally described as an arc mode gas discharge because of the high current and power densities required to excite the upper states of the laser transitions.

The principal ultraviolet lines of this laser, at approximately 351 nm and 364 nm, issue from doubly ionized argon (ArIII). The excitation process for these lines requires high tube currents and/or solenoidal magnetic field strengths to produce highly energetic electrons. These high energy electrons are required to populate the upper energy states of the ultraviolet emitting lines.

Figure 3-7 shows the output power of a typical ultraviolet and visible argon laser as a function of the input power to the capillary discharge. As can be seen in this figure, the excitation processes in argon lasers are inefficient, especially for the ultraviolet emission lines. This positive column ultraviolet argon laser has a capillary diameter of 2.3 mm (0.09″) and cannot be operated at optimum output because of the excessive tube current required. Threshold current for lasing occurs for visible laser lines at very low tube currents (typically, about 5 amps with input power about 1 kW). For ultraviolet laser lines, the threshold occurs at tube currents of about 30 amps with input power about 10 kW. Although the output power increases rapidly above threshold, the high threshold current and substantial input power level establish the need for special materials to dissipate waste heat from the discharge at nearly all operating levels.

The output power of argon lasers is normally regulated using the tube current, since the tube voltage is nearly independent of tube current. Although gas pressure is often not regulated in visible argon lasers, this is done in UV versions because of the rapid argon pressure loss occurring at the higher tube currents needed for UV operation.

All commercially available UV argon lasers use "cold disc" laser tube configurations. Visible argon lasers use either cold disc or beryllia capillary laser tubes. Both of these constructions are necessary to dissipate the large quantities of waste heat generated in the discharge capillary, especially in UV configurations of these lasers.

Carbon Dioxide Lasers

Carbon dioxide lasers were originally demonstrated in 1964 and emit in the deep infrared near 10,600 nm. Carbon dioxide lasers are most commonly excited

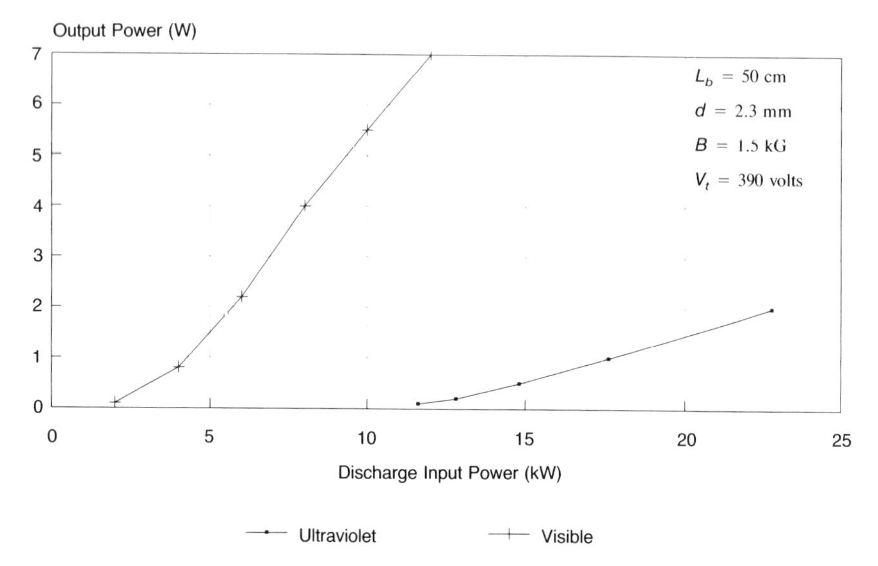

Figure 3-7. Argon laser output power: ultraviolet and visible.

either by direct electrical discharge or an rf source. They are among the highest efficiency of all laser types, with typical wall plug efficiencies between 5% and 10%. For this reason, they have received considerable development as a source of radiation for cutting, welding, and sintering of materials, and have come into widespread usage worldwide.

Present Status of Lasers for RP&M

Ultraviolet photopolymer resins dominate stereolithographic and other rapid prototyping applications. These resins are presently exposed using helium-cadmium lasers with output power levels up to about 40 mW at 325 nm and argon lasers with output power levels up to about 400 mW divided between 351 nm and 364 nm.

RP&M systems that employ direct laser sintering of particulates or cutting of patterns use exclusively CO_2 lasers. DTM uses a 25 W CO_2 laser for sintering particulate materials together in patterns similar to SL. Helisys uses a 50 W CO_2 laser to cut sheet material into patterns, which are then stacked and adhered to form parts.

The key advantages of helium-cadmium lasers are low power consumption, long lifetime, and low acquisition, installation, and operating costs. The disadvantage is low output power. Improvement of the UV output of helium-cadmium lasers has been dramatic. As recently as 1986, the highest UV output from a commercial helium-cadmium laser was about 2 mW. Today, these lasers are available with output up to about 50 mW. Values of over 100 mW at 325 nm have been demonstrated in the laboratory.

The advantage of argon lasers is higher UV output power. The UV output from available lasers exceeds one watt. Higher power consumption, shorter lifetime, and higher acquisition, installation, and operating costs are the disadvantages.

CO_2 lasers are very high in efficiency, have high power capability and low acquisition, installation, and operating costs. Although not useful for exposure of photopolymer resins, these lasers are effective in providing the focused heating necessary for sintering and cutting applications. The disadvantage of these lasers is their long wavelength and inability to focus as tightly as shorter wavelength lasers. Table 3-2 describes the characteristics of available helium-cadmium, argon, and CO_2 lasers. Helium-cadmium lasers are used exclusively in SLA-250 systems, and argon lasers are used exclusively in SLA-500 systems.

Table 3-2
Laser characteristics

Laser Type	HeCd	Argon	CO_2
Emission Wavelength (nm)	325	351, 364	10600
Power Output Range	20-40 mW	100-500mW	25-50 W
Power Input Range (kW)	0.6–0.8	10–20	0.3-0.6
Cooling Requirements	Air (100 CFM)	Water (20GPM)	Air
Laser Head Size (cm)	15x18x84	18x13x114	8x8x50
Power Supply Size (cm)	33x10x36	43x15x48	20x20x30
Laser Acquisition Cost (k$)	15	45	15
Laser Installation Cost (k$)	0	15	0
Operating Cost Per Month			
-Laser Tube Wearout (k$)	3	12	3
-Electricity Cost (k$)	0.05	1	0.05

3.4 Laser Beam Propagation and Focusing

To the casual observer, a laser appears to emit a thread-like beam of light with a uniform cross section. A detailed map of the transverse or cross-sectional intensity distribution of a typical laser beam shows, however, that the beam may consist of a single beam called a Gaussian or TEMoo beam, or a collection of sub-beams closely spaced together or concentric with each other and usually overlapping each other.[3,4,5,6] These beams are generally referred to as multi-mode beams. Laser beam writing applications such as SL do not require TEMoo beams since the focused spot size does not need to be extremely small.

Figure 3-8 shows the theoretical transverse intensity distribution of 12 of the lowest order cylindrical mode possibilities of a laser.[7] The intensity, I, at any radius, r, and azimuth, θ, from the optical axis is proportional to:

$$I \propto \left(\frac{2m \, ! \, cos^2 \, (n \, \theta)}{\pi(m \, + n)!(1 + \delta) \, W_o^{\ 2}} \right) (2u^2)^m \left[L_m^n \, (u) \right]^2 e^{-2u^2} \tag{3-4}$$

where $u \equiv r/W_o$ *and* $\delta = 1$ *if* $m = n = 0$, $\delta = 0$ *if* $m, n > 0$, W_o is the $1/e^2$ radius of the Gaussian beam, and L_m^n is the associated Laguerre polynomial of order m and index n. Not shown are the ''starred'' or ''donut'' modes. In actual lasers, the transverse intensity distribution is rarely as perfect as illustrated in *Figure 3-8*. Typical transverse intensity distributions are a superposition of modes up to and including the highest mode allowed by the limiting aperture within the laser tube. This superposition of modes causes a smoothing effect on the intensity profile. The transverse intensity distribution of any of these modes is independent of position along the laser beam axis.

The lowest order mode ($m=0$, $n=0$) is the TEMoo or Gaussian mode. This and higher order, TEMmn, modes can be produced in relatively endless variety in a given laser type by controlling the optical losses within the laser cavity between the laser mirrors. These losses are introduced by means of an optical limiting aperture within the laser tube: *the smaller the aperture the closer to TEMoo performance*; the larger the aperture the higher the degree of multimode structure.

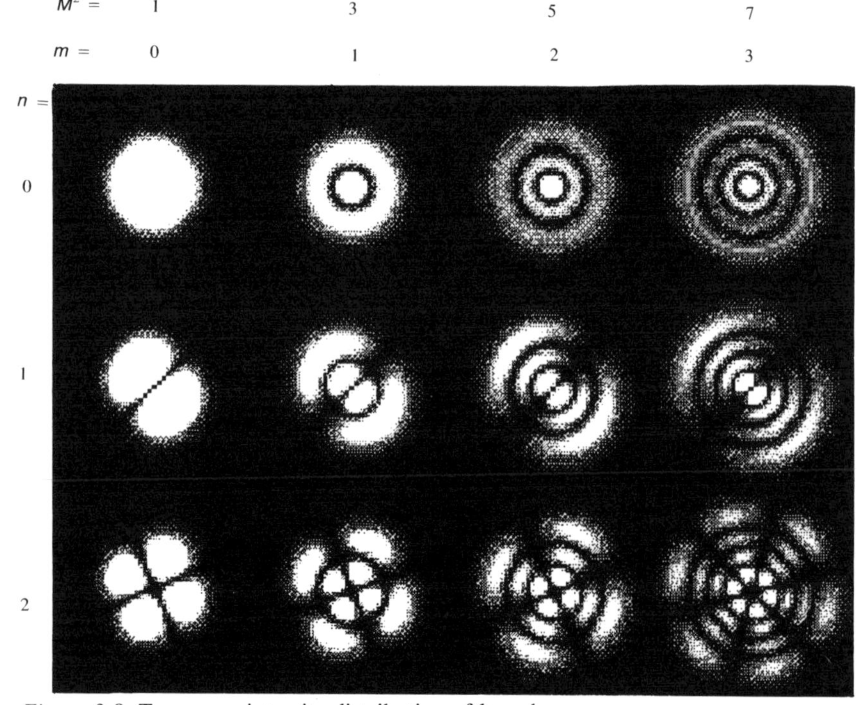

Figure 3-8. Transverse intensity distribution of laser beams.

Illustrated in *Figure 3-8*, for each mode type, is the beam diameter ratio, M = Dmn/Doo. This ratio describes the actual $1/e^2$ diameter of a multimode laser beam compared to the diameter of the "embedded" TEMoo beam. The TEMoo beam would occur if the limiting aperture were adequately small to allow it to dominate. The ratio M, called the *mode purity parameter*, can be determined by measuring the $1/e^2$ radius, W, at the exit of the laser head and the full angle divergence of the laser beam, Φ, and using the equation:

$$M^2 = \pi W\Phi/2\lambda \tag{3-5}$$

where λ is the wavelength of the laser beam. The mode purity parameter, M, describes both the approximate intensity distribution of a laser beam as illustrated in *Figure 3-8* and its full angle divergence. *Figure 3-9* shows photographic enlargements of two ultraviolet helium-cadmium laser transverse intensity distributions: *Figure 3-9A* with $M^2 = 3.32$ and *Figure 3-9B* with $M^2 = 8.84$.

The mode purity parameter for a laser is determined by the aperature ratio, d/W_{oo} where d is limiting aperture diameter and W_{oo} the TEM_{oo} radius at that limiting aperture. *Figure 3-10* shows M versus the aperture ratio. As the aperture ratio of a laser tube design is made smaller, the TEMoo mode purity increases (M approaches 1). Unfortunately, the output power of the laser decreases due both to increased diffraction losses within the laser and decreased utilization of the optical gain volume in the gas discharge capillary. The laser tubes in *Figure 3-9A* and *Figure 3-9B* have an aperture ratio of 5.26 and 7.65 respectively.

Two advantages of using lasers with multimode beams are as follows:

1. A multimode laser can provide several times the output power of a Gaussian mode laser of equal size or length because the larger mode diameters make more effective use of the gain medium of a laser tube.
2. The output power of a multimode laser is less sensitive to mechanical misalignment of its resonator mirrors.

Figure 3-11 shows a comparison of the beam divergence at the focus of a lens for the case of a multimode and TEMoo mode laser. The higher the value of M, the higher the divergence away from a waist. Illustrated here is the important but not well understood fact that *a multimode laser can be focused to the same diameter spot as a Gaussian mode laser*. However, the multimode laser will have a larger divergence and require a higher numerical aperture focusing lens.

The general equation for the radius, W_o, of a beam focused by a lens with focal length f is given by:

$$W_o = Wf/[X^2 + Z_R^2]^{1/2} \tag{3-6}$$

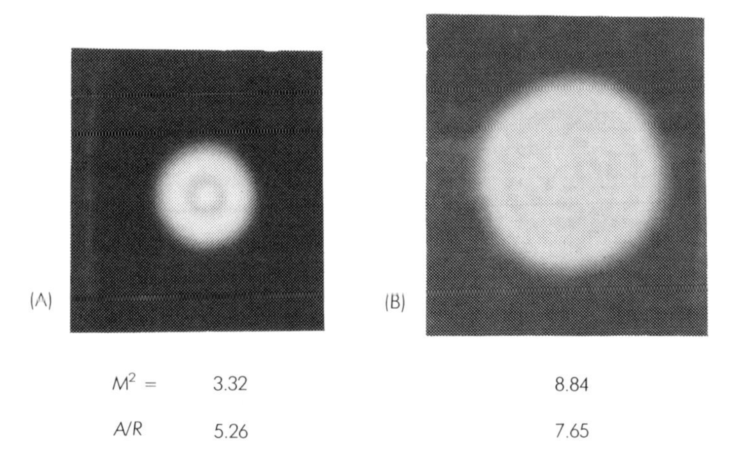

Figure 3-9. Photographs of two transverse intensity distributions.

where W is the primary waist radius of the beam before the lens, X is the axial distance between the primary waist W and the focal point of the focusing lens, and Z_R is the Raleigh range around the primary waist defined as:

$$Z_R = \pi W^2/M^2 \lambda \tag{3-7}$$

The Raleigh range is the axial distance away from a waist for which the beam radius expands to 1.4 times the waist radius or, equivalently, the beam cross-sectional area doubles. Equation 3-6 reduces to a familiar simplified equation for laser beam focusing for the case where $X = 0$:

$$W_o = \Phi f = 2\lambda M^2 f/\pi W \tag{3-8}$$

In the case of SL, the focused spot radius, W_o, is limited by the diameter of the galvanometer mirror aperture, the distance to the photopolymer surface, and the mode purity parameter, M. The equation governing this relationship is:

$$W_o = 2\lambda M^2 L/\pi D_L \tag{3-9}$$

where L is the distance from the galvanometer aperture to the surface of the liquid photopolymer and $D_L = 2W$ is the diameter of the galvanometer mirror limiting aperture. For the case of the SLA- 250: $L = 652$ mm (25.7"), $D_L = 4.1$ mm (0.16"), $M^2 = 3.2$. Therefore, the spot radius at the photopolymer surface, $W_o = 0.1$ mm (0.004"). A TEMoo laser with $M = 1$ would be able to provide a focused spot radius of 0.03 mm (0.0012") using the same 4.1 mm galvanometer limiting aperture. However, because of the scan speed limitations, and the fact that the maximum exposure scales as $P_L/W_o V_s$, the output power would need

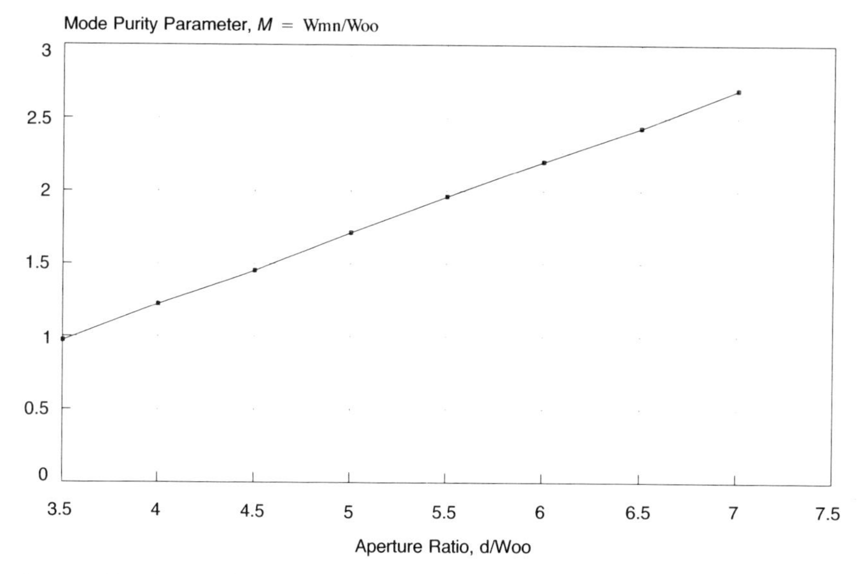

Figure 3-10. Mode purity parameter versus aperture ratio.

to be reduced by a similar factor of 3 to maintain the same exposure and the same cure depth.

In the case of direct laser sintering or pattern cutting using a CO_2 laser, the previous equations also apply. The wavelength difference between the helium-cadmium laser emission at 325 nm and the CO_2 at 10,600 nm has significant systems implications. For a CO_2 laser to form a focused spot equal in size to that of a helium-cadmium laser, the beam divergence will be 30 times greater than

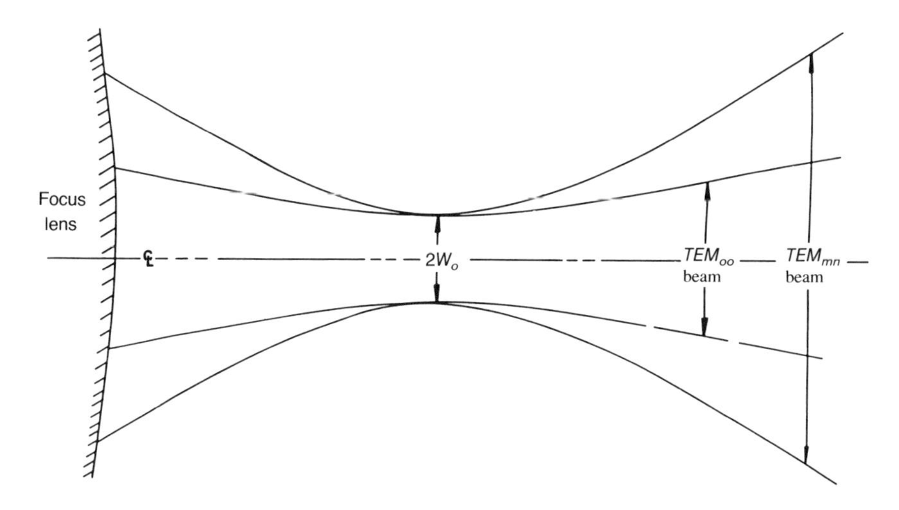

Figure 3-11. Comparison of beam divergence of TEMoo and TEMmn laser beams.

that of the helium-cadmium laser. This implies the need for a much larger galvanometer mirror with its implied slower scan speed.

3.5 Immediate Future of RP&M Lasers

Over the next five years, part building speed will almost certainly improve as a result of reductions in recoating time and increases in shutter and scanner speed. The resulting requirement for laser power is not expected to change significantly.

Helium-Cadmium Lasers

The task for helium-cadmium laser manufacturers will be to increase ultraviolet laser output power to the 100 mW range. This will require either scaling up the size of present lasers using positive column laser discharge designs, or developing lower cost methods of separating cadmium isotopes. Recently, helium-cadmium output power at the 100 mW level has been announced, but these lasers are not presently available commercially.

The focus for helium-cadmium lasers will also be improvements in ruggedness, reliability, and lifetime. Ultraviolet helium-cadmium lasers have demonstrated lifetimes in excess of 12,000 hours. However, consistent lifetime performance over 10,000 hours will require improvement in manufacturing processes and materials.

Finally, there is the matter of cost reduction. Helium-cadmium laser design and construction is amenable to substantial cost reduction as manufacturing volume increases. Because of the low power density of helium-cadmium lasers, the cost of basic construction materials and cooling subsystems is inherently less than argon lasers.

Argon Lasers

The task for argon laser manufacturers in the future will be to scale down the size, input power consumption, and waste heat of these devices. They also need to improve the reliability, ease of installation, and operation compared to existing systems. Output powers of 500 mW are likely not to be needed for SL. Optimizing a laser at the 100 mW level is now appropriate. High power densities require liquid cooling for the laser tube, which results in high installation costs for water and power hookups. Recirculating cooling towers are often required. Therefore, as argon laser designs are scaled down to the 100 mW level, more effective hybrid cooling systems become practical. A new 100 mW ultraviolet argon laser was recently announced which consumes less than 10,000 watts of input power and uses a liquid-to-air heat exchanger for heat dissipation.

Another focus for the future of argon lasers will also be improvements in ruggedness, reliability, and lifetime. Due to the high power densities required to produce ultraviolet output, argon lasers have a formidable task of improving inherent lifetime. Present ultraviolet argon lifetimes are in the order of 1000 to 2000 hours. Improvements to 5000 hours will be necessary for SL, where systems are often operated in excess of 400 hours per month.

Argon laser cost will also decrease as production volumes increase. The cost of materials is high whether the laser tube utilizes a cold-disc design or beryllium oxide to form the discharge capillary. Power supply costs will always remain substantial simply because of the high power levels.

Carbon Dioxide Lasers

The development of carbon dioxide lasers is not driven by RP&M presently. The market for these lasers is much larger in medical applications such as surgery and dermatology and in industrial markets for welding, cutting, and marking of materials. The present range of performance capabilities is adequate to cover the needs of RP&M for the foreseeable future.

3.6 Long-Term Future of Lasers for RP&M

The longer term future trends of lasers for SL are difficult to predict because of the close dependence of laser requirements on photopolymers. Part building speed in these systems also depends on the viscosity and surface tension dependent recoating time. In addition, the speed of future scanners and shutters will also strongly affect the power output requirements of lasers. Helium-cadmium and argon lasers will continue to be the dominant lasers used for SL. However, three other types of lasers offer some prospect for application to SL.

In the ultraviolet, the possible alternative lasers are frequency tripled NdYAG emitting at 355 nm; and neon-ion emitting at 332 nm. Both of these systems are available in the marketplace as research devices. Neither, however, could be considered viable lasers for SL in the near future. Currently, these systems are very expensive compared to either helium-cadmium or argon lasers. The neon-ion laser is, in addition, much larger than present devices for the same output power, and consumes many times more power than existing UV argon lasers.

A final possibility would be frequency doubled solid state diode lasers. Currently, these diodes emit 420 nm with output levels near 1 mW. It will probably be some years before UV emission is commercially available for SL applications.

3.7 Summary

The present and projected laser power needs for SL can be provided with existing technology laser products. There are future opportunities for evolutionary developments of lasers principally to improve laser reliability, cost, and performance. Unless there is a significant breakthrough in scanner speed, laser power requirements of about 100 mW will be satisfied by helium-cadmium and argon laser technology in the future.

The present and projected laser power needs for direct laser sintering and cutting can be provided with existing CO_2 lasers. Evolutionary development of these lasers is driven by other market applications which will provide products more than adequate for RP&M.

ACKNOWLEDGMENTS

For useful discussions, comments, and the computer generated mode plots, the author gratefully acknowledges the assistance of Ron Zimmerman and Kevin Tice, both of Omnichrome Corporation, and Albert Campbell of Directed Energy, Inc.

REFERENCES

1. Weber, M.J. ed., *Handbook of Laser Science and Technology*, Volume II "Gas Lasers," CRC Press, 1988.
2. Arecchi, F.T. and Schulz-Dubois, E.O., eds., *Laser Handbook*, Volume I, "Design of Gas Lasers."
3. Goldsborough, J.P., North Holland Publishing Co, Amsterdam, 1972.
4. Boyd, G.D. and Kogelnik, H., Bell System Tech. J, 41, 1347, 1962.
5. Freiberg, R.J. and Halstead, A.S., *Applied Optics*, 8, 355, Feb. 1969.
6. Kruger, J.S., *Electro-Optical Systems Design*, Sept. 1972.
7. Siegman, A.E., *New Developments in Laser Resonators*, SPIE/OE LASE '90 Conference, Los Angeles, 1990.

chapter 4

Fundamental Processes

Mathematical analysis has therefore necessary relations with sensible phenomena;...it is a pre-existent element of the universal order, and is not in any way contingent or fortuitous; it is imprinted throughout all nature.

—Jean Baptiste Joseph Fourier
Theorie Analytic de la Chaleur
Paris, France (1822)

4.1 Background

As noted in the preface, StereoLithography was the first of the rapid prototyping systems. Further, of the various Rapid Prototyping & Manufacturing (RP&M) systems described in this book, and especially in Chapter 16, six are based on a technology which is either identical to, or very similar to StereoLithography (abbreviated as SL hereafter to avoid confusion with the STL software format). Finally, SL is the most mature of the RP&M technologies, with a considerable body of research and development results forming the basis for much of our present level of understanding in this field.

SL technology is fascinating in its breadth. Taken in its entirety, SL involves major contributions from polymer chemistry (as discussed in Chapter 2), laser physics (Chapter 3), applied mathematics (Chapter 4), computer software (Chapter 5), CAD modeling (Chapter 6), viscous fluid dynamics, material science, optics, as well as electrical and mechanical engineering. To gain a solid understanding of this new technology, the reader should become familiar with the basic physical principles involved.

This chapter describes a number of these fundamental processes. It is by far the most mathematical of the book. The reader who is able to follow the analysis will gain a deeper understanding of the basic phenomena, as well as important insights which should ultimately lead to better part building.

By **Paul F. Jacobs, Ph.D.**, *Director of Research and Development, 3D Systems, Inc., Valencia, CA*

Nonetheless, for the benefit of those who do not wish to go into the full details of the mathematics, the key equations have been boxed, and the important conclusions stemming from those equations have been highlighted.

The best method to begin describing a new technology is not always immediately evident. The approach taken here involves posing some very fundamental questions, and then attempting to develop mathematical models which hopefully will either answer those questions or at least provide some useful insights.

During the past three years, this author has become only too aware of the fact that many of the phenomena in this technology are very complicated and involve many different variables. Thus, it is absolutely imperative that analytical results be tested by careful and repeatable experiments. Throughout this chapter, we shall provide experimental data to allow the reader to assess the validity of the various models.

Some examples of a few of the most fundamental questions are as follows:

1. When an actinic laser beam is scanned in a straight line across the surface of a photopolymer, how does the resulting exposure (energy per unit area) depend upon the laser and the resin parameters?
2. What is the cross-sectional shape of a single cured "line?"
3. How does the depth of the cured "line" depend on the laser exposure and the resin parameters?
4. How does the width of the cured "line" depend on the laser and resin parameters?
5. How does the polymer build up strength with increased exposure?

In the remaining sections of this chapter, we shall present some of the mathematical foundations behind the fundamental physical processes of this new technology. We shall also present relevant experimental data. In most cases, this data will tend to confirm the mathematical models. In other cases, the data may point to deficiencies in the model or models, suggesting areas where future work is still needed.

While we have made considerable progress during the past few years, many questions remain unanswered. It is probably true that the key to future advances and the ability to reap the potentially enormous benefits of RP&M lies in our ability to gain an even better understanding of these fundamental processes.

4.2 The Line Spread Function of a Scanned Gaussian Laser Beam

Consider a Gaussian laser beam being scanned in a straight line at constant velocity, V_s, over the surface of a photopolymer, as shown in *Figure 4-1.*

Now, consider some arbitrary point, Q, within the resin. We shall adopt a coordinate system, as shown in *Figure 4-1*, such that:

1. The x-y plane is coincident with the resin surface.
2. The x axis is coincident with the centerline of the scanned laser beam, with positive x in the direction of scan.
3. The z axis is normal to the resin surface, with positive z directed downward into the resin.
4. The projection of Q onto the resin surface shall be Q',
5. The origin is selected such that the x and z coordinates of point Q' are zero ($Q'(x,y,z) = Q'(0,y,0)$).

Thus, the centerline of the laser beam is on the x axis and is scanned from $x = -\infty$ to $x = +\infty$, at constant velocity, V_s. We shall also assume that the absorption of laser radiation within the resin follows the Beer-Lambert Law. Under these conditions, the irradiance (radiant power per unit area), $H(x,y,z)$ at any point within the resin, is related to the irradiance incident on the resin surface, $H(x,y,0)$ by the relation

$$H(x,y,z) = H(x,y,0) \exp -(z/D_p) \tag{4-1}$$

where D_p is the "penetration depth" of the resin, defined as that depth of resin which results in a reduction in the irradiance to a level equal to 1/e of the surface irradiance, where e = 2.7182818..., the base of natural logarithms. Thus, for example, at a depth $z = D_p$, the irradiance would be about 37% of the surface irradiance.

Furthermore, for a Gaussian laser

$$H(x,y,0) = H(r,0) = H_o \exp -(2r^2/W_o^2) \tag{4-2}$$

where the quantity W_o is the so-called $1/e^2$ Gaussian half-width, as shown in *Figure 4-2*. Note that when $r = 0$, $H(0,0) = H_o$. Also, note that when $r = W_o$, $H(W_o,0) = H_o\, e^{-2} = 0.13534\, H_o$.

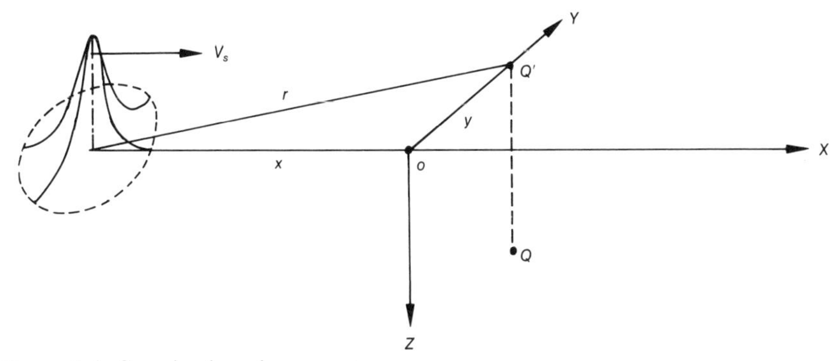

Figure 4-1. Gaussian laser beam scan.

We may determine the constant H_o by recognizing that the integral of the surface irradiance over the entire x, y plane, from $r = 0$ to $r = \infty$, must be equal to the laser power, P_L, incident on the resin surface. Thus,

$$P_L = \int_{r=o}^{r=\infty} H(r,0)\, dA \tag{4-3}$$

where $dA = 2\,\pi\, rdr$

Substituting for $H(r,0)$ from equation 4-2,

$$P_L = \int_o^{\infty} H_o \exp\,(-2r^2/W_o^{\,2})\; 2\,\pi\; rdr \tag{4-4}$$

Let us now define the dimensionless variable

$$u \equiv 2r^2/W_o^{\,2} \tag{4-5}$$

Differentiating,

$$du = 4rdr/W_o^2 \qquad \text{or} \qquad rdr = (W_o^{\,2}/4)\; du \tag{4-6}$$

Further, regarding the limits of integration, it is clear from equation 4-5 that

$$\text{when } r = 0,\; u = 0, \qquad \text{and} \qquad \text{when } r = \infty,\; u = \infty$$

Substituting equations 4-5 and 4-6 into equation 4-4,

$$P_L = \frac{\pi}{2}\, W_o^{\,2}\, H_o \int_o^{\infty} \exp(-u)\; du = \frac{\pi}{2}\, W_o^{\,2}\, H_o\, [-\exp(-u)]_0^{\infty}$$

$$= \frac{\pi}{2}\, W_o^{\,2}\, H_o \tag{4-7}$$

Thus, solving for the peak surface irradiance at $r = 0$ and $z = 0$, we obtain the result:

$$H_o = 2P_L/\pi\; W_o^{\,2} \tag{4-8}$$

Substituting this result back into equation 4-2, we obtain:

$$H\,(r,0) = [2P_L/\pi W_o^{\,2}\,]\, \exp\,(-2r^2/W_o^{\,2}) \tag{4-9}$$

Now, from the geometry of *Figure 4-1*, and the Pythagorean Theorem,

$$r^2 = x^2 + y^2 \tag{4-10}$$

Also, during the laser scan, y = constant, and

$$V_s = dx/dt \tag{4-11}$$

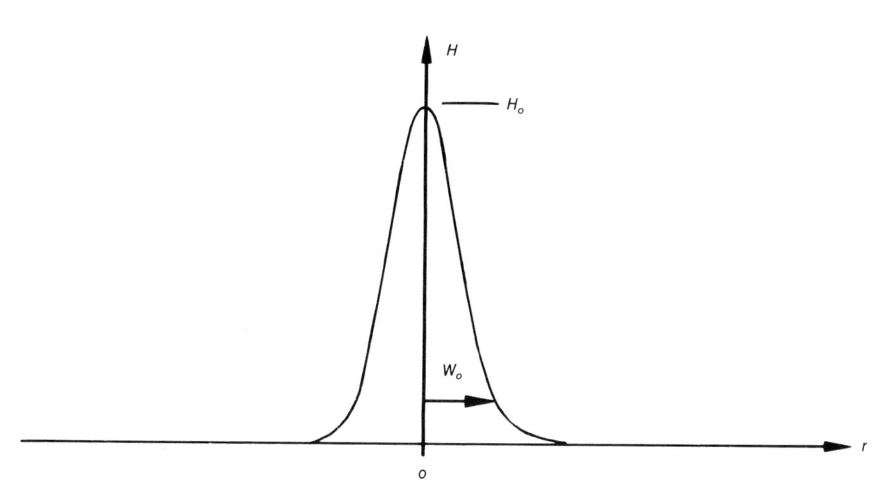

Figure 4-2. Gaussian half-width.

By definition, the exposure, E, which has the units of energy per unit area, at any point Q', is simply the integral of the irradiance at Q' over time. Since the beam initially comes from far to the left, passes by Q', and finally moves far to the right of Q', then on the resin surface, $z = 0$, we may write:

$$E(y,0) = \int_{t=-\infty}^{t=\infty} H\,[r(t),\, 0\,]\, dt \qquad (4\text{-}12)$$

However, from equation 4-11,

$$dt = dx/V_s \qquad (4\text{-}13)$$

Differentiating equation 4-10, with y = constant,

$$dx = rdr/x = rdr/[r^2 - y^2]^{1/2} \qquad (4\text{-}14)$$

Substituting from equations 4-9, 4-13, and 4-14 into equation 4-12, we obtain an integral expression for the exposure on the resin surface,

$$E(y,0) = [2P_L/\pi\, W_o^2\, V_s] \int_{-\infty}^{\infty} \exp(-2r^2/W_o^2)\, rdr/[r^2 - y^2]^{1/2} \qquad (4\text{-}15)$$

At this point, it is worth discussing in some detail the limits of integration used in equation 4-15. For mathematical convenience, the limits in x are taken from $x = -\infty$, to $x = +\infty$. Obviously, an infinite scan length at a finite scan velocity would result in an infinite scan time for a single line! This would hardly be compatible with the whole concept of "rapid prototyping."

Fortunately, however, the laser exposure on the resin surface is proportional to exp $(-2r^2/W_o^2)$. This function falls off very rapidly with distance. For this reason, let us define the concept of a "zone of influence," such that within this zone a differential area $dA = dxdy$ will receive 99.99% of its total exposure; while outside of this zone the contribution to the total exposure will amount to a maximum of only 0.01%. This tiny discrepancy (1 part in 10,000) is almost always much less than the error associated with ignoring the higher order mode content when we assume the laser beam to be purely Gaussian.

Defining $r = R$ as the radius of the zone of influence, then

$$\exp(-2R^2/W_o)^2 = 0.0001 \qquad \text{or} \qquad \exp(2R^2/W_o^2) = 10{,}000$$

Taking natural logarithms of both sides of the latter equation,

$$2\,R^2/W_o^{\,2} = \ln(10{,}000) = 9.2103$$

Dividing by 2 and taking square roots, we obtain the result:

$$\boxed{R = 2.146\ W_o} \qquad (4\text{-}16)$$

Thus, the entire "zone of influence" extends from $-R$ on the left to $+R$ on the right of any point, or a span of only about 4.3 W_o. Since W_o is typically about 5 mils for most SL systems, the entire zone of influence amounts to only about 22 mils, or about 11 mils on each side of a given point.

We may now define the "characteristic exposure time," t_e, by the expression

$$t_e \equiv 2R/V_s \approx 4.3\ W_o/V_s \qquad (4\text{-}17)$$

which is the time it takes for the laser beam to traverse the entire zone of influence when scanned at a constant velocity, V_s. For current SLA systems, V_s generally ranges from 10 to 100 inches per second, with a maximum of about 300 inches per second on the SLA-500. Taking $W_o = 0.005$ inches, this leads to values of the characteristic exposure time that are in the range:

$$70 \text{ microseconds} < t_e < 2 \text{ milliseconds}$$

Thus, while the limits are expressed as $x = -\infty$ to $x = +\infty$, in actual practice $x = -11$ mils to $x = +11$ mils will account for about 99.99% of the total exposure, and, as seen above, this can be accomplished in times which are short relative to the time required to draw a typical "line."

Recalling from equation 4-10 that $r^2 = x^2 + y^2$, then

$$\exp(-2r^2/W_o^2) = \exp(-2x^2/W_o^2) * \exp(-2y^2/W_o^2) \quad (4\text{-}18)$$

Substituting equations 4-14 and 4-18 into equation 4-15, we obtain

$$E(y,0) = [2P_L/\pi W_o^2 V_s] \exp(-2y^2/W_o^2) \int_{-\infty}^{\infty} \exp(-2x^2/W_o^2)\, dx \quad (4\text{-}19)$$

where we have brought the $\exp(-2y^2/W_o^2)$ factor outside the integral sign since it is not a function of x.

At this point, we shall define another dimensionless variable very similar to the variable u previously defined by equation 4-5. This time we define the quantity v as follows:

$$v^2 \equiv 2x^2/W_o^2 \quad (4\text{-}20)$$

from which $v = (2^{1/2}/W_o)\, x$ (4-21)

Differentiating, $dv = (2^{1/2}/W_o)\, dx$

and transposing, $dx = (W_o/2^{1/2})\, dv$ (4-22)

Now, substituting equations 4-20 and 4-22 into equation 4-19, we obtain:

$$E(y,0) = [\sqrt{2}\, P_L/\pi W_o V_s] \exp(-2y^2/W_o^2) \int_{-\infty}^{\infty} \exp(-v^2)\, dv \quad (4\text{-}23)$$

Furthermore, since the integral in equation 4-23 is symmetric (the contribution from minus infinity to zero is equal to that from zero to infinity) we may, therefore, write the result in the form:

$$E(y,0) = [2\sqrt{2}\, P_L/\pi W_o V_s] \exp(-2y^2/W_o^2) \int_{o}^{\infty} \exp(-v^2)\, dv \quad (4\text{-}24)$$

From Reference 1, it can be shown that:

$$\int_{o}^{\infty} \exp(-v^2)\, dv = \sqrt{\pi}/2 \quad (4\text{-}25)$$

Substituting the result from equation 4-25 into equation 4-24 we obtain the "surface exposure equation:"

$$E(y,0) = \sqrt{\frac{2}{\pi}}\, [P_L/W_o V_s] \exp(-2y^2/W_o^2) \quad (4\text{-}26)$$

Finally, returning to equation 4-1, to include the absorption function $\exp(-z/D_p)$, we obtain the "volumetric exposure equation" for any arbitrary point, Q, at a distance y from the laser scan axis, and at a depth z below the resin surface:

$$E(y,z) = \sqrt{\frac{2}{\pi}} [P_L/W_o V_s] \exp(-2y^2 / W_o^2) \exp(-z/D_p) \quad (4\text{-}27)$$

It is worth noting that this result shows that *the line spread function of a scanned Gaussian laser beam is itself Gaussian in the surface coordinate orthogonal to the scan direction*. This is perhaps a more precise statement than the often quoted, but not always fully understood expression that "the line spread function of a Gaussian is a Gaussian."

4.3 The Parabolic Cylinder

For photopolymers, when the exposure is less than a critical value, E_c, the resin remains liquid. When the exposure is greater than E_c, the resin polymerizes. When the exposure is equal to E_c, the polymer is at the "gel point," corresponding to the transition from the liquid phase to the solid phase.

Thus, we may solve for the locus of points $y = y^*$ and $z = z^*$, which are just at the gel point, or equivalently, the outer boundary of that portion of the resin which has been at least partially polymerized. This locus provides a mathematical description of the three dimensional shape of a single laser cured "line," which is the fundamental "building block" for all SL parts.

Thus, setting $y = y^*$ and $z = z^*$ when $E(y,z) = E_c$, we obtain from equation 4-27,

$$E_c = \sqrt{\frac{2}{\pi}} [P_L / W_o V_s] \exp -(2y^{*2}/W_o^2 + z^*/D_p) \quad (4\text{-}28)$$

or, after some simple algebra,

$$\exp(2y^{*2}/W_o^2 + z^*/D_p) = \sqrt{\frac{2}{\pi}} [P_L/W_o V_s E_c] \quad (4\text{-}29)$$

Taking natural logarithms of both sides of equation 4-29,

$$2 y^{*2}/W_o^2 + z^*/D_p = \ln \{ \sqrt{\frac{2}{\pi}} [P_L/W_o V_s E_c] \} \quad (4\text{-}30)$$

Noting that W_o, D_p, V_s, P_L, and E_c are all constants, then equation 4-30 may be written in the form

$$a\,y^{*2} + b\,z^{*} = c \tag{4-31}$$

where a, b, and c are all positive constants. Equation 4-31 is the general equation in three dimensions for a parabolic cylinder whose axis is parallel to the laser scan axis.

Thus, scanning an actinic Gaussian laser beam at a constant velocity across a photopolymer obeying the Beer-Lambert Law will result in a single cured "line" which is actually a parabolic cylinder (a parabolic cross-section extruded in space along the x axis).

Figure 4-3 shows a schematic view of a cured "line" having the parabolic cross-section described above.

4.4 The Working Curve Equation

Let us define the maximum cure depth for a single laser cured line by the symbol C_d. From equation 4-31, since the first term involves the quantity y^* squared, and the constant a is positive, then all nonzero values of y^* must result in a positive value for the first term. Thus, it is clear that the largest value of z^* will occur when $y^* = 0$. This can also be seen from *Figure 4-3*, where the "deepest" part of the parabola occurs directly under the scan axis, which, by definition of our coordinate system, requires that $y^* = 0$.

First, however, setting $y^* = 0$ and $z^* = 0$ in equation 4-27, the maximum centerline laser exposure, E_{max}, incident on the resin surface is given by the relation:

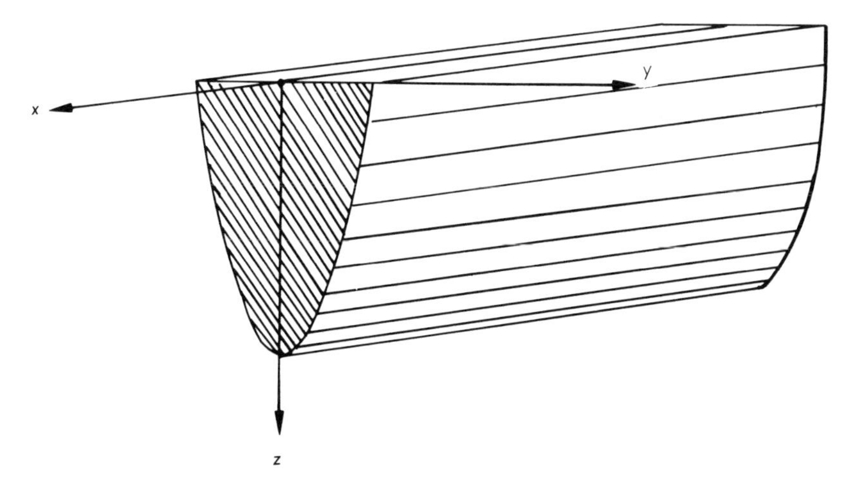

Figure 4-3. Schematic view of a cured line.

$$E(0,0) = E_{max} = \sqrt{\frac{2}{\pi}}\,[P_L/W_o\,V_s] \tag{4-32}$$

We may now solve for the maximum cure depth of a single laser cured line. Setting $y^* = 0$, when $z^* = C_d$, and substituting equation 4-32 into equation 4-30, we finally obtain the "working curve equation":

$$\boxed{C_d = D_p \ln\,(E_{max}/E_c)} \tag{4-33}$$

This equation is absolutely fundamental to SL. Equation 4-33 states, in mathematical form, the following five basic points:

1. The cure depth is proportional to the natural logarithm of the maximum exposure on the centerline of the scanned laser beam.
2. A semilog plot of C_d versus E_{max} should be a straight line. This plot is known as the working curve for a given resin.
3. The slope of the "working curve" is precisely D_p, the penetration depth of that resin, at the laser wavelength.
4. The intercept of the working curve, specifically the value of the exposure at which the cure depth is zero, is simply E_c, the critical exposure of that resin, at the laser wavelength.
5. Since D_p and E_c are purely resin parameters, then both the slope and the intercept of the working curve are independent of laser power.

Figure 4-4 shows an experimentally measured working curve for Ciba-Geigy resin XB 5081-1. It is important to note that this figure is a semilogarithmic plot; with cure depth on the ordinate and the logarithm of the maximum laser exposure on the abscissa.

It is evident from inspection of *Figure 4-4* that C_d is indeed directly proportional to the logarithm of the exposure, as indicated by the fact that the experimental data conform very closely to a straight line. The slope, indicated graphically, corresponds to a value of the resin penetration depth, $D_p = 0.0071 \pm 0.0003''$ (7.1 $\pm$0.3 mils, or 0.180 $\pm$0.008 mm).

It is worth noting that D_p is determined by the increase in cure depth, δC_d, corresponding to an increase in exposure by a factor equal to $e = 2.7182818\ldots$ Since a factor of e is not very easy to determine graphically, a much easier method to determine D_p is to find the increase in cure depth ΔC_d, for a factor of 20 increase in exposure, which is easy to determine from a semilog plot. Noting that the natural logarithm of 20 is 2.9957323..., this is within 0.15% of 3. Thus, to the same close level of approximation:

$$D_p \equiv \Delta C_d/2.9957 \approx \Delta C_d/3 \tag{4-34}$$

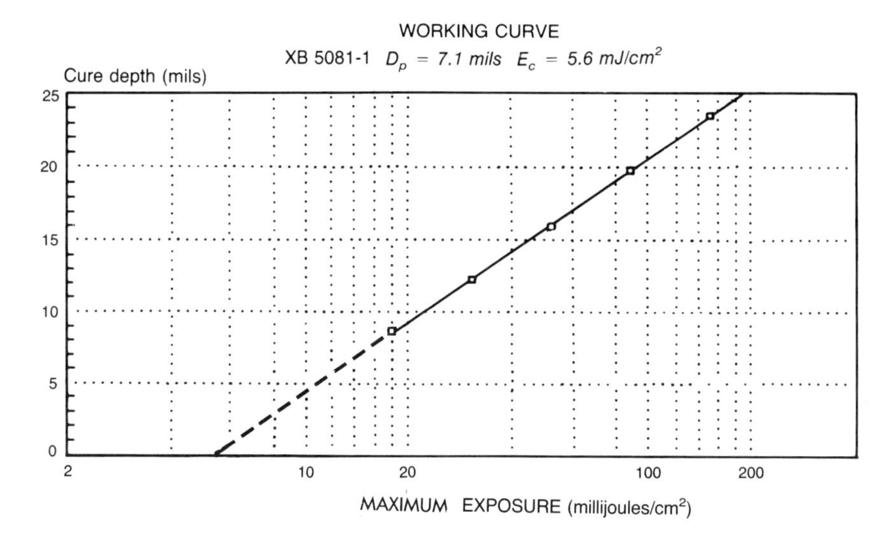

Figure 4-4. Experimentally measured working curve.

Finally, note that the straight line trend of the data shows an abscissa intercept, or critical exposure, $E_c = 5.6 \pm 0.3$ mJ/cm^2.

The data obtained in generating a working curve are generally confined to a cure depth range given by $D_p < C_d < 4D_p$. Thus, E_c must be determined by linear extrapolation. For values of $C_d < D_p$, the physical properties of the partially cured photopolymer are so weakly developed, as described in detail in sections 4.9 and 4.10 of this chapter, that it is impossible to make accurate, repeatable measurements of cure depth.

Also, for cure depths greater than about $4D_p$, the working curve will often begin to exhibit nonlinear, or strictly speaking nonlogarithmic behavior. The following two trends have been observed:

Superlogarithmic Behavior. Here C_d increases more rapidly than a simple logarithmic dependence upon laser exposure. Two possible explanations have been advanced. The first involves "optical bleaching." Here it is assumed that the partially solidified photopolymer may have a reduced absorption coefficient, at the laser wavelength, relative to the liquid resin (in this hypothesis, cured polymer would become more transparent). This would, therefore, result in less absorption per unit length, and a corresponding increase in D_p. Since D_p is the slope of the working curve, then the result would be a *positive curvature* to the working curve.

A second possible explanation involves the phenomena of "optical self-focusing." It is well known that resin shrinks as it is polymerized. Thus, the density of the resin increases, and hence the index of refraction increases as well. Therefore, optical self-focusing, similar to the phenomena occurring in fiber optics, may then obtain.

In this hypothesis, the laser photons would tend to be "steered" or "self-focused" due to an index of refraction profile which is lower near the outer portions of the cured photopolymer and higher near the center. In fiber optics this is done quite intentionally to confine the optical radiation within the fiber. In a laser cured polymer, optical self-focusing may be the natural consequence of parabolic iso-exposure contours, as defined by equations 4-27 and 4-31.

The result of these parabolic contours would be higher levels of exposure, and hence greater levels of polymerization, leading to higher densities and, therefore, higher index of refraction values on the centerline and, conversely, lower values at the outer edges. This in turn could cause some of the laser photons, that would otherwise have been incident on the outer portions of the polymerized region, to be directed towards the centerline. In this case, the local centerline laser exposure would exceed the theoretical value predicted by equation 4-27, for $y^* = 0$, and $z^* = C_d$, with a corresponding increase in cure depth. *Figure 4-5A* shows a schematic representation of a superlogarithmic working curve.

At this time, it is not clear if either of these hypotheses is the true explanation for superlogarithmic behavior. Nonetheless, once the linear portion of the working curve has been exceeded, the value of the cure depth is no longer reliably predictable. For this reason, it is very important to confine the various cure depths used in actual part building to the linear regime of the working curve.

Sublogarithmic Behavior. Here C_d increases less rapidly than the simple logarithmic dependence on laser exposure. A plausible explanation for this phenomena involves "optical scattering." In this case, any microscopic imperfections, occlusions, voids, etc. will act as scattering centers. The resulting

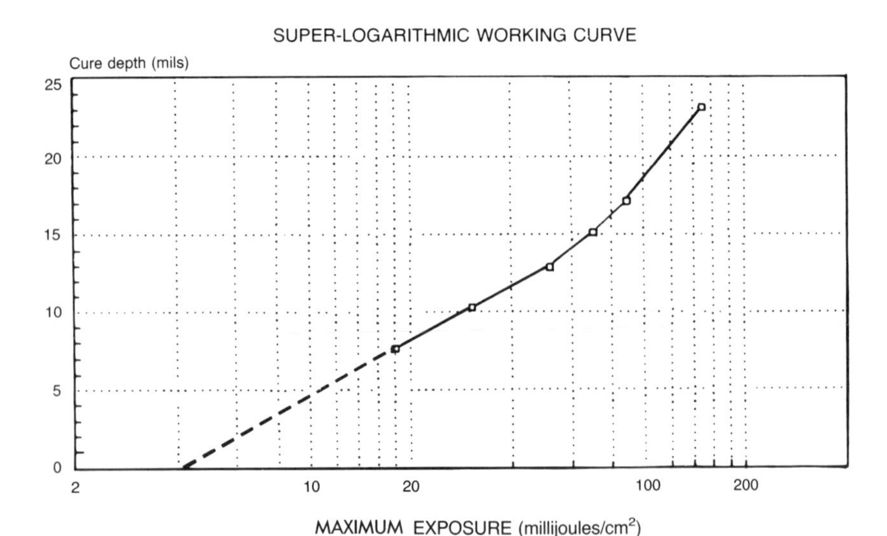

Figure 4-5A. Schematic representation of a superlogarithmic working curve.

scattering will tend to reduce the peak exposure at $y = 0$, and $z = C_d$, relative to that predicted from equation 4-27. This will then correspondingly reduce the actual cure depth relative to the value of C_d from the working curve. The result would be a *negative curvature*, as seen in *Figure 4-5B*, for a sublogarithmic working curve.

In either case, it is absolutely critical that the two resin constants D_p and E_c be determined only from the linear portion of the working curve, and that actual cure depths used in building SL parts also be confined to this region.

4.5 The Cured Linewidth Function

Returning to equation 4-30 and *Figure 4-3*, we may also note that the parabolic cross-section of a single laser drawn line will reach its maximum cured width at the resin surface ($z^* = 0$). It follows from equation 4-30 that $y^* = y_{max}$ when $z^* = 0$. Substituting, we obtain:

$$2\, y_{max}^2 / W_o^2 = \ln \left[\sqrt{\frac{2}{\pi}} \,(P_L / W_o\, V_s\, E_c) \right] \quad (4\text{-}35)$$

Defining the maximum cured linewidth, L_w, by the expression:

$$L_w \equiv 2 y_{max} \quad (4\text{-}36)$$

Now, substituting the value for E_{max} from equation 4-32, and L_w from equation 4-36, we obtain after some algebra:

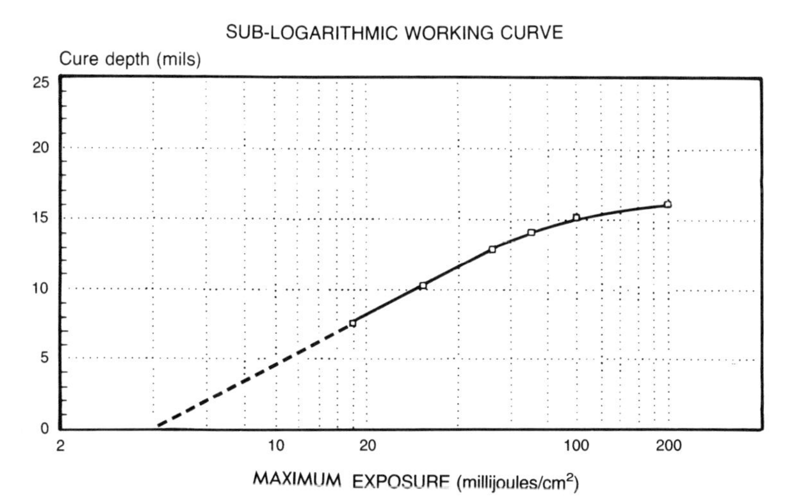

Figure 4-5B. Schematic representation of a sublogarithmic working curve.

$$L_w = 2^{1/2} W_o \{\ln [E_{max}/E_c]\}^{1/2} \quad (4\text{-}37)$$

Finally, substituting for the ln $[E_{max}/E_c]$ factor from equation 4-33, we obtain the "cured linewidth function":

$$L_w = W_o [2 C_d / D_p]^{1/2} \quad (4\text{-}38)$$

However, since W_o is defined as the Gaussian half-width of the laser irradiance distribution (at the $1/e^2$ point), while L_w is defined as the full-width of a laser cured line, the author has found it very useful to define the full-width of the Gaussian laser beam by the expression:

$$B \equiv 2 W_o \quad (4\text{-}39)$$

Substituting equation 4-39 into equation 4-38 we obtain the important result:

$$\boxed{L_w = B \sqrt{C_d/2D_p}} \quad (4\text{-}40)$$

This cured linewidth function shows that for a Gaussian laser scanned in a straight line across a photopolymer obeying the Beer-Lambert Law of absorption:

a. The cured linewidth, L_w, is directly proportional to the laser spot diameter, B.
b. The cured linewidth is also proportional to the square root of the ratio of the cure depth to the resin penetration depth (C_d / D_p).

Thus, curing deeper will also result in increased linewidth. This becomes very important when attempting to perform "linewidth compensation" algorithms in software. Fortunately, however, the dependence is only as the square root of C_d, so the variation is not very great.

For example, if as noted earlier we stay in the linear portion of the working curve, then $D_p < C_d < 4D_p$, which is equivalent to the range $1 < C_d / D_p < 4$. From equation 4-40, this corresponds to:

$$2^{-1/2} < L_w/B < 2^{1/2}$$

Hence, in the linear region of the working curve, the cured linewidth will always be within the range of about 0.7 B to about 1.4 B. For a typical laser employed in SL, we may take $B = 9$ mils. For Ciba-Geigy resin XB 5081-1, with $D_p = 7.1$ mils, this would correspond to a range of predicted linewidths, as shown in Table 4-1.

Table 4-1
Predicted Linewidths for XB 5081-1

C_d	C_d/D_p	L_w
5 mils	**0.704**	**5.3 mils**
10 mils	**1.408**	**7.6 mils**
15 mils	**2.113**	**9.3 mils**
20 mils	**2.817**	**10.7 mils**

This amounts to a variation of over 5 mils in the width of a single line. Thus, if one wishes to produce truly accurate parts with SL technology, the variation in L_w with C_d must be taken into account.

The analysis presented in sections 4.2 through 4.5 is based on three assumptions:

1. The laser irradiance distribution is *Gaussian*.
2. The photopolymer obeys the *Beer-Lambert Law* of absorption.
3. The photopolymer has a *threshold exposure*, E_c, corresponding to the transition from the liquid phase to the solid phase at the gel point.

Of these, the first assumption is probably most often violated. Helium-cadmium (He-Cd) lasers, as discussed in Chapter 3, are often multimode and may even occasionally have a ''doughnut-like'' irradiance distribution. Argon-ion (Ar+) lasers are often closer to exhibiting true Gaussian beam character, but even these lasers will tend to have some higher order mode structure. Experience with lasers having non-Gaussian irradiance distributions has shown that departure from Gaussian behavior has very little effect on the working curve, and hence on the value of C_d, but will have a somewhat greater influence on the value of L_w.

This result should not be considered very surprising since the value of $z^* = C_d$, determined from equation 4-30, is the result of setting $y^* = 0$. This, in effect, washes out the entire radial structure of the laser irradiance distribution function, and leaves C_d purely dependent upon the maximum exposure value. Furthermore, E_{max} depends on the line spread function, not the laser irradiance distribution itself. Since linespread functions are integrals, by their very nature they tend to ''smooth out'' details of the original distribution function.

Therefore, despite considerable differences between purely Gaussian, single mode laser irradiance distribution functions and those of lasers with significant higher order mode content, the resulting line spread functions show relatively little change. This is certainly fortuitous, since both obtaining and maintaining purely Gaussian mode He-Cd or Ar+ lasers is both expensive and difficult.

However, the same cannot be said for the cured linewidth function. While careful experiments have generally confirmed the fundamental character of equation 4-40, including a basically square root dependence upon C_d, it has been observed that actual cured linewidth measurements generally exceed the values predicted by equation 4-40.

Further, the extent by which the actual cured linewidths exceed the theoretical values for a Gaussian laser appears to be related to the mode structure of the beam. Quasi-Gaussian lasers, such as a properly tuned Argon-Ion laser, will generally produce measured linewidths not exceeding one mil greater than the corresponding theoretical value of L_w.

As the beam structure becomes more "doughnut like" (with a helium-cadmium laser nearing the end of its tube life) the discrepancy in cured linewidth may increase to as much as two mils. As we shall see in Chapter 8, this difference can be significant when using some of the more advanced building techniques, such as WEAVE™ or STAR-WEAVE™.

4.6 Mechanical Properties

The determination of the mechanical properties of photopolymers as a function of exposure is absolutely fundamental to a better understanding of the SL process. In section 4.2 we have already determined that the laser exposure varies as a function of the horizontal position, y, from the beam scan centerline, as well as with depth, z, below the resin surface. Since the exposure in a cured line is inhomogeneous, and the mechanical properties of the partially cured photopolymer are a function of the exposure, it follows that the mechanical properties of a cured line would also be inhomogeneous.

The subject of the remaining sections of this chapter is the dependence of Young's modulus of elasticity, Y, defined as the initial slope of the photopolymer stress versus strain curve as a function of exposure. An understanding of the function $Y(E)$ would allow the prediction of the so-called "green strength" of a laser cured SL part. Green strength is critical to successful part building, and involves the composite average strength of a multilayered part consisting of nonhomogeneous layers.

Further, knowing $Y(E)$ also allows prediction of the strength of these parts after post curing. Finally, this information has already proved to be valuable in optimizing techniques to reduce shrinkage generated distortions in both part building and post curing, as discussed in Chapter 8.

Unfortunately, all previous measurements of photopolymer modulus had been performed on samples generated by unilateral exposure. As with a single laser exposed line, this results in test samples that have more exposure on one side than the other. Since E is a function of depth into the resin, as can be seen from equation 4-27, then $Y(E)$ is also a function of depth, and hence is nonuniform throughout the test sample. As a result of these nonuniformities, variations in test results from sample to sample were large, errors were large, and the inevitable "averaging" tended to mask any detailed functional dependence of Y upon E.

For this reason, it was important to develop an altogether new method of delivering more uniform exposure throughout a test sample. The remainder of this chapter will describe:

1. The theoretical and experimental background for this new and more accurate method (see Reference 2).
2. The mathematical basis for a resin property referred to herein as the "photomodulus coefficient," K_p.
3. Detailed experimental data describing actual $Y(E)$ functions for five different SL resins.

4.7 Bilateral Exposure of a Thin Sample

Consider a thin slab of photopolymer, as shown in *Figure 4-6*, having a constant thickness, w, and exposed uniformly from both sides. Let E_o be the exposure at $x = 0$, coming from the left, and E_o also be the exposure at $x = w$, coming from the right. As before, we shall assume the resin obeys the Beer-Lambert Law of absorption with a penetration depth, D_p, as defined earlier.

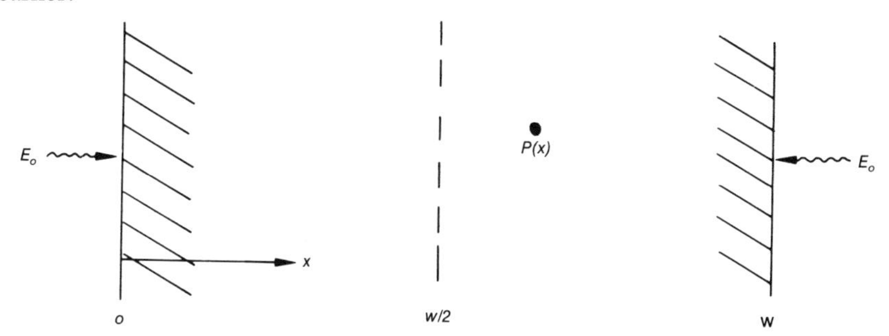

Figure 4-6. Bilateral exposure of a thin slab of photopolymer.

The total exposure at any point, P, a distance x from the left edge of the slab, is the sum of the exposure arriving at P from both the left and the right. Thus, from the Beer-Lambert Law,

$$E(x) = E_o \exp[-x/D_p] + E_o \exp[-(w-x)/D_p] \qquad (4\text{-}41)$$

Define the dimensionless variables:

$$U \equiv E/E_o \qquad s \equiv x/w \qquad F \equiv w/D_p$$

Substituting into equation 4-41, we obtain:

$$U(s,F) = \exp(-sF) + \exp(-F)\exp(sF) \tag{4-42}$$

After some algebra, it can be shown that this is equivalent to:

$$U(s,F) = \exp(-F/2)\{\exp[F(1/2-s)] + \exp\text{-}[F(1/2-s)]\} \tag{4-43}$$

Define

$$G \equiv F\,(1/2 - s) \tag{4-44}$$

Substituting from equation 4-44 into equation 4-43, we obtain:

$$U(G,F) = 2\exp(-F/2)\cosh(G) \tag{4-45}$$

where the hyperbolic cosine function is defined (see Reference 3) by the relation,

$$\cosh(G) \equiv \{\exp(G) + \exp(-G)\}/2 \tag{4-46}$$

Thus, the dimensionless exposure behaves like a catenary function. The departure from exposure uniformity is directly analogous to the sag of a rope in a uniform gravitational field. Here the finite absorption per unit length of resin plays the same type of role as the finite weight per unit length of the rope.

Figure 4-7 is a plot of the dimensionless exposure $U = E/E_o$, as a function of the normalized depth into the slab $s = x/w$, for different values of the dimensionless thickness ratio $F = w/D_p$.

Some observations are appropriate:

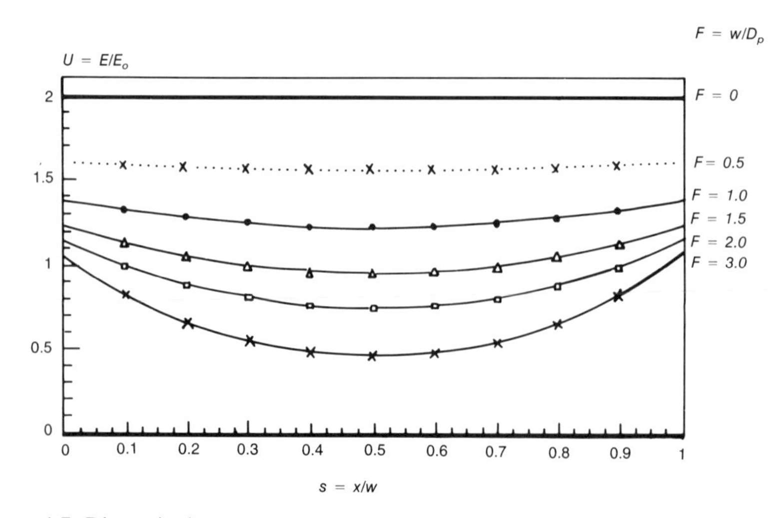

Figure 4-7. Dimensionless exposure.

1. The functions are all symmetric about the midpoint of the slab, at a position $s = 0.5$, or $x = w/2$.
2. In the limit as F approaches zero, we find that the dimensionless exposure is everywhere constant at $U = 2$. This is the so-called "transparent slab" limit, which may occur for either a slab of zero thickness or an infinite penetration depth resin. In either case, nothing is being absorbed, and hence the dimensionless exposure at any point within the slab is simply unity from one side plus unity from the other side.
3. For any value $F > 0$, there is some "sag." That is, the exposure at the center of the slab is less than it is at either $x = 0$ or $x = w$.
4. For $0 < F < 0.9$, the maximum variation in E throughout the slab $[E(0)–E(w/2)]$ is calculated to be less than 10% of the maximum value, E(0). Therefore, provided that the sample thickness is less than 90% of the resin penetration depth (w is less than 90 % of D_p), then the variation in the sample exposure will always be smaller than 10%. This may be referred to as the "ninety/ten rule."
5. Using the ninety/ten rule, with bilateral exposure, we can ensure that the samples generated for tensile testing will be quite uniform, and the resulting uncertainties in resin modulus will be much smaller than with prior experimental techniques.

4.8 The Photomodulus Model

The discussion that follows presents an almost classical interplay between theory and experiment. At the time this work began in early 1990, no mathematical model had been developed which would allow prediction of the $Y(E)$ function. For this reason, the experimental results led the way. Once the initial results for $Y(E)$ became available, using the new experimental technique detailed in section 4.9 and Reference 2, three important trends became evident:

1. $Y(E)$ was zero for $E < E_c$.
2. For experimental exposure values at, and just beyond, E_c, the slope, dY/dE, of the resulting modulus versus exposure curve was steeply positive.
3. With further increase in E, the slope decreased and the value of Y asymptotically approached the maximum modulus, Y_{max}, for large values of E.

Inspection of the data led to the formulation of a model for the development of modulus with exposure (referred to hereafter as photomodulus) based on a classical, first order, linear relaxation differential equation. Concisely stated, *the rate of change of photopolymer modulus of elasticity with actinic exposure is proportional to the modulus deficit.*

Mathematically, this may be written in the form,

$$dY/dE = K_p\,[(Y_{max}-Y)/Y_{max}] \tag{4-47}$$

where K_p is a constant of proportionality

$$Y_{max} = \lim_{E\to\infty} [Y(E)]$$

E = actinic exposure incident on an element of the photopolymer.

Let us define the following dimensionless variables:

$$\Phi \equiv (Y_{max}-Y)/Y_{max} \qquad dY = -Y_{max}\,d\,\Phi \tag{4-48}$$

$$\Theta \equiv K_p\,E/Y_{max} \qquad dE = (Y_{max}/K_p)\,d\,\Theta \tag{4-49}$$

Substituting equations 4-48 and 4-49 into equation 4-47, we obtain the linear, first order, ordinary differential equation

$$d\,\Phi/d\,\Theta = -\Phi \tag{4-50}$$

Rearranging and integrating,

$$\int d\,\Phi/\Phi = \ln(\Phi) = -\int d\,\Theta = -\Theta + \beta \tag{4-51}$$

where β is a constant of integration.

The single boundary condition for this first order equation is

$$Y = 0 \qquad \text{when} \qquad E = E_c \tag{4-52}$$

Physically, this corresponds to the fact that up to $E = E_c$, the resin is a liquid, and hence has zero Young's modulus. In the dimensionless variables this is equivalent to $\Phi = 1$ when $\Theta = \beta$. Substituting $E = E_c$ into equation 4-49, we may now evaluate the constant of integration:

$$\beta = K_p\,E_c/Y_{max} \tag{4-53}$$

Note that β, which is also dimensionless, is purely a function of the three resin parameters K_p, E_c, and Y_{max} . After some simple algebra, we obtain the following from equation 4-51:

$$\Phi = \exp -(\Theta-\beta) \tag{4-54}$$

Returning to physical variables, we finally obtain the so-called "photomodulus equation":

$$Y/Y_{max} = 1 - \exp\left[-\beta \left(E/E_c - 1\right)\right] \qquad (4\text{-}55)$$

This solution actually describes a family of curves, as shown in *Figure 4-8*, where each curve is uniquely described by the value of the resin constant β.

Further, note that β is directly proportional to E_c. Thus, attempts to produce resins with low values of E_c, so that they will be efficient with respect to laser energy, must also recognize that low E_c values may seriously compromise the mechanical properties of the resulting photopolymers.

It is also worth reviewing a number of additional properties of this solution. Specifically, the photomodulus equation, (4-55), predicts the following:

1. $Y = 0$ for $E = E_c$. This is a statement of the fact that the resin remains liquid for exposures below E_c. Also, equation 4-55 has no physical meaning for $E < E_c$. The resin is still a fluid and the Young's modulus of a fluid is meaningless. Further, from a purely practical standpoint, we are really only interested in the mechanical properties of the photopoly-

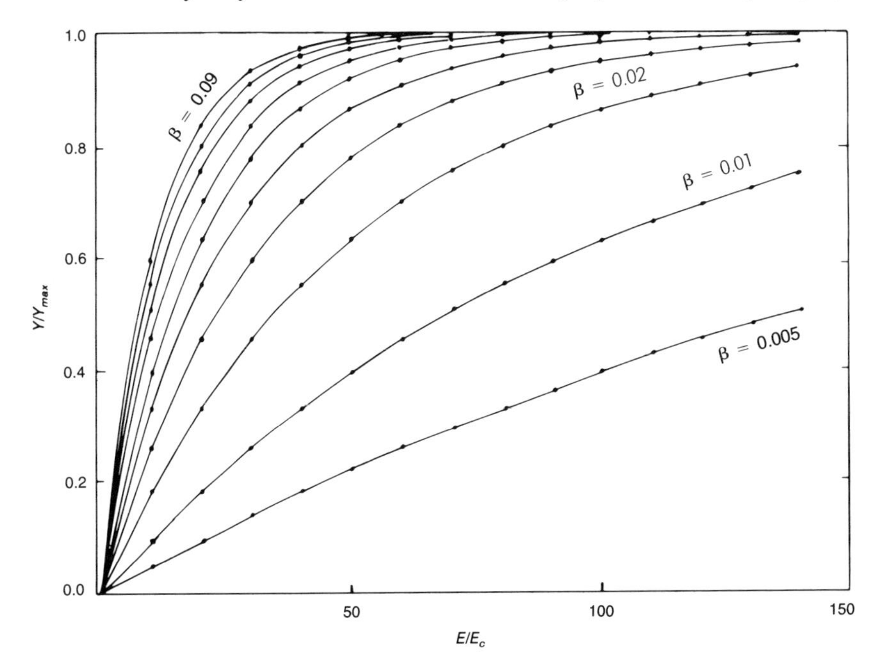

Figure 4-8. The family of curves derived from the photomodulus equation.

mer after it has been at least partially solidified, and this only occurs for exposure values greater than E_c.

2. For E just above E_c we can expand the exponential function in equation 4-55, for small values of the argument. Thus, if we take $E/E_c = 1 + \epsilon$, where $\epsilon << 1$, then $(E/E_c - 1)\,\beta = \epsilon\,\beta$, and

$$\exp\text{–}(\epsilon\,\beta) = 1 - \epsilon\,\beta + (\epsilon\,\beta)^2/2! \;\; -(\epsilon\,\beta)^3/3! + \ldots \quad (4\text{-}56)$$

Provided that $\epsilon\,\beta << 1$, we may neglect higher order terms. This leads to the result $1 - \exp\text{–}\epsilon\,\beta = \epsilon\,\beta$. Substituting this result back into equation 4-55, and recalling that $\epsilon = [E/E_c - 1]$, and that $\beta = K_p\,E_c/Y_{max}$, we obtain

$$Y/Y_{max} = [E/E_c - 1]\,K_p\,E_c/Y_{max} \quad (4\text{-}57)$$

or, equivalently

$$\boxed{Y = K_p\,(E - E_c)} \quad (4\text{-}58)$$

We now see that K_p is simply the slope of the modulus versus exposure function in the region just beyond E_c. *Thus, K_p can be interpreted as the initial rate of increase in Young's modulus of a photopolymer with increased actinic exposure just beyond E_c.* For this reason, we have chosen to call K_p the "photomodulus coefficient."

The units of K_p are modulus of elasticity divided by exposure. The modulus, Y, is expressed in units of Newtons per square millimeter, or force per unit area. Exposure is expressed in units of millijoules per square centimeter, or energy per unit area. Thus, the dimensions of K_p are that of a reciprocal length (10^5 m^{-1}). For the resins studied, the characteristic length, $S \equiv 1/K_p$, ranges from 1.1 to 6.7 microns.

It is worthwhile noting that S is much larger than the laser wavelength, λ (roughly 0.32 to 0.36 microns) or the length of a so-called "long chain" polymer molecule, d_m, in a configuration typical of SL resins, which would be in the range of 1 to 7 nanometers, or 0.001 to 0.007 microns.

Conversely, S is much smaller than either the laser spot diameter, B, or the radius R of the "zone of influence" for photopolymer exposure, or D_p, the photopolymer penetration depth, all of which are within the range of about 150 to 300 microns. Finally, all of these quantities are much smaller than L, a typical part dimension. Ordering the characteristic lengths of SL, we find

$$d_m << \lambda << S << D_p < B < R << L$$

A possible physical interpretation of S might be as an average characteristic size for local polymerization "islands" just prior to general macroscopic cross-linking between these initially separated domains. In this view, K_p would then be a measure of the spacial frequency of these domains. A high value of K_p would correspond to a photopolymer initially linked together from small, closely spaced domains, while a low value of K_p would correspond to a more loosely spaced structure.

These conclusions, regarding the formation of photopolymer "domains," as well as their characteristic dimensions, are in general agreement with the findings of Kloosterboer[4], and give additional weight to the physical significance of the photomodulus coefficient.

3. In the limit as $E/E_c >> 1$, it is clear that the exponential term of equation 4-55 tends toward zero, and that Y/Y_{max} approaches unity. Physically this means that for exposures very much greater than E_c, the Young's modulus of the photopolymer will asymptotically approach the fully cured value.

The validity of the photomodulus equation 4-55 must be determined by direct comparison with experimental results for a variety of SL resins. The remaining sections of this chapter describe the experimental methods used and the results obtained.

4.9 Experimental Method

Early attempts to produce uniformly exposed thin tensile specimens included cutting samples from preformed films and exposing thin, uncured resin layers through dog-bone shaped masks. However, these attempts did not result in reproducible data. The best experimental repeatability was achieved with samples produced through laser raster scanning techniques using an SLA-250.

The laser drawing parameters were carefully chosen to ensure maximum uniformity of exposure and, therefore, minimum inhomogeneity of the resulting test sample modulus. The laser beam diameter, B, as previously defined, was maintained at approximately 10 ± 0.5 mils by carefully positioning the test fixture along the z axis, centered in the laser beam focal waist.[5]

The hatch spacing, h_s, (the lateral distance between adjacent laser scan centerlines) was chosen to be 3 mils. The value $h_s = 3$ mils was selected because it produced samples that had no visible drawing artifacts. The lack of artifact structure is generally a good indicator of sample uniformity for these resins. Spacings of $h_s = 4$ mils, or greater, did result in visible artifacts.

To produce samples of nearly uniform modulus, it was necessary to devise a method of tensile sample preparation that provided for bilateral exposure. From the ninety/ten rule described earlier, provided the specimen thickness, w, is kept below 90% of the resin D_p, then the maximum possible variation in exposure, from center to edge, will be less than 10% throughout the sample. Furthermore, this corresponds to less than a 5% difference in exposure between any arbitrary point within the sample and the average exposure for all points.

Consequently, a special experimental fixture had to be developed that would ultimately produce uniform thin samples of photopolymer, having thicknesses less than 90% of D_p, for the particular resin being tested.

To accomplish these goals, an SLA-250 was fitted with the special clamping apparatus shown schematically in *Figure 4-9*.

Uncured resin was inserted with an eyedropper between two glass microscope slides separated by a shim of known and constant thickness. The shim, which resembled the general character of an automotive gasket, had a cutout in the shape of a dog-bone to allow the laser to expose the sample in the correct region. Next, the two glass slides, the shim, and the liquid resin were clamped with sufficient force to ensure that the remaining resin sample thickness would be very nearly equal to the thickness of the shim.

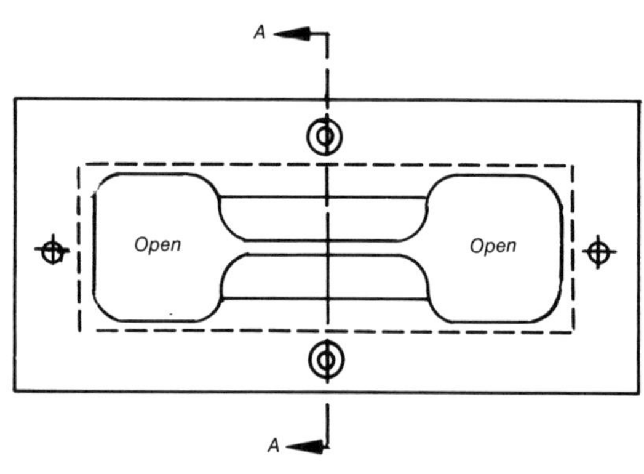

Section *A-A*:

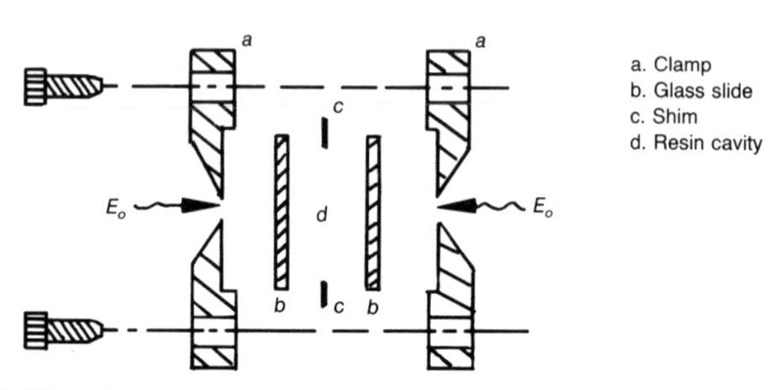

Figure 4-9. Clamping apparatus.

Of course, care must be taken not to exert so much clamping force that the glass slides either deform or break. Optimum clamping force was determined, by trial and error, to be that value which led to the most consistent sample thickness values. A very thin cut in the perimeter of the shim allowed for resin "bleed off." This is important in being certain that enough resin has been added to preclude any voids.

The thickness of the shim was typically chosen to be 1 mil less than the value of D_p for the resin being tested. This resulted in values of F in the range from 0.83 to 0.87. These values more than satisfied the ninety/ten rule, and resulted in maximum exposure variations of about 8% throughout the sample. This corresponds to only about a $\pm 4\%$ variation about the mean value.

Next, the sample was exposed from one side. To do this, the SLA laser was intentionally scanned in a "raster" fashion with $h_s = 3$ mils. When the initial scanning had completely exposed the dog-bone shaped test sample, the entire test fixture, shown in *Figure 4-10*, was flipped over to expose the sample to a second, otherwise identical exposure from the opposite side.

Sample registration in the *x-y* plane is achieved from one exposure to the next by locating the clamping fixture onto a special jig having precision registration pins, as shown in *Figure 4-10*. The pins are positioned parallel to the scanning axis, which also happens to be the tensile test axis of the specimen. This ensures accurate registration, and essentially eliminates translational position errors.

To eliminate possible angular alignment errors between the laser scan axis and the centerline of the jig, the entire jig is mounted on a rotating platform. The

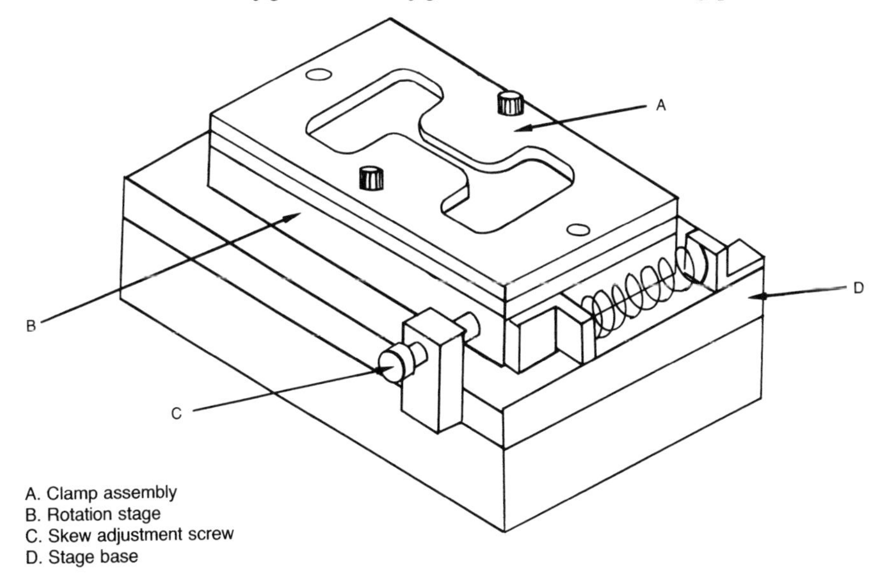

Figure 4-10. Test fixture.

rotation adjustment is made with a precision screw mechanism having a very fine pitch and positive spring-loaded restraint to guard against backlash. In this way, one can carefully iterate, on a few sacrificial samples, until any tendency towards "skew" has been effectively minimized. Obviously, quite careful attention to sample registration is necessary to achieve a high level of exposure uniformity.

Earlier, in equation 4-27, we determined the exposure distribution for a single laser cured line. However, in this case we are, in effect, creating a virtually continuous "skin" by raster scanning numerous closely spaced lines. Since we have selected a hatch spacing $h_s < W_o$ (3 mil separation of scan centerlines for a beam with a Gaussian radius of about 5 mils), then the exposure at any point involves the superposition of the exposure contributions from neighboring lines.

A detailed computer analysis[6] has determined the composite exposure distributions resulting from a series of closely spaced raster scans of a Gaussian laser as a function of the ratio of h_s/W_o. *Figure 4-11* shows one of the resulting computer-generated plots for the specific case where $h_s/W_o = 0.6$, appropriate to the fabrication of photomodulus samples.

The results of this analysis have shown that provided $h_s < W_o$, then the average exposure, E_{av}, is in close agreement with the result

$$E_{av} = P_L\, t_d/A_s \qquad (4\text{-}59)$$

where P_L = laser power (milliwatts)

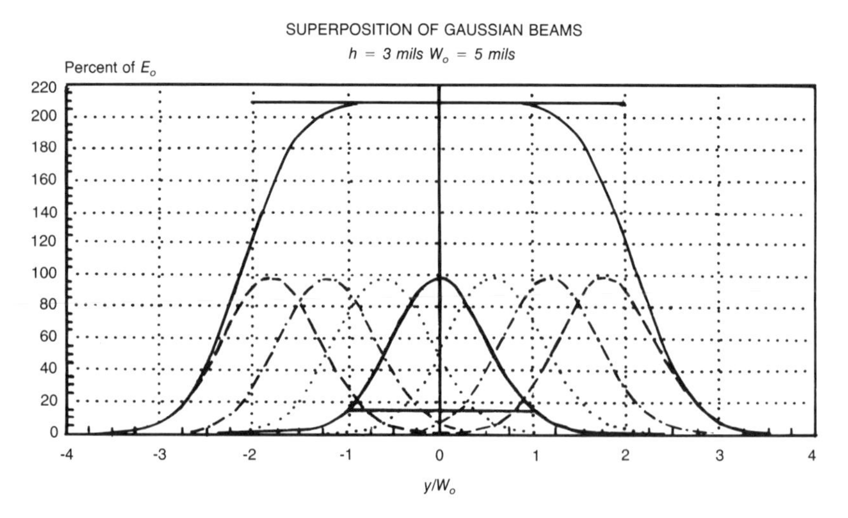

Figure 4-11. Composite exposure distributions.

t_d = total laser drawing time (seconds)

A_s = sample surface area (square centimeters)

Note that since the sample is exposed sequentially from both sides, then t_d refers to the total bilateral drawing time.

It is important to be aware of the practical exposure limits associated with using the laser scanning technique to produce tensile specimens. If the average exposure is below E_c, no resin is cured. If $E_c < E_{av} < 2E_c$, the resulting sample typically does not have sufficient mechanical integrity to survive even very gentle handling.

Finally, if E_{av} exceeds about 50 E_c (typically 200 to 300 mJ/cm^2), then ''line spreading'' occurs. At these high exposures, generally beyond the linear portion of the working curve, the cured linewidth increases as discussed in section 4.5. This leads to uncertainties in the effective gauge cross-sectional area, and hence uncertainties in the determination of the stress versus strain data, and ultimately in the calculation of Young's modulus. For these reasons, the great majority of the data taken lie in the range from $2E_c$ to about 50 E_c.

Next, numerous tensile samples are prepared for a given resin type. Each sample is given a different exposure within the practical exposure limits discussed. Some repetitions at the same exposure are done intentionally, to check repeatability. Once the test samples have been exposed from both sides, in the special tooling described, they are then carefully removed from the location between the two glass microscope slides, cleaned of excess liquid resin, and their gage cross-sectional areas are measured. Low contact pressure digital micrometers are used to measure the sample thickness. Measurements are made at a number of locations. If the differences between measurements are significant, the sample is rejected. An optical microscope is used to measure the gauge widths.

The samples are then tensile tested using a Lloyd Instruments Model 500 test apparatus. The exposure, E_{av}, is determined using a calibrated laser power meter, an accurate stopwatch (drawing times with these small hatch spacings and relatively high exposures are sufficiently long that human reaction times are negligible), and careful measurements of the sample surface area.

The stress versus strain data for each sample is collected in the form of a load versus extension curve, as shown in *Figure 4-12*.

The slope of the linear region of the load versus extension curve is then determined graphically, and Young's modulus for that test sample is calculated by correcting for the sample cross-sectional area and gauge length.

4.10 Experimental Results

This section describes the results of experimental determinations of Young's modulus of numerous test samples as a function of actinic exposure for five

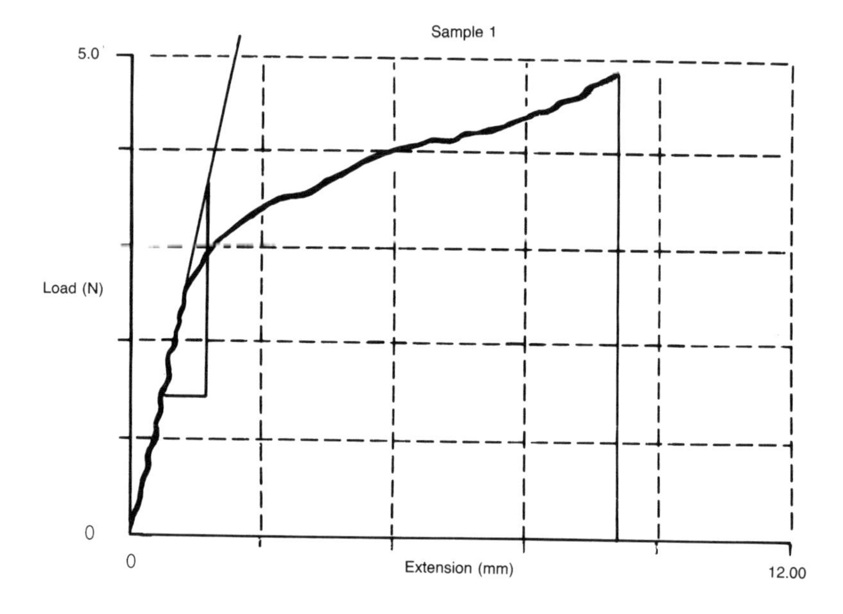

Figure 4-12. Load versus extension curve.

different SL resins. These results were determined using the sample preparation technique described in section 4.9.

Figure 4-13 shows a normalized plot of modulus (Y/Y_{max}) versus exposure (E_{av}/E_c) for Ciba-Geigy XB 5081-1 resin. The data shown as open circles indicate experimentally measured Young's modulus values for specific test samples. The figure also shows the theoretical curve predicted by the photomodulus equation 4-55, where the dimensionless resin parameter, $\beta = K_p E_c/Y_{max} =$ 0.064 for this resin. This value corresponds to a photomodulus coefficient, K_p $= 5.7 \times 10^5\ m^{-1}$, *as determined from the initial slope,* dY/dE, of the nonnormalized modulus versus exposure curve for XB 5081-1.

From *Figure 4-13*, it is clear that the experimental data correlates quite well with the theoretical curve. Furthermore, at exposures just above E_c, we see that the data is closely approximated by the simple linear expression $Y = K_p (E-E_c)$, as described earlier by equation 4-58. Finally, at high exposures, Y does indeed asymptotically approach Y_{max}.

Figures 4-14 through *4-17* show the experimental data, as well as the theoretical photomodulus curves, for the four additional Ciba-Geigy SL resins: XB 5131, XB 5134-1, XB 5139, and XB 5143 respectively. In all five cases, excellent correlation exists between the experimental data and the curves predicted from the photomodulus equation. Furthermore, this correlation exists over almost an order of magnitude of variation in the value of the resin constant β; from $\beta = 0.007$ for XB 5134-1, to $\beta = 0.064$ for XB 5081-1.

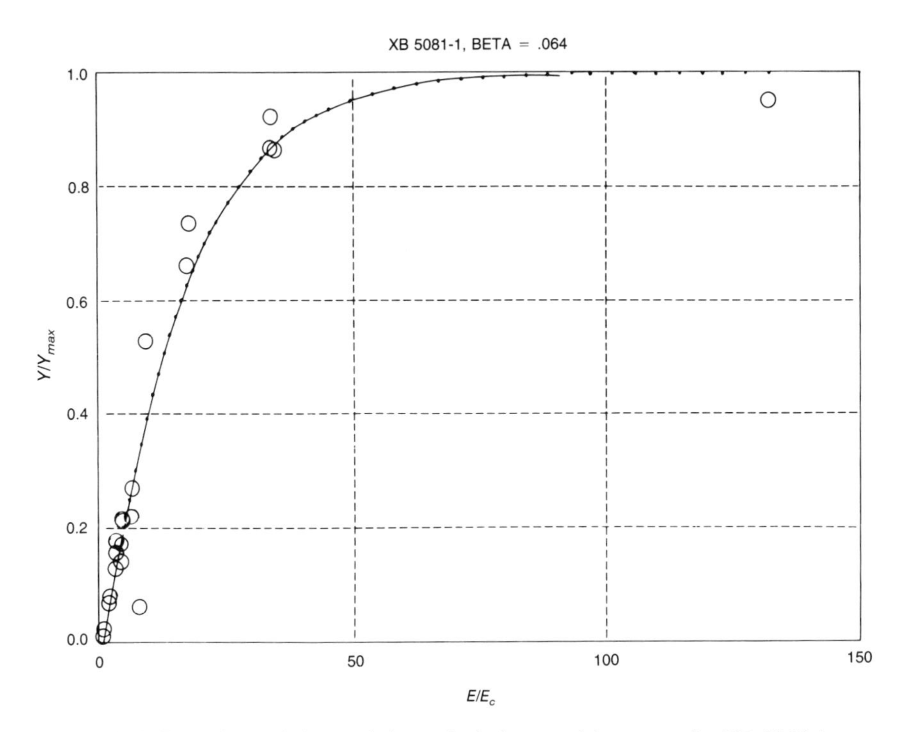

Figure 4-13. Experimental data and theoretical photomodulus curve for XB 5081-1.

The scatter that exists in each data set is most likely due to small errors in determining cross-sectional areas, damage caused by over-handling of the samples, and any mechanical discontinuities in the sample gauge region, such as voids or microcracks. Nonetheless, considering the difficulty of the measurements, the scatter in the data is quite reasonable.

Table 4-2 summarizes the values of D_p, E_c, K_p, and β for the five SL resins tested to date.

Table 4-2
Summary of Resin Values

Resin	D_p (mils)	E_c (mJ^2/cm)	K_p (10^5 m^{-1})	Beta
XB 5081-1	7.1	5.6	5.7	.064
XB 5131	6.9	5.0	8.7	.058
XB 5134-1	6.6	7.6	1.6	.007
XB 5139	6.2	5.1	5.9	.015
XB 5143	5.7	4.3	4.0	.045

As is evident from inspection of the data presented, the theoretical model predicts the function $Y(E)$ over a wide range of exposure values and resin types.

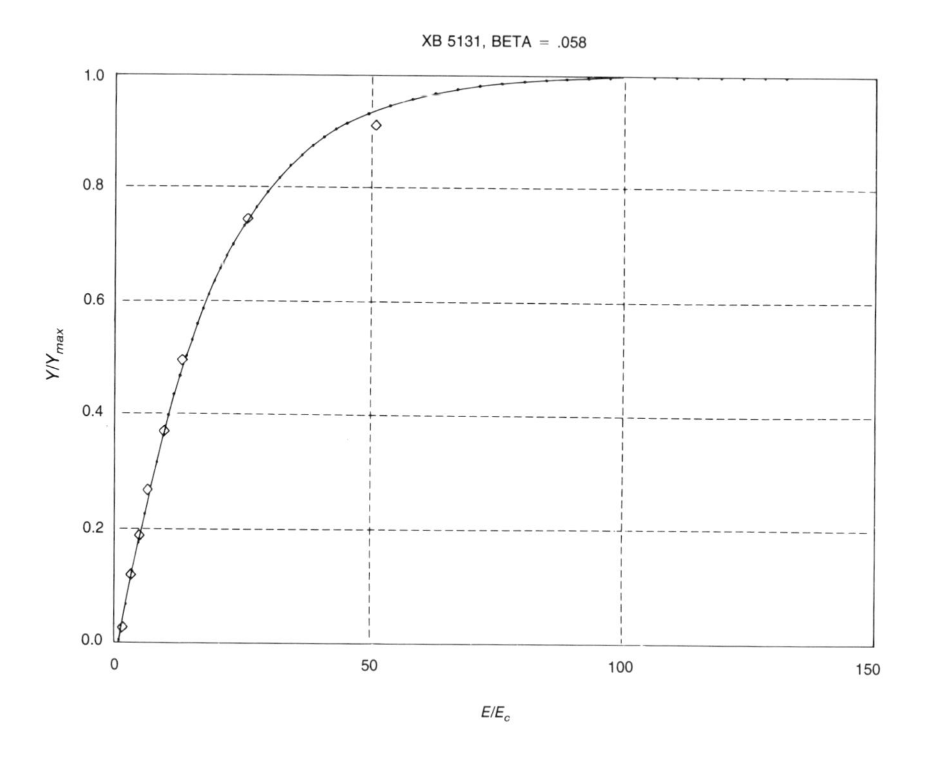

Figure 4-14. Experimental data and theoretical photomodulus curve for XB 5131.

Thus, once the three basic resin constants E_c, K_p, and Y_{max} have been determined for a particular resin, the entire photomodulus behavior may be predicted with good accuracy. This greatly reduces the amount of experimental data needed to fully characterize the modulus of elasticity of a given resin.

Also, there is excellent reproducibility of the experimental results. Independent testing performed at Ciba-Geigy (Fribourg, Switzerland) produced virtually identical results to those obtained at 3D Systems (Valencia, California). Both groups determined the value of $K_p = 1.6 \times 10^5 \ m^{-1}$, for resin XB 5134-1. Thus, the goal of improving repeatability and reducing errors due to the nonuniformity of the samples has apparently been met.

Further, once $Y(E)$ is known for a given resin, this information can be used with the exposure distribution corresponding to a particular SL part building method to calculate both the "green strength" as well as the strength of the fully postcured part.

Finally, this model, coupled with cure shrinkage data as a function of actinic exposure, has already been used to optimize part building techniques. Specifically, these results were instrumental in the development of both WEAVE™ and STAR-WEAVE™, which are described in detail in Chapter 8.

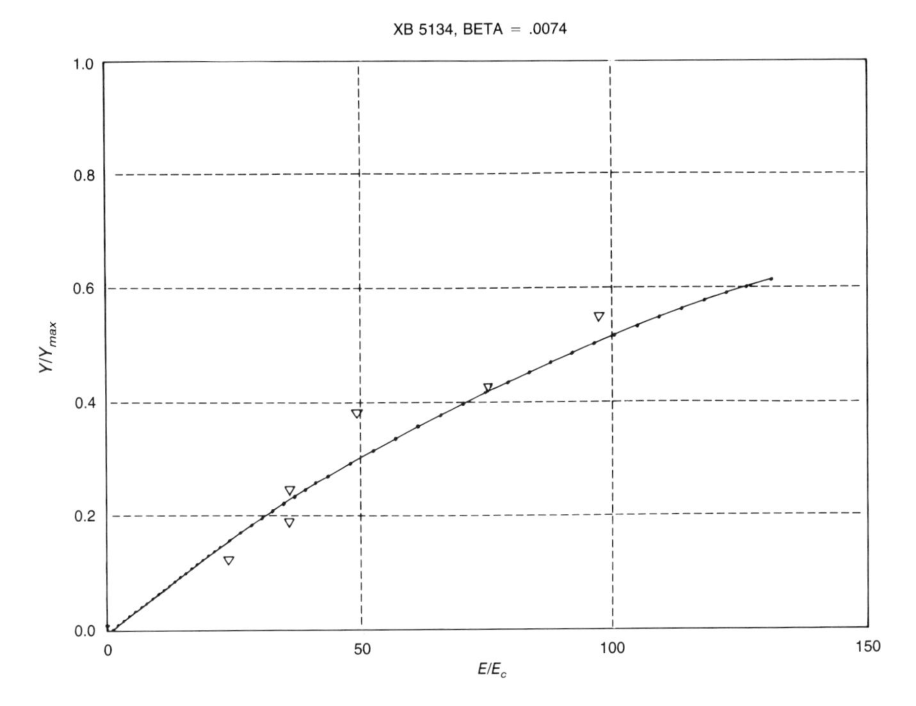

Figure 4-15. Experimental data and theoretical photomodulus curve for XB 5134-1.

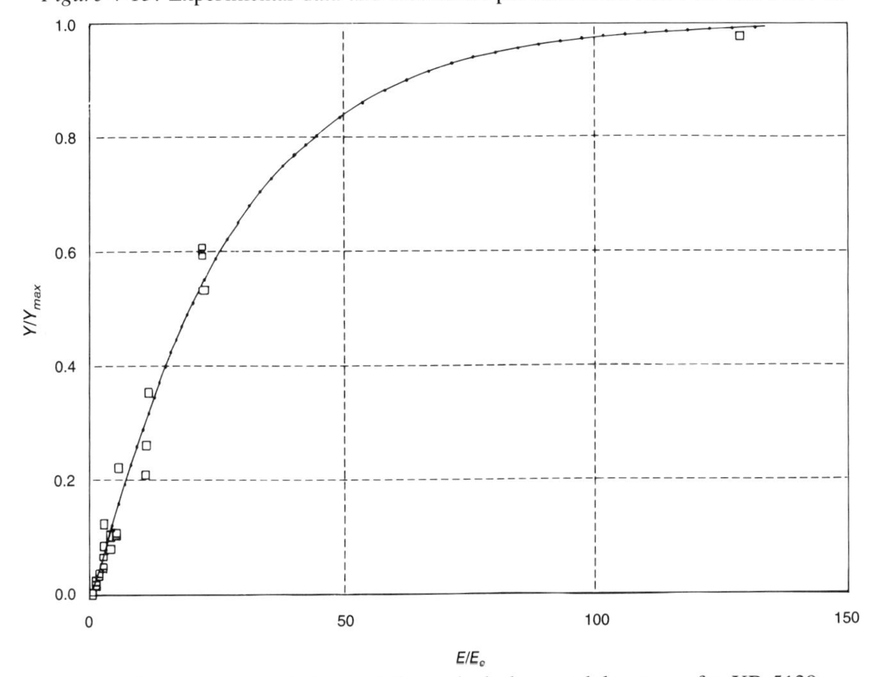

Figure 4-16. Experimental data and theoretical photomodulus curve for XB 5139.

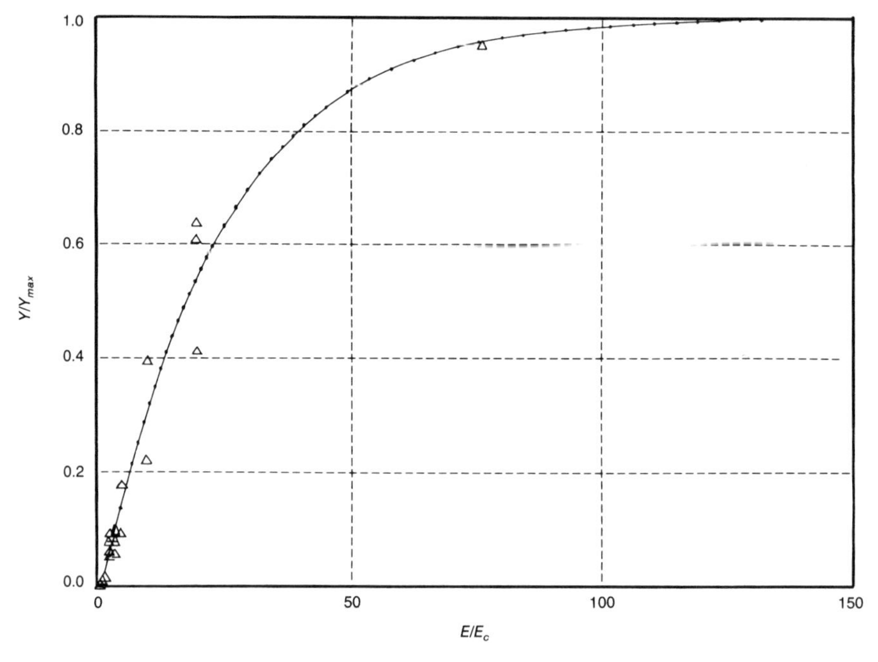

Figure 4-17. Experimental data and theoretical photomodulus curve for XB 5143.

REFERENCES

1. Sneddon, I.N., *Special Functions of Mathematical Physics and Chemistry*, Oliver and Boyd, Edinburgh, 1961, pp. 13-14.
2. Evans, H. and Jacobs, P.F., *The Development of Photopolymer Modulus with Actinic Exposure*, Conference Proceedings, Second International Conference on Rapid Prototyping, University of Dayton, Dayton, OH, June 23-26, 1991, pp. 69-85.
3. Sokolnikoff, I.S. and Sokolnikoff, E.S., *Higher Mathematics for Engineers and Physicists*, McGraw-Hill Inc., New York, 1934, pp. 216-220.
4. Kloosterboer, J.G., *Network Formation by Crosslinking Photopolymerization and Its Applications in Electronics*, Published in *Advances in Polymer Science*, Volume 84, *Polymers in Electronics*, pp. 1-61, 1988.
5. Siegman, A.E., *An Introduction to Lasers and Masers*, McGraw-Hill Inc., New York, 1971, pp. 312-314.
6. Nguyen, H., Internal communication, 3D Systems, Valencia, CA, June 1991.

chapter 5

Software Architecture

Wallach was a scholar of computer architectures. He knew by heart the works of the Michelangelos, the Frank Lloyd Wrights, and the Gaudis of the computer profession. He imagined himself standing in front of a roomful of experts when Eagle was all done. They would question him about the architecture, inserting sharp little knives in his flank and then twisting them. 'Why didn't you do it this way, Steve, when it's obviously better?'

—Tracey Kidder
The Soul of a New Machine
The Atlantic Monthly Press, 1981

5.1 Software as the Common System Link

This chapter provides an overview of the software architecture utilized in RP&M systems, focusing primarily on StereoLithography. The chapter explains how software is involved in the part building process, how the software has evolved, and finally some future software concepts. For any RP&M system, the natural first questions are "what objects can I build with it," and "how well can I build those objects?" Because almost all aspects of the process involve software control, a basic understanding of software functioning is mandatory. The answers involve multiple disciplines, including chemistry, physics, and mechanical and electrical engineering. Defining and synchronizing all the processes required to produce a high-quality part is a very complex task.

By **Grady O. Floyd**, *Software Project Manager, Research and Development Group, 3D Systems, Inc., Valencia, CA.*

Data Gathering

Upon deciding to build any object, a less natural question arises: "How shall I represent the object I desire to construct?" Even with a perfect RP&M system, the object can only be as accurate as the data which represents it. Typically, RP&M systems receive their data from computer-aided design (CAD) systems. CAD packages come in many forms:

1. **Two dimensional**–similar to a single blue print. Data is stored in the form of 2D points, 2D lines, 2D circles, 2D splines, text, etc.
2. **Two-and-a-half dimensional**–several 2D drawings linked together. Data storage is similar to item 1, with the addition of data constructs which define how the various 2D drawings link together.
3. **Three-dimensional wireframe**–similar to a toothpick sculpture where only the edges of the part are represented. Data is stored in the form of 3D points, 3D lines, 3D circles, 3D splines, "3D" text, etc.
4. **Three-dimensional surfaces**–similar to building an object with flexible cardboard strips where the cardboard has "zero thickness" and "no mass." Data is stored in the form of ruled surfaces, surfaces of revolution, Bezier surfaces, Non-Uniform Rational B-Spline (NURBS) surfaces, etc.
5. **Three-dimensional solid**–similar to representing the object with modeling clay which is altered to produce the desired object. The first class of this type of CAD package is referred to as Constructive Solid Geometries (CSG), where data is stored as a collection of Booleaned 3D primitives such as spheres, parallelpipeds, polyhedra, cylinders, cones, toroids, etc. Reference 1 explains the Boolean operation.

   A second class of three-dimensional solid packages are enhanced wireframe packages where patches are formed by hooking the edges together forming "faces." The topology is defined by the local normals. Normals point away from the part mass. A third class is a hybrid of class 1 and class 2. A fourth class is Voxel Representations, where mass properties are associated with each pixel.

Because the output of an RP&M system is a tangible object, the best representation of the object comes from those CAD systems which utilize the three-dimensional solid modeling approach. Unfortunately, these software packages represent a relatively small percentage of the total CAD systems in use today. This has slowed the transition to RP&M systems.

The RP&M system requires data in a particular format. The original standard for RP&M system input is the StereoLithography file (STL) format. This standard is based upon a mesh of connected three-dimensional triangles whose verticies are ordered to indicate which side of the triangle contains the mass. Most CAD companies with three-dimensional solid modeling packages provide the ability to output STL files. Currently, STL files have become the defacto standard for input into all types of RP&M systems. 3D Systems has also announced development agreements with several companies to accept other data formats in addition to STL files. Companies such as Cubital, DTM, and Stratasys also accept other input formats. Some of the industry standard file formats are as follows:

Initial Graphics Exchange Specification (IGES, version 5). IGES is a standardized format for most 2D and 3D CAD system data. Generally, CAD systems do not work directly with IGES models. Most CAD systems have their own internal data formats. To change data from one CAD format to another, a neutral IGES file is usually created, then the final CAD system data is generated from the neutral file. Because two translations are involved instead of one, data translation problems are approximately twice as likely to occur. More importantly, because each system may have different data constructs, certain types of information may not be exchangeable, and that information is then lost. Problems of this nature apply to most of the other standardized file formats as well. IGES is defined by the U.S. Department of Commerce Document NISTIR 4412, and is distributed through the National Computer Graphics Association (NCGA) in Fairfax, Virginia.

Numerical Control (NC) G-codes. Numerical control G-codes comprise the standard format used in the NC market. Most CAD/CAM systems offer outputs of "cutter location" and "sequence of operations" data. This data is output in the ANSI defined APT cutter location (CL) data format and is processed into G-codes. These codes are then used to drive the programmable mill or lathe. The codes are basically a series of "go to X,Y,Z" commands nested between "set spindle speed" and "change tool" commands. More complex machines support more G-codes. For example, a five-axis milling machine generally supports more G-codes than a three-axis machine.

Association of German Automotive Manufacturers–Surfaces Interface (VDA-FS, version 2). VDA-FS is the standardized CAD format for European auto makers. The data is similar to the internal representations of data in CATIA, which is a popular CAD package worldwide. VDA also has a format called VDA-IS, which is portable to IGES. VDA-FS is explained in the German DIN 66301, and related DIN documents.

Hewlett-Packard Graphics Language 2 (HP/GL-2). HP/GL-2 is a standard data format for graphic plotters. Data types are all two-dimensional, including lines, circles, splines, text, etc.

Computerized Axial Tomography (CAT) Scan. CAT scan data is a particular approach for medical imaging. This is not standardized data. Formats are proprietary and somewhat unique from one CAT scan machine to another. The scan generates data as a grid of three-dimensional points, where each point has a varying shade of grey indicating the density of the body tissue found at that particular point. Thus, a tumor would have a different shade of grey in the CAT scan than the surrounding healthy tissue. Data from CAT scans have been used to build skull, femur, knee, and other bone models on SL systems. Some of the reproductions were used to generate implants, which have been successfully installed in patients.

Nuclear Magnetic Resonance Imaging (NMRI or MRI). NMRI is similar to CAT scan data, but the data is gathered by a different process.

As more data input formats become acceptable to the RP&M software architectures, the easier it will be to integrate RP&M systems into existing CAD or medical sites. However, each additional format requires further analysis software to ensure that the data is sufficiently complete to accurately build the desired object.

Data Analysis

This section addresses data analysis, focusing on problems encountered with STL files and providing a framework to understand three-dimensional data problems in general. While errors in the data may not reveal themselves on the CAD system's computer monitor, flaws become readily apparent when the part is constructed.

STL files are comprised of a mesh of connected "three-dimensional" triangles, much like the Eiffel Tower. Although a triangle is, strictly speaking, a two dimensional object, the "three dimensional" terminology applies to the *X*, *Y*, and *Z* coordinates which form the triangle's endpoints. *Figure 5-1* shows how triangles can be used to represent rectangles and approximate curves. *Figure 5-2* shows an object represented by triangles, often referred to as a tesselated object.[2] Triangles in an STL file must all mate with other triangles at the verticies; this is known as the "vertex to vertex" rule. Furthermore, triangles must be properly oriented to indicate which side of the triangle contains mass.

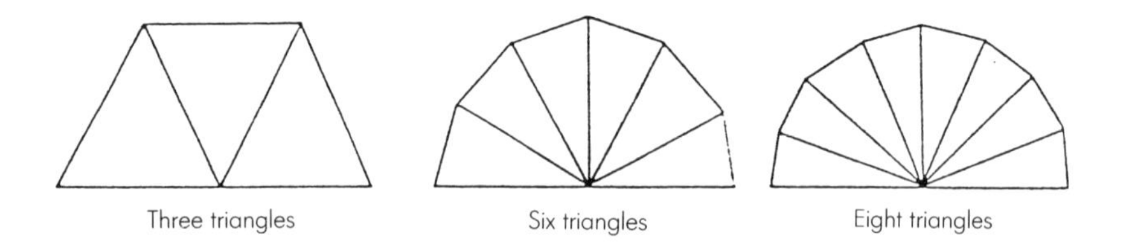

Figure 5-1. Triangles approximating half-circles.

Some methods used to construct a part may be particularly sensitive to data anomalies. One such item is triangle connectivity. A lack of connectivity in the three-dimensional triangle matrix which comprises the STL file leaves ambiguous gaps in the object's representation. What a human can look at and instinctively repair is not always obvious to a computer program. Gaps between triangles are further convoluted in the slicing process, in which a two-

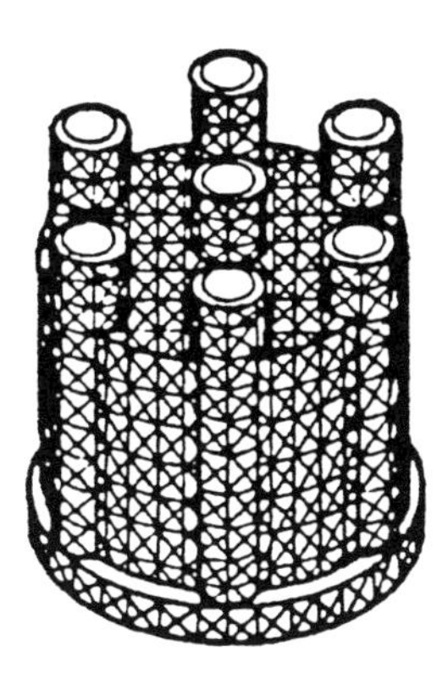

Figure 5-2. Tesselated distributor cap.

dimensional plane is cut through a three-dimensional STL file. To discuss the plane cut operation, a few concepts must first be established:

1. The "universe" in which a part exists is the maximum volume of a specific RP&M system. On an SLA-500 this is 20″ × 20″ × 24″. On an SLA-250, this is 10″ × 10″ × 10″. A part that exceeds these dimensions would surpass the boundaries of the RP&M universe. In practice, situations of this nature are prevented through software parameter checks.
2. The slicing technique uses a ray-tracing algorithm which systematically scans through the particular Z level of the RP&M universe. A ray trace usually starts at one edge of the universe, gathers data as it moves across the space, and ends when it reaches the opposite boundary of the universe. The type of data gathered involves the coordinates where the solid (mass) sections of the part do and do not exist in the plane cut.
3. Thus, if a gap in the triangles exist, the plane cutting algorithm may determine that a solid mass extends to the edge of the RP&M universe. *Figure 5-3* shows a coffee mug with a plane cut and the resulting two-dimensional cut. *Figure 5-4* shows a coffee mug with a gap and the resulting incorrect two-dimensional cut. The incorrect cut is the result of the gap in the triangle and the method used in the ray trace algorithm, which determines solid versus nonsolid regions. The ray trace algorithm described in *Figure 5-4* has been simplified for clarity.

A second category of STL file problems involves incorrect normals where two adjacent triangles indicate that the mass of the object is on opposite sides. *Figure 5-5* shows correctly and incorrectly oriented normals on adjacent triangles. Normals are computed by the right hand rule. The numbers inside the triangle vertices indicate the order in which the coordinates were listed in the file.

As incorrect normals propagate, Möbius strip conditions can begin to exist, where surfaces transform into each other. An Escher rendition of a Möbius strip, appropriately covered with software bugs, is shown in *Figure 5-6*.

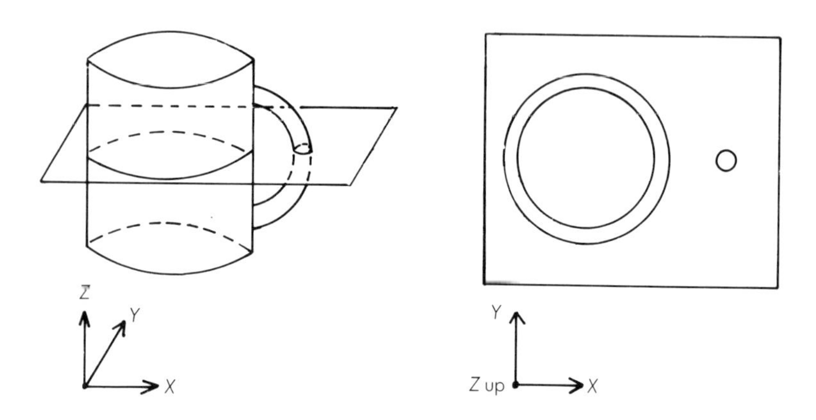

Figure 5-3. Coffee mug with a plane cut.

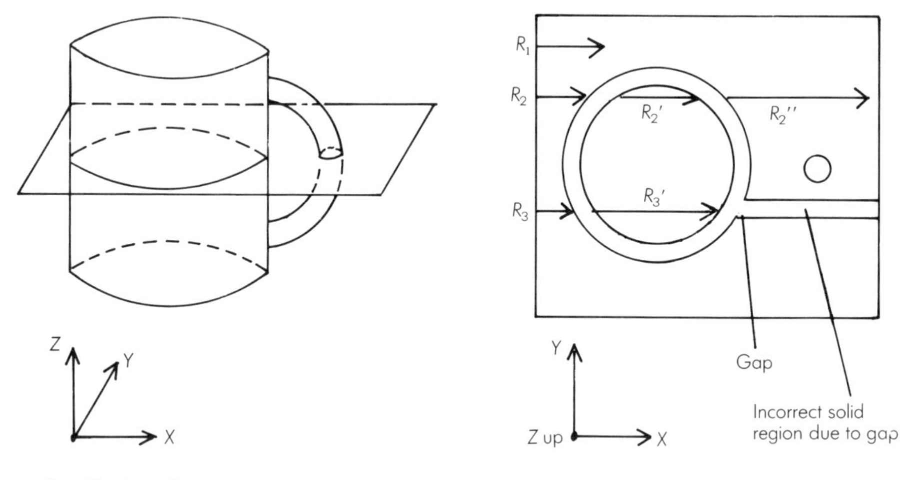

R_1 = Ray trace 1

R_1 never finds a solid and stops at the end of the "universe."

R_2 = Ray trace 2

R_2 finds the coffee mug edge, a nonsolid region in the middle of the cup, R_2', a second solid region after R_2', and then the final nonsolid region R_2''

R_3 = Ray trace 3

R_3 is similar to R_2 but because of the gap, it never encounters a nonsolid region.

Figure 5-4. Coffee mug with a gap in the STL file.

Problems such as gaps between triangles, incorrect normals, and Möbius strip conditions are usually the result of minor problems in the CAD to STL translators aggravated by particular part geometries. 3D Systems' software

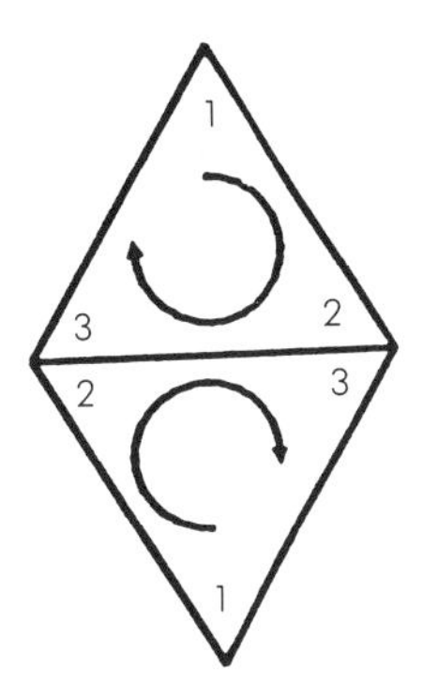

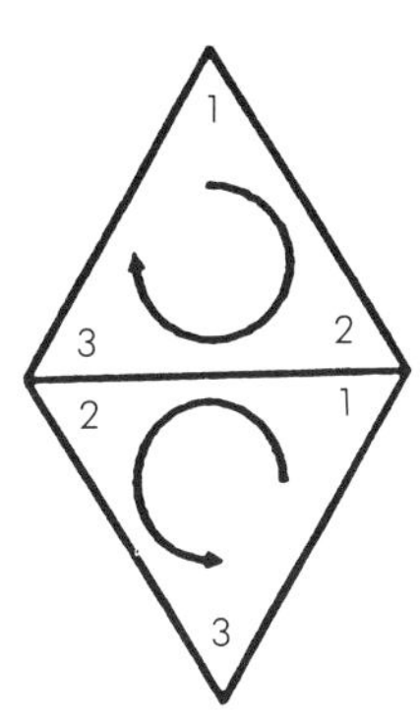

Figure 5-5. Correct and incorrect triangle orientation.

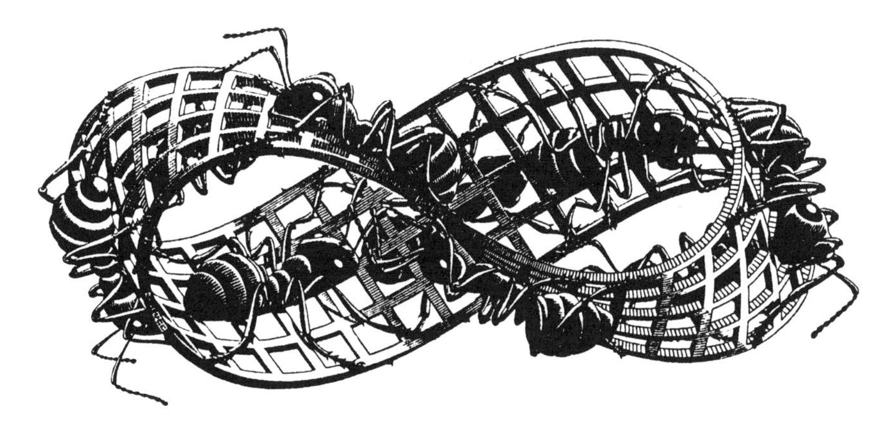

Figure 5-6. Möbius strip II, by M.C. Escher. [3]

includes a program (3DVerify) to identify and/or correct problems in STL files. In addition, file viewing programs are provided on both the DOS and the workstation platforms to help further identify file anomalies. Naturally, not all incorrect STL files are capable of being repaired, and not all repairs are necessarily what the user desires.

Part Preparation

A third question is "how well can my system recreate the object?" This process execution and control question is intrinsically tied to the part preparation software. The final object is the result of a combination of the data which represents it, the part preparation parameters submitted to the building cycle, the precision with which the machine can execute the part building commands, and the postprocessing techniques utilized.

Part preparation software has improved significantly in recent years, and is frequently required to adapt to new building methods. In the early days of SL, all aspects of creating the final object were subject to question. Research to determine those build methods that produced the best results was just beginning. Parameters which might yield an excellent part with some geometries failed with others. Large flat areas and ''trapped volumes'' were of particular concern.

Part preparation begins by determining how the object(s) should be positioned in the vat. The criteria is dependent on what the RP&M user wants. Usually part accuracy is most important, aesthetics are next, and build time is somewhat less critical. Deciding which surfaces need the best finish, fitting the maximum number of parts on the platform, minimizing the number of part supports, attaching proper supports to any cantilevered areas of the part, and eliminating trapped volumes are all considerations in optimal part orientation. Some of the parameters requiring definition include unit type (inch or millimeter), layer thickness, hatch spacing, number of sweeps, the recoater speed and blade gap, and the cured linewidth and resin shrinkage compensation factors. Fortunately, commonly used part building parameters can be grouped and saved in building styles. Default values can be employed in many cases. Users may elect to establish their own styles which are optimal for their particular needs.

Part Building

Various parameters are utilized by the machine for data processing, motion control, and motion synchronization. Items such as units and layer thickness affect data processing, while drawing and recoating styles affect motion control and synchronization. RP&M systems require software and hardware feedback mechanisms to prevent the system from causing damage. Because RP&M systems are required to repeatedly create an object with exacting dimensional tolerances, every motion must be tightly controlled.

Most part building improvements require software changes. Enhancements that might make the product more marketable also require software changes. Thus, the needs of technical product improvement versus marketability must be balanced. Before coding a software change, requirements and specifications must be agreed upon. If the project is approved for delivery, then coding, internal alpha testing, external beta testing, and final delivery must also occur. Training documents, training procedures, manuals, and other literature are required.

The ability to consistently deliver quality software plays a large part in the success that any RP&M system will achieve. Likewise, complete software documentation is necessary for users to understand what the parameters control and how they interact. Ideally, the software architecture should be mentally intuitive. Controversy will always exist as to how it should be improved, since one person's intuition is not always identical with anothers. Software is the common link through almost every aspect of the RP&M system.

5.2 Software Evolution

To better understand existing RP&M software, it is helpful to discuss its evolution. In this section, the history of file formats, conventions, and part building improvements will be covered. The marketing, engineering, and manufacturing forces which shaped the existing software will be examined. Software development methods will also be covered.

SLA-1 Development Software

From 1984 to 1986, before the SLA-1 was developed, SL parts were built on prototypes which directed the focused ultraviolet radiation from a mercury lamp through a fiberoptic tube positioned by a plotter's *X/Y* traveller. Command line software read a data file which drove the traveller. After the data in the file was read, a secondary command line instruction was utilized to move an elevator. The process was then repeated for each part layer. Small turbine fans, test boxes, and other diagnostic and demonstration parts were built. The original Stereo-Lithography patent applications were made based on these results.

As fundamental questions involving laser and photopolymer interactions began to be solved, grouping of discrete software tasks was initiated. Various part preparation parameters became associated with the slicing algorithm, and others with certain part geometries. Motion control software was assembled into appropriate libraries, such as control of the plotter or the elevator.

By 1987, precision control of the *X/Y* mirrors, which direct the laser beam onto the surface of the vat, had been accomplished. Version 1.21 and 1.30 of the "laser controller system" was a DOS PC-based set of programs.[4] The Laser.exe program consists of a foreground software interrupt handler and a background instruction queuing system. The Stereo.exe program is responsible for geometric correction of the mirrors using a file generated in the mirror calibration process. Stereo.exe also receives data relating to drift correction of the mirrors during calibration and part building. To gather the drift correction data, mirrors perform a search pattern to locate two fixed laser beam profilers. Any changes are fed into the Stereo program, which then performs translate or zoom functions upon the calibration results to keep the coordinate locations drawn in the vat consistent from layer to layer.

The STL File Format

The STL file format opened the door for the RP&M market by allowing CAD data to reach SL systems. The STL file is a faceted representation of the exterior surfaces of the part. Faceted representation means that each surface is defined by a set of small planar triangular facets. Most CAD systems already used faceted representation to generate shaded image displays, so providing an STL file interface was relatively straightforward.

As mentioned earlier, many of the initial CAD to STL translators had a number of bugs. It was several years before the 3DVerify program was written to identify and repair bad STL files. In the meantime, much effort was devoted to improving the translators and to providing diagnostic tools to locate the problematic areas within faulty STL files. The View program on DOS and the 3DView program on the Silicon Graphics Workstation were the first such programs that allowed users to view their files from all angles and thereby detect obvious errors. Translation, rotation, zoom, and triangle query functions provided the ability to closely examine questionable regions. Also, after the Slice program was executed, the user could view the cross sections that had been generated. The ability to view slice layers one at a time, or in groups, was of significant value in preventing unsuccessful part building attempts.

The First Production Release

The first beta release (release 2.62) of SL software went to five customers in January of 1988: Baxter Healthcare (Illinois), Eastman Kodak (New York), Pratt and Whitney (Florida), General Motors (Michigan), and AMP (North Carolina). The release had been put together in time for demonstration at the International Exposition for Automated and Computer Integrated Manufacturing (AUTO-FACT) trade show in October of 1987. The first production SLA-1 was delivered to Precision Castparts Corp. (Oregon) during April of 1988. The programs that controlled the SLA-1 were written in several different languages, and followed no particular protocols as to screen conventions, data input, or error processing. Most were written by physicists, chemists, or electrical engineers to test particular aspects of the system. Some last minute changes were made to improve consistency, but the primary concern was building parts.

STL files were sliced into cross-sectional data (SLI files), and multiple SLI files were merged into build files. At this stage of development, the software required a very patient and methodical user to achieve successful results. Little in the system was automated, and determining the correct part preparation parameters often required having a calculator next to the SLA. Because the software was the last thing the user saw before starting his or her part, any failure in the part was first assumed to be a software problem. However, failures were caused by many aspects of the system including software, hardware, and suboptimal part preparation conventions. Laser failures were much more frequent than today, and no official methods existed to restart parts or rapidly diagnose machine or part building failures.

Meanwhile, in the research and development lab, more sophisticated methods of recoating parts were under evaluation. It was determined that "dip and dunk" was no longer an acceptable method for recoating parts of arbitrary geometries, and, thus, the SLA-1 was discontinued.

SLA-250 Software Development

The primary difference between the SLA-250 and the SLA-1 was the addition of the recoater blade. The last SLA-1 was built in November of 1988, and the first SLA-250 was shipped to Japan in March of 1989. The SLA-1 was entirely sufficient for part geometries with "critical circles" of less than one or two inches in diameter. A "critical circle" is the maximum distance on any cross section's geometry that resin must travel to flow off the part. *Figure 5-7* shows some simple examples of "critical circles." As the radius of the circle increases, the time to recoat (t_r) using only "dip and dunk" increases proportionally. "Critical circles" are less important when a recoater blade is used, as this method provides significantly more uniform recoating in much less time.

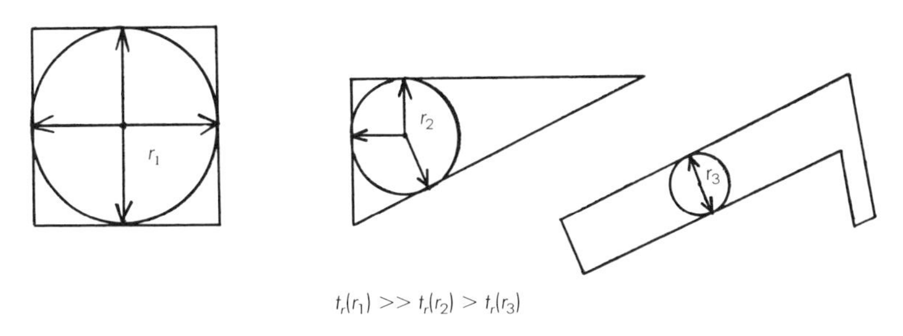

Figure 5-7. Critical circle examples.

Clearly, a recoating mechanism was vital to allow StereoLithography, at least in principle, to build any part geometry independent of "critical circle" limitations. The SLA-250 was a significant improvement over the SLA-1 in this regard, in part due to numerous software enhancements.

The basic hardware of the original SLA-250 consisted of a remote slice computer, which did the slicing of STL files into cross sections, linked by Ethernet to a controller computer which resided on the actual SLA. During this period, controller code from different languages was enhanced and rewritten. Pascal was chosen because the majority of programs were already in that language, and libraries existed to read and write STL, SLI, and L, R, and V files, handle screen communications, and control the motion of the elevator and other devices. Code was further enhanced to support the recoater blade. A sweeper parameter file (Sweep.prm) was added. All slice computer code remained in "C." The history of software releases for the SLA-1, the SLA-250, and the SLA-190 is as follows:[5]

October 1987: Release 2.62 (SLA-1)

- First beta release to customers (SLA-1 beta units were delivered to customer sites in January 1988).
- Slice program on the slice computer.

- Laser and Stereo "Terminate and Stay Resident" (TSR) programs on the Controller allowed precision control of the mirrors through software interrupts.
- Merge, Prepare, Build, and Utility programs on the Controller.

April 1988: Release 3.00/3.06 (SLA-1)

- First production release.
- Improvements and bug fixes on all programs.
- More consistent user interfaces.
- SLA-1 training manual.

September 1988: Release 3.20 (SLA-1)

- Infinite loop problem fixed in Slice.
- Calibration improvements.
- Major improvements to View program on DOS.
- Automatic laser on/off code.
- Bug fixes and more.

November 1988: Release 3.20 Upgrade 1 (SLA-1)

- Laser controller board Rev. D support.
- 287 Coprocessor support increases throughput.
- Beam analysis program improvements.
- Bug fixes, especially to Prepare.

February 1989: Micom to Excelan Upgrade Kit (SLA-1)

- Supported improved Ethernet capabilities.

SLA-250 Introduced

March 1989: Release 3.40 (SLA-250)

- Support the SLA-250 recoater blade and other new hardware.
- Support reading the thermistors and setting the heater thermostat.
- Inch/millimeter part preparation introduced.
- Add the /Restart option to the Build program.
- Workstation slicing supported on SLP-20 (Silicon Graphics 4D-20) as an alternative to the UNIX PC slice computer for customers desiring faster slice execution times.

September 1989 : Release 3.60 (SLA-250)

- Automatic features:
  –auto multiple parts.
  –auto zoom in Build viewports.
  –auto dip depth.

- Binary slice files (SLI) allowed more data and thus more complex parts to be built.
- Support new Silicon Graphics operating system IRIX 3.1 (SLP-20).
- Range checking at all prompts.

November 1989: Release 3.60 Upgrade 1 (SLA-250)

- Support new NCR 386 controller and 387 coprocessor.
- Software supports either the WYSE 286 or NCR 386 controllers from the same executable files.
- Automatic postdip delay.
- Laser centering software.
- Support InterActive UNIX 2.0.2 on PC slice computer.

December 1989: Release 3.60 Upgrade 1 (SLP-20 only)

- Provide 3DView program for workstations.
- Provide SliMerge program for merging multiple SLI files into one larger SLI file on the workstation, plus other utility programs.

Early and Mid-1990: Release 3.60 Upgrades 2 and 3 (SLA-250)

- Bug fixes.

SLA-190 Introduced. Similar to SLA-250, the SLA-190 had no recoater blade and a smaller vat. The first SLA-190 was delivered to Allied Signal (Illinois) during November of 1990.

September 1990 : Release 3.80 (SLA-190 only)

- Automatic step size and step period.
- Simplified DOS slice.
- Easy to use DOS menuing system.
- No slice computer required.

December 1990: Release 3.60 Upgrade 4 (SLA-250)

- WEAVE™ building method supported, which dramatically improved part accuracy.
- Provide automatically generated material files through the GenMat program.

May 1991: Release 3.81 (SLA-190 and SLA-250)

- Doubled throughput of SLA-250 by reducing laser processing delay time and providing more accurate, faster leveling.
- Merging program execution speed increased by over tenfold.
- Overcure and cure depth parameters replace complicated step size and step periods.

- Resin files introduced for ease of selection; material files and banjotop diagnostic part no longer required.
- Easy to use DOS menuing systems on the SLA-250.

October 1991: Release 3.82 (SLA-190 and SLA-250)

- STAR-WEAVE™ building method supports the new tooling resin and further improves part accuracy.
- Desktop DOS software for slicing, merging, preparing, and simulating build.
- Speedier menu systems.
- Add/Remove resin software improved.
- Advanced Slice interface on DOS.
- 3DVerify program to verify and/or repair STL files which marginally did not meet the STL specifications.

SLA-500 Software Development

During 1988, in one of the back rooms of 3D Systems, a colossal new machine was under construction. This was to be the machine that would build entire automotive engine blocks in a single build cycle. The preliminary specifications showed an SLA-500 with a Silicon Graphics workstation linked to a 386 DOS PC controller. Two digital signal processing (DSP) boards within the controller would handle geometric correction and motion control of a new set of scanning mirrors. Part preparation utilized a spreadsheet method of input and generated a single build file called the "bff" (build file format) file.

Over time, the architecture of the SLA-500 proved to be much more user friendly than that of the SLA-250. With pull-down menus and pop-up windows, the spreadsheet approach to part preparation was slated to eventually replace the SLA-250 part preparation programs. The first SLA-500 was delivered to Pratt and Whitney (Connecticut) in December of 1989. *Figure 5-8* is a simplified drawing of the SLA-500 BuildStation.[6]

The history of the SLA-500 software is as follows:

June 1990: Release 1.01/1.02

- First production release to customers.
- Software supports Silicon Graphics workstation.
- Software supports DSP boards in Controller.
- Software supports a faster and more accurate set of mirrors.

December 1990: Release 1.21

- WEAVE™ build style supported.
- Support of metric units introduced.
- Laser linewidth compensation supported.
- Automatic laser power settings.

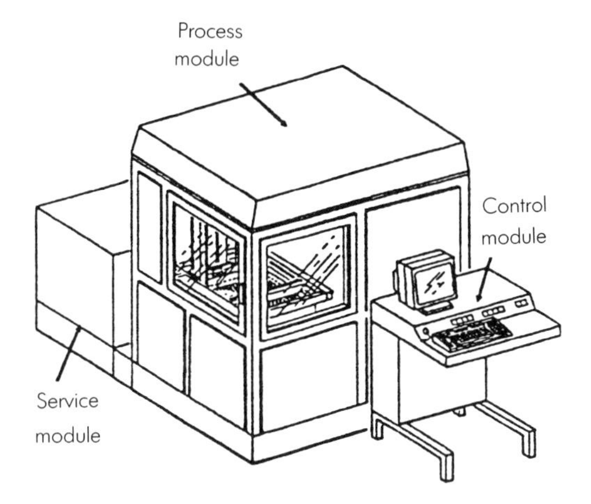

Figure 5-8. Simplified SLA-500 BuildStation diagram.

- Overcure parameters.
- Improvements to the restarting software.

June 1991: Release 1.30 (Upgrade to 1.21)

- Support new full area laser beam detector.
- Field service engineer required to install the hardware.
- Resin files introduced for ease in selecting resin types.
- Material files and banjotop diagnostic parts no longer required.

September 1991: Release 1.30A

- Diagnostic code to inspect for and remove any software viruses.

December 1991: Release 1.40

- STAR-WEAVE™ building method further improves part accuracy.
- Enhancements to the View program.
- Enhancements to the Partman (Spreadsheet) program.
- Productivity enhanced by more efficient leveling algorithms.
- 3DVerify program accessible through the spreadsheet to verify and/or repair STL files which did not meet the STL specifications.

Figure 5-9 shows a time line of product introductions and software releases.

Evolution of Software Development Methods

One of the most significant improvements in SL software was an improvement in the development methods. Whereas shared libraries, revision control systems (RCS), and computer augmented software engineering (CASE) tools were already in use by 1989, the methods to decide exactly what each software release would contain continued to be a debate. Also, during late 1989 and 1990, customers pointed out that 3D Systems did not deliver sufficiently frequent

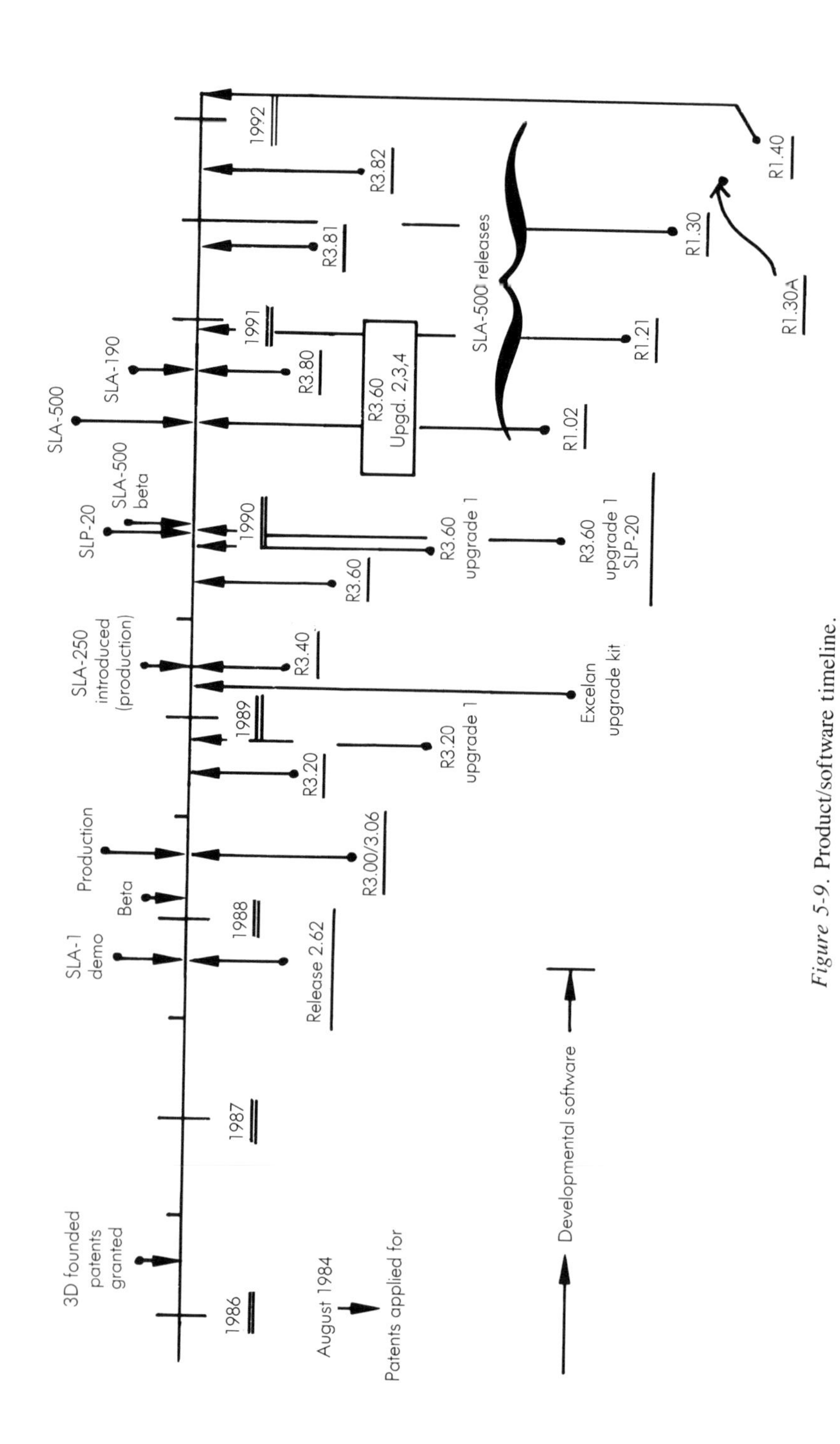

Figure 5-9. Product/software timeline.

software releases to address enhancements and necessary bug fixes. This led to a desire to perfect the products in the next software release. Ironically, this approach only slowed progress further and made software release offerings even less frequent.

In recognition of this circular problem, 3D Systems instituted the first formalized long-term software plan in early 1991 to promote *incremental product improvement*, reduce the technical risk in software releases and upgrades, and bring software products to market more quickly.[7] This method helped end the "just one more fix" phenomena, which had been a problem in the past. By freezing design requirements and detailed specifications earlier in the product cycle and adhering to a long-term software plan, software releases began arriving at customer sites with more regularity, better quality, and more significant improvements.

A compressed requirements sheet was instituted to allow maximum information exchange without excessive paperwork:

1. Introduction–Definition/Goals/Objectives.
2. Applicable Documents.
3. Environmental Description–Platforms/Operating Systems.
4. Information Description–Flow/Content/Structure.
5. Functional Description–Control/Processing/Performance.
6. Testing–Error Processing.
7. Maintainability–Compatibility/Portability/Reusability.
8. Other Requirements–Enhancements.

A more modern approach to the software specification and development cycle was instituted in accordance with Edward Yourdon's approach.[8] The newer modeling techniques necessitated more time in the specification stage, but greatly reduced the time required to code the software and verify its functionality.

5.3 Current Software

This section explains the usage, architecture, and philosophy of the current SLA-190, SLA-250, and SLA-500 software. It examines the external and internal aspects of the software architecture. This section does not explain how to build parts (see Chapters 7 and 8). The discussion begins with the universal elements of 3D Systems' software architecture, and then moves into the current SLA-190/SLA-250 software, release 3.82 upgrade 1 (R382U1), and the current SLA-500 software, release 1.41 (R141). Next, production and calibration issues are covered. Third party software and how it fits into the existing architecture is the final topic of this section.

The difference between building parts with the R382U1 and R141 involves part preparation. With R382U1, parameters are entered in several separate stages: before the slice program is run, before the merge program is run, before the prepare program is run, and sometimes before the build program is run. With R141, all slicing, drawing, and recoating parameters are input to a spreadsheet before any subprograms are executed. After entering all parameters, the user begins the computations by selecting a button on the workstation screen. At that time, the appropriate subprograms such as Slice and Converge are called. If a small change is requested, only the necessary programs are reexecuted to enact the change.

All part data comes from CAD models, medical applications, mathematical model generators, or some other form, regardless of which SLA will build the part. Most parts enter the system as an STL file, which is then sliced into cross sections. If the data comes in an already sliced form such as contour data, the slicing stage is bypassed and only layer comparison algorithms are performed to determine flat, near flat, and steep regions. The rules for categorizing each region depend on how much horizontal and vertical separation is involved in drawing the vectors and layers. Building parameters are critical to a part's build cycle, and all parameters are set through the software. The following section lists all the main SL parameters and a brief description of their function. Additional parameters are accessible if the individual programs, such as Slice and 3DVerify, are executed directly from a command line instead of being indirectly called from the menu systems. The main purpose of the menu systems is to increase the software's user friendliness by allowing sets of parameters to be grouped into "slice styles," "recoating styles," and "build styles." In R1.41, slice styles are incorporated into build styles.

In the following section, parameters which are exclusive to the SLA-190 and SLA-250's software release 3.82 Upgrade 1 are denoted by a single asterisk (*), and all parameters exclusive to the SLA-500's software release 1.40 Upgrade 1 are denoted by a double asterisk (**). Parameters not marked apply to all the SLAs. Parameters that are set by manufacturing, field service, or application engineers at either assembly and/or installation are not covered in this list. Default values and ranges for each parameter are listed. Because the parameter's defaults and ranges are approximately the same for both software releases, only one set of values are listed.

1. *Units of the input file*. Usually these were CAD units. Default = millimeters. Range = inch or millimeter.
2. *Layer Thickness of the final part*. If the file contains contour data (in cross-sectional form), this parameter is read from the input file. All machines have the ability to assign variable layer thickness for regions of the part which may require more or less detail. Decreasing layer thickness improves resolution by decreasing the "stair-stepping" effect on parts but usually increases the build time. This is shown in *Figure 5-10*. Default = 0.25 mm and range = 0.0635 to 0.50 mm.

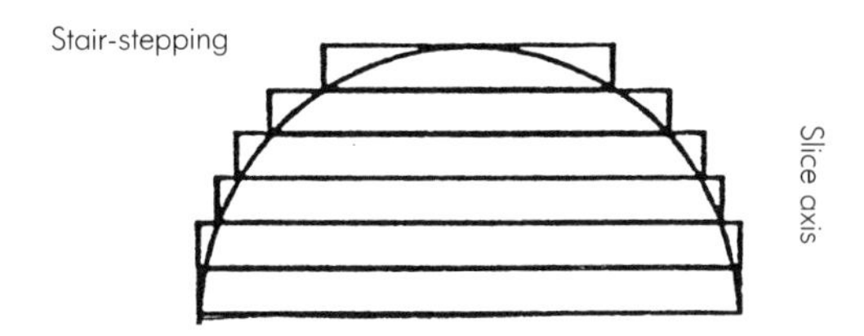

Figure 5-10. Stair-stepping effect.

3. *Linewidth compensation of the final part.* This function operates at the borders of the part to allow for the cured laser linewidth. The laser must be offset by half the linewidth toward the part mass to draw accurate borders. The default is usually overridden after the machine is installed and the linewidth compensation determined. This is displayed in *Figure 5-11*. Default = 0 (disabled) and range = 0 or 0.025 to 0.50 mm.
4. *X hatch spacing controls the separation distance between each X hatch vector, which forms the internal structure of the final part.* Default = 0.28 mm and range = 0 (deactivates drawing *X* hatch vectors) or 0.025 to 25.0 mm. Values in the upper region of the acceptable range are sometimes used with equally large *Y* hatch spacings to build support structures using the bottom borders of the part projected downward to the build platform.
5. *Y hatch spacing.* Same as item 4, but in the *Y* direction.
6. *60/120 hatch spacing* controls the separation distance between the 60° and 120° cross-hatch structure. This parameter applies to the Tri-Hatch build style. Default = 0 (deactivates drawing 60/120 crosshatch vectors) and range = 0 or 0.025 to 25.0 mm.
7. *X skin spacing controls the separation distance between each X skin vector.* Skin vectors are used to seal the part. They are equivalent to putting the final sheet metal on an airplane frame. Default = 0.125 mm and range = 0 (deactivates the *X* skin vectors) or 0.025 to 25.0 mm.
8. *Y skin spacing.* Same as item 7, but in the *Y* direction.
9. *Border overcure*** *controls the cure depth of all the border vectors.* Actual border cure depth = layer thickness + border overcure. Default = dependent on selected build style. Range = negative layer thickness to 0.508 mm.
10. *Hatch overcure*** *controls the cure depth of all the hatch vectors.* Actual hatch vector cure depth = layer thickness + hatch overcure. Default = dependent on selected build style. Range = negative layer thickness to 0.508 mm.
11. *Fill overcure*** *controls the cure depth of all the skin fill vectors.* Actual skin fill vector cure depth = layer thickness + fill overcure. Default = dependent on selected build style. Range = negative layer thickness to 0.508 mm.

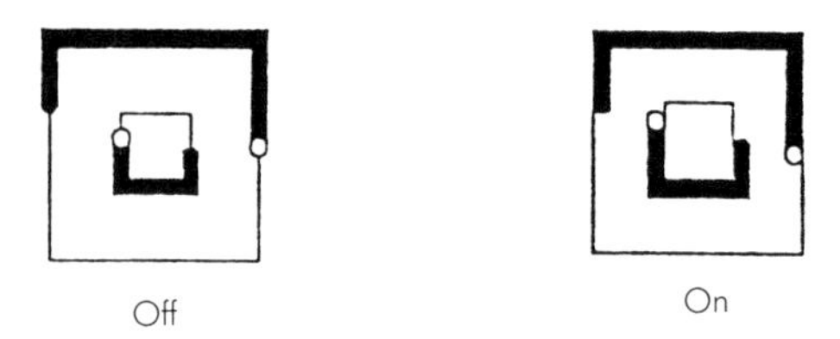

Figure 5-11. Linewidth compensation.

12. *Minimum surface angle (MSA) controls the angle at which a surface is considered "steep" versus "near flat."* Near flat indicates that the angle to the horizontal is too small to "self-seal" without requiring a skin. Default = 50° (with a default layer thickness of 0.25 mm). Range = 0° to 90°.
13. *Slice resolution specifies the number of slice units per CAD unit.* It divides the CAD space into smaller slice units creating a three-dimensional grid. Slice resolution must be the same for all parts in a build cycle. Default = 200/mm. Range = 40/mm to 400/mm.
14. *STAR-WEAVE™ is a building method.* WEAVE™ is a subset of STAR-WEAVE™ with the three STAR parameters turned off. The STAR acronym stands for the three switches: STagger, Alternate, and Retract, settings which apply to hatch vectors. Default = OFF, OFF, OFF (WEAVE™ is the default build method). Range = ON or OFF for each switch. WEAVE™ and STAR-WEAVE™ are discussed in detail in Chapter 8.
15. *Slice axis determines which CAD axis is perpendicular to the cross sections.* It is strongly recommended that all parts be sliced along the *Z* axis to eliminate confusion when positioned on the platform. Default = *Z*. Range = *X*, *Y*, and *Z*.
16. *Build Styles*** *group the slicing and building parameters into convenient style files (STY) which are accessible through the spreadsheet program.* The parameters in the style file are CAD units, slice resolution, layer thickness, MSA, border overcure, hatch overcure, fill overcure, slice axis, and hatch type, which is a grouping of the hatch, fill, and 60/120 vector spacing parameters. Hatch type also contains all WEAVE™ or STAR-WEAVE™ parameters.
17. *Scale factor** *changes the size of the STL file before it is sliced.* Default = 1 (no change in size). Range = 0.001 to 50. Note that R141 has a menu option which allows the user to scale STL files separately in the *X*, *Y*, or *Z* directions and rename the rescaled file.
18. *Slice argument files** *group the slicing parameters into a convenient parameter file* (.ARG on DOS, .UII on UNIX), which is accessible

through the DOS or UNIX menu system. The parameters in the slice argument file are CAD units, scale factor, slice resolution, layer thickness, *X* and *Y* hatch spacing, 60/120 hatch spacing, *X* and *Y* skin fill spacing, MSA, linewidth compensation, slice axis, and STAR switches (*Figures 5-12* and *5-13*).

19. *Part offsets are used to position the part on the building platform.* Offsets are an X, Y, Z transformation applied to the object. In R140U1, this is done through the spreadsheet, in R382U1 through the Merge program. Default = 0, 0, 0. Range is checked by software to ensure that the part still fits in the vat.

    Note: The following parameters apply to the recoating of the part. Since the SLA-190 does not have a recoater blade, a global variable in R382U1 software determines the machine type and the applicable recoating parameters.

20. *Postdip delay (PD) specifies the number of seconds that the elevator stays in the "deep dip" position.* Resins with higher viscosities and/or parts with large flat areas need longer post dip delays. Most parts build fine with PD = 1. Default = 4 seconds. Range = 1 to 99 seconds.
21. *Z-wait (ZW) specifies the number of seconds that the elevator pauses prior to the laser drawing.* This time allows for resin surface relaxation and depends upon resin viscosity and surface tension. SLA-250 default = 30 seconds. Range = 0 to 999 seconds.

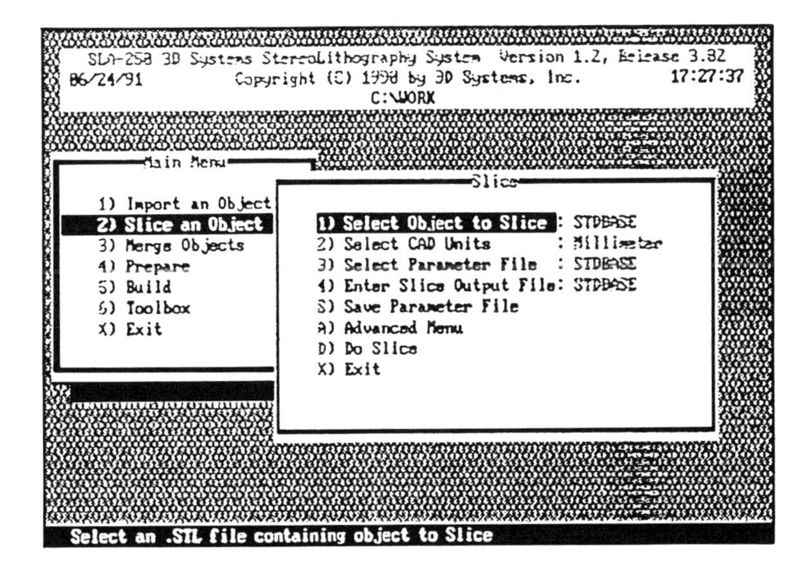

Figure 5-12. DOS slice screen.

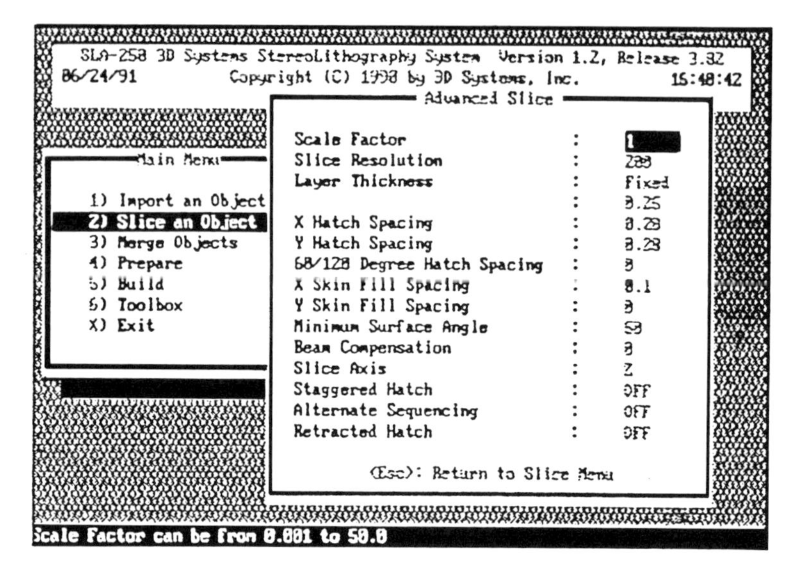

Figure 5-13. Advanced slice screen.

22. *Z acceleration and velocity** *(ZA, ZV) control the motion of the elevator during the build cycle.* The ZA and ZV values may need to be reduced in particularly fragile areas of the part. Default = 0.20, 0.20. Range for both = 0.05 to 1.00 in machine-specific units.
23. *Number of sweeps (NS) is the number of recoater blade sweeps.* Odd numbered values are recommended to avoid cumulative fluid dynamic effects near the leading edge of the part, relative to the recoater blade. Default = 1. Range = 1 to 7 sweeps.
24. *Sweep period for sweep n* (P_n, where n = 1 to 7) *specifies the sweep period in seconds for the recoater blade to move the length of the vat.* On the SLA-500 (R1.41), more sophisticated recoater blade control is available. R382U1 default = 5 seconds. R382U1 range = 3 to 30 seconds.
25. *Part-to-blade gap for sweep n* (G_n, where n = 1 to 7) *specifies the blade clearance of the nth sweep*, expressed as a percentage of the layer thickness. Peculiar part geometries may require different combinations of recoater blade periods and gaps (see Chapter 7). Default = 100% (the gap between the part and the recoater blade = 100% of the layer thickness). Range = 100% to 999%.
26. *Recoating styles*** *group the recoating parameters into convenient recoat style files* (.RCS), which are accessible through the spreadsheet program.
27. *Overcure** *(OC) controls the cure depth of either border or hatch vectors.* Actual cure depth = layer thickness + overcure. Default = resin and vector type dependent. Range = negative layer thickness to 0.50 mm.

28. *Absolute cure depth* (CD) controls the cure depth of up- and/or down-facing skin fill vectors.* Actual cure depth = absolute cure depth. Default = resin and vector type dependent. Range = 0 to 1.0 mm.
29. *Ranges divide the part into user-selected vertical regions.* One may elect to change parameter values from range to range to improve part quality. Ranges are frequently used to assign different parameters to part supports than those assigned to the actual parts. On R1.41 spreadsheets, build styles and recoat styles can be different from range to range. On R382U1, selected parameter groups and default parameter groups may be assigned to different ranges.
30. *X and Y and Z mini-scale is used to account for resin shrinkage.* These shrink factors are assigned at the BuildStation on the SLA-190/250/500 and may also be assigned on the SLA-190/250 desktop PC software. Default = 0% (part size unchanged). Range = -5.000 to +5.000%.
31. *Number of copies allows the user to build the maximum number of selected parts which will fit in the vat.* Spacing can be user selected or assigned automatically.
32. *Resin data files are selected by choosing the resin type, and they contain the E_c and D_p of the resins.* With R382U1, the resin file also contains the recommended overcure and cure depth values. Default = user selectable. *Figure 5-14* shows an example of the R382U1 "Select Resin" screen. Range = user definable.

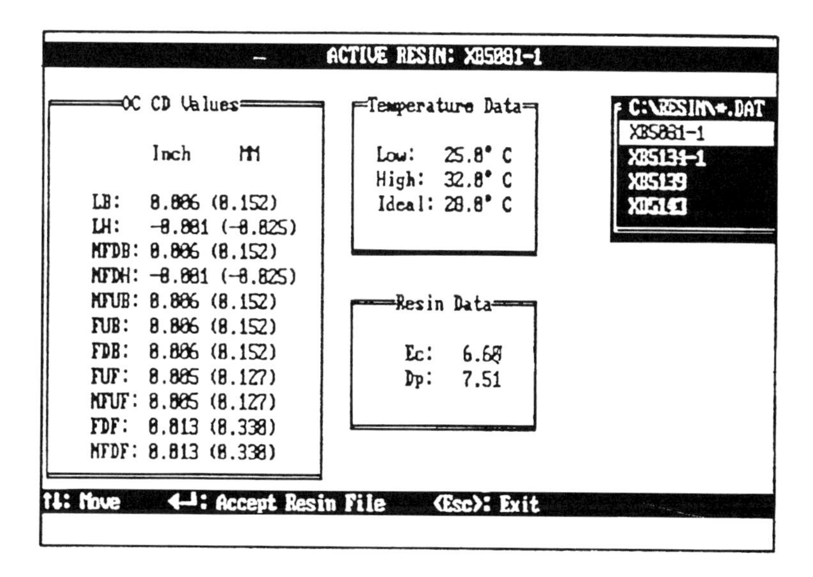

Figure 5-14. Select resin screen. (Note: At this time, the standard D_p and E_c values for XB 5081-1 are 7.1 and 5.6, not the values displayed in this example.)

33. *Build options selectable at or before part building time include the following:*

    A. Simulate building without operating the elevator or laser. Build simulation can be done on a desktop PC.
    B. Restart a partially completed part.
    C. Start and/or end a part at a particular layer.
    D. Pause the part at a particular layer.
    E. Deactivate the mirror drift correction process.
    F. Do not draw in the vat, but move the elevator and show the part layer drawing on the controller screen.
    G. Do not move the elevator, but laser draw the part in the vat.
    H. Run the build process in text mode only.
    I. Invoke writing to various diagnostic files.
    J. Mirror the part, or swap the *X* and *Y* axes.

Release 3.82 Upgrade 1

The release 3.82 software architecture was designed to increase total machine productivity by providing the ability to run the entire software package, including the slicing, merging, preparing, and simulated building operations on almost any DOS 386 or 486 PC. With previous releases slicing could only be done on the turnkey 386 ''Slice'' PC with a UNIX operating system or the SLP-20/25 workstation. Likewise, merging, preparing, and simulated building could only be done on the SLA itself. By allowing all these operations to be performed on a desktop PC, the SLA is freed up to build parts unattended 24 hours a day. This provides a substantial advance in productivity. It also allowed the software to propagate to almost any desktop which contained a suitable PC.

Release 3.82 also provided the ability to verify or repair any STL files which did not completely meet the STL specifications. This program had been used internally at 3D Systems for quite some time in its developmental stage to help analyze files received from customers who had called the 3D Systems' support hotline.

Release 3.82 enabled the STAR-WEAVE™ building method and hence the release of a new machinable resin, Ciba-Geigy XB 5143. This resin was tougher than previous resins and could be drilled, tapped, milled, and turned. Finally, the release improved the advanced slice user interface, while the main menu system was enhanced and its speed increased.

The difference between upgrades and releases is as follows:

- Upgrades are used to deliver functionality and bug fixes for individual software programs stemming from a baseline release.
- Releases contain all the executable programs and data files to run the SLA.

Upgrades can be designed, coded, and tested faster enabling quick response to customer needs. A strong release can allow a series of upgrades to follow without hampering the development of subsequent major releases. Upgrades are cumulative, meaning that the entire contents of upgrade 1 are included in upgrade 2, and so on. For example, a customer wishing to receive the benefits of upgrade 1, upgrade 2 and upgrade 3 to release "A" need only install the baseline release "A," and then upgrade 3.

Release 3.82 upgrade 1 contains:[9]

- Independent *X*, *Y*, and *Z* miniscale factors to account for part shrinkage. These miniscale factors are accessible through the Prepare program. Formerly, all three shrinkage factors were *not* independent. This feature allows "fine tuning" of the part building process and can provide still further improvements in accuracy.
- A recompiled 3DView program for the SLP workstation, which allows the software to run on any of the Silicon Graphics workstation family.
- Further improvements to the add, remove, and level resin sections of the software.
- Several minor bug fixes.

The SLA-250 release 3.82 buildstation software menu tree is detailed in *Figure 5-15*. *Figures 5-16A* and *5-16B* are data flow diagrams for release 3.82 with user interactions shown.

Generally, CAD models become STL files through a CAD interface translation program. If the input file is contour or other data it may come into the system in a different data file format, but still becomes an SLI file before proceeding through the system at this time. The Slice program may be directly accessed through the command line or through a user interface.

Sliced files are merged together to form build files (.Layer, .Range, .Vector, and .Parameter files) using the Merge program. The Prepare program may be used to further alter the build files if necessary.

Parts are built by reading the build files, the machine parameter file, the resin file, the calibration file, and other files. The grey area represents the laser beam calibration operation which is carried out whenever the system is installed, moved, or the laser is replaced.

The checkpoint data file (CheckPt.DAT) is used to restart the part, and the system state file (3DState.DAT) is used to store the state of all components in the SLA. The profile module (LibProfile) is used to obtain profiler locations and feed them into the Stereo.exe program to correct for mirror drift, if necessary.

Releases 1.40 and 1.41

Release 1.40 streamlined and added function to part preparation software on the workstation and provided new capabilities and improved performance on the

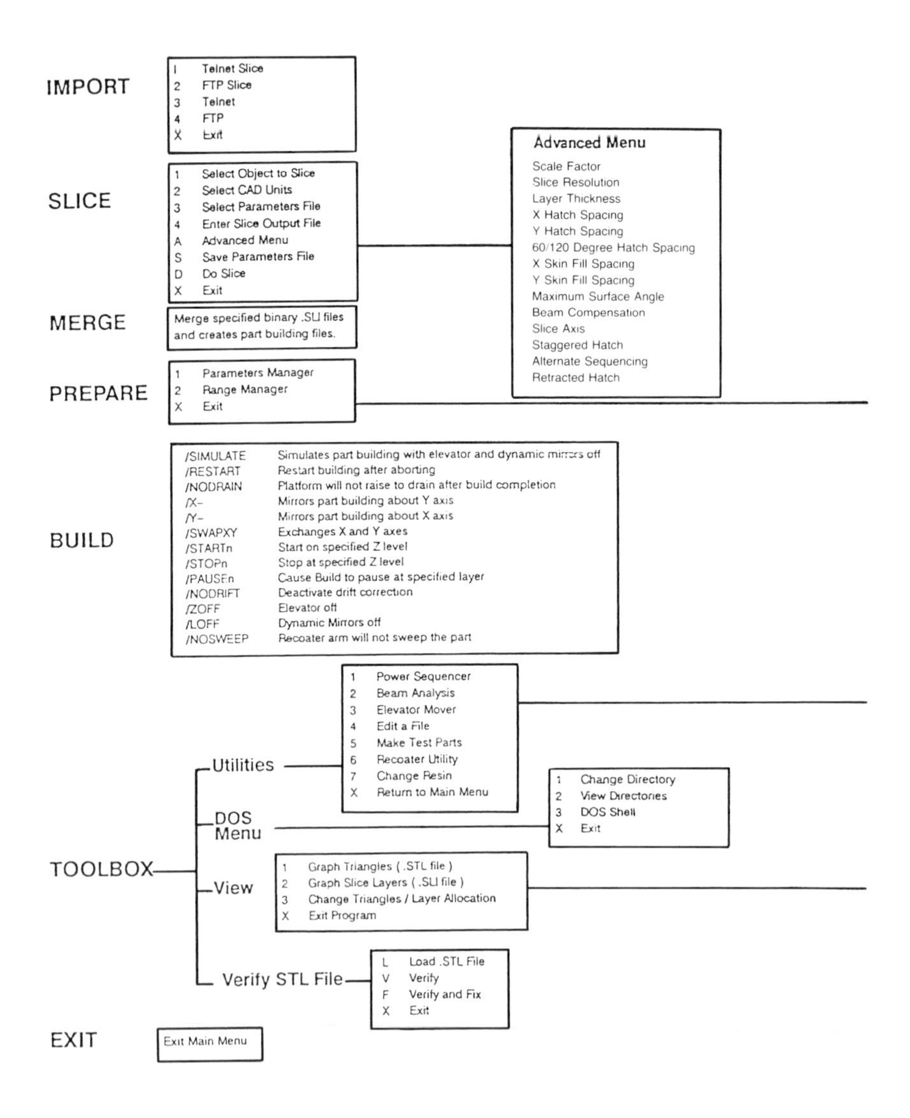

Figure 5-15. SLA-250 menu tree, release 3.82.

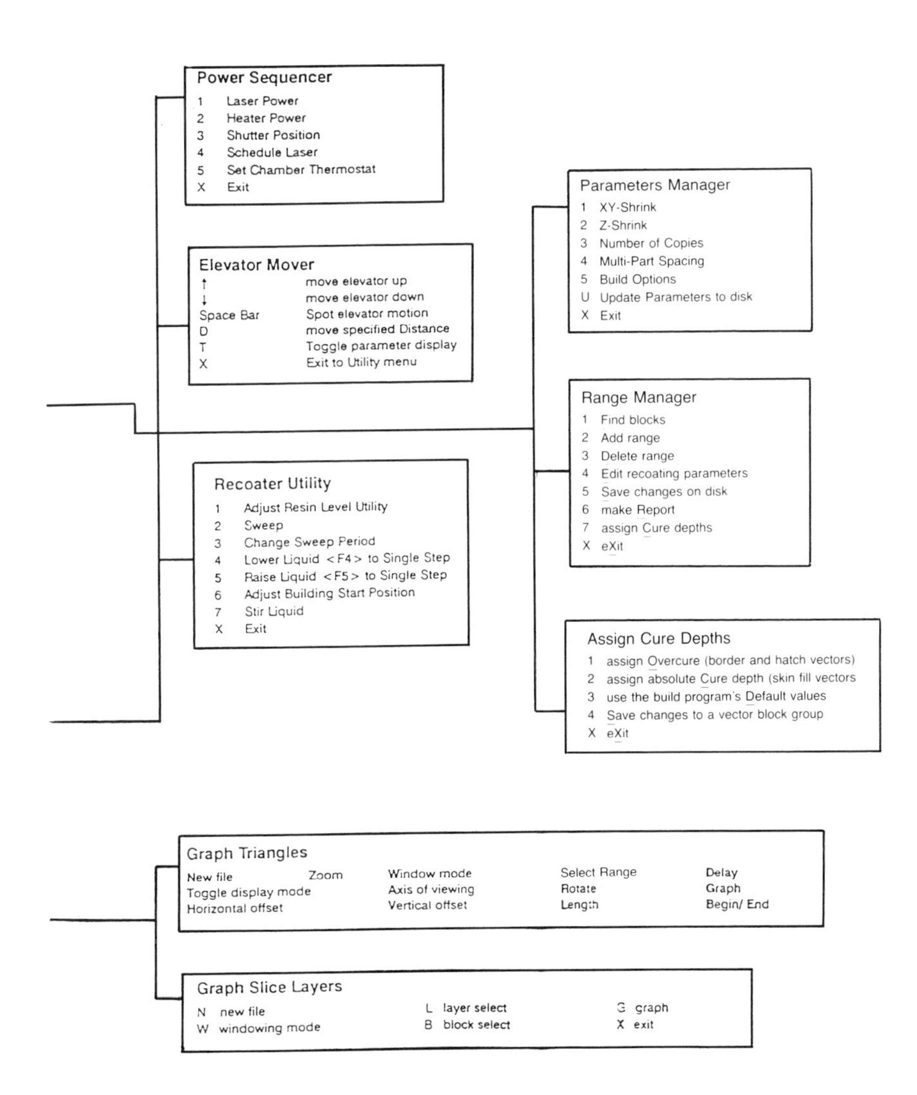

Figure 5-15 Continued.

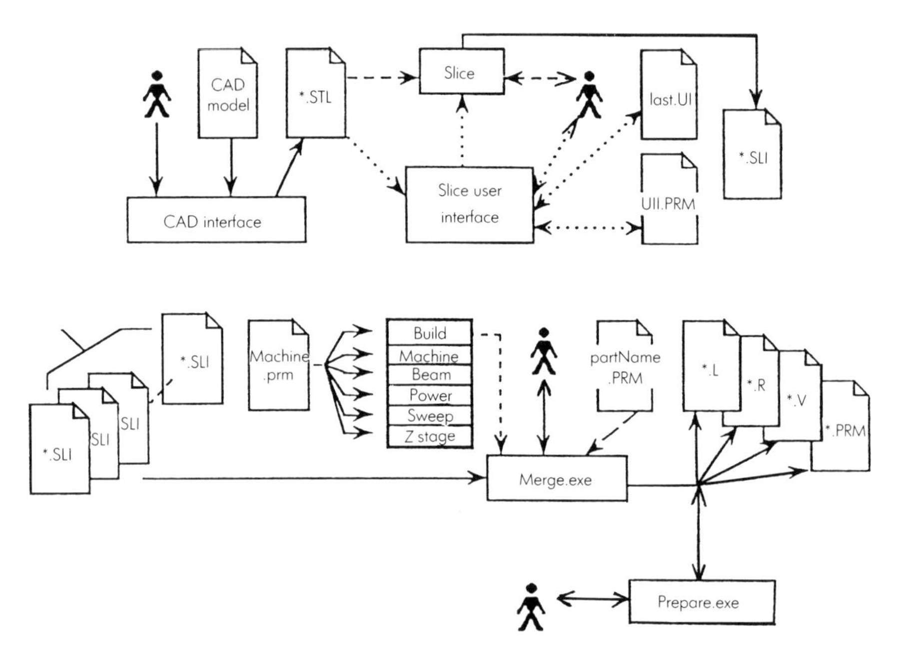

Figure 5-16A. SLA-250 data flow, release 3.82.

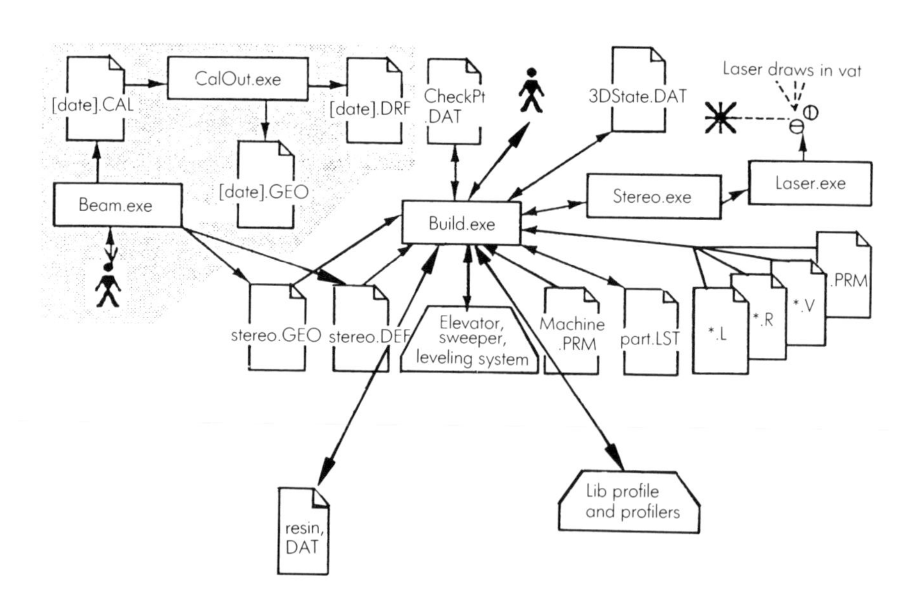

Figure 5-16B. SLA-250 data flow, release 3.82.

controller. The workstation software contains a larger and easier to read spreadsheet, a compact user interface in the View program, linewidth compensation on a component basis, an improved program to analyze STL files (3DVerify), and supports STAR-WEAVE™. The BuildStation software contains access to additional machine operations through the User Utilities and incorporation of Z scaling, an enhanced and easier to use part restarting routine, a faster resin level adjust routine, and a vat loading and unloading program.

Release 1.41 contains:[9]

- View as a replacement to 3DView for the SLA-250 customers, plus other enhancements to the View program for SLA-500 customers. To support SLA-250 customers, the display of vector files was added. This upgrade converged the View program across the 3D Systems product line.
- Provide "Indigo" support for the SLA-500 customers. The Indigo is Silicon Graphics' newest and least expensive workstation. The Indigo allows customers a *reduced price* of SLA ownership accompanied by a simultaneous *increase in performance*.
- Proprietary improvements to the mirror control system.

The SLA-500 also contains an extensive on line help file. *Figure 5-17* shows some of the help topics and the introduction to the help file. *Figure 5-18* displays the SLA-500 workstation menu tree.

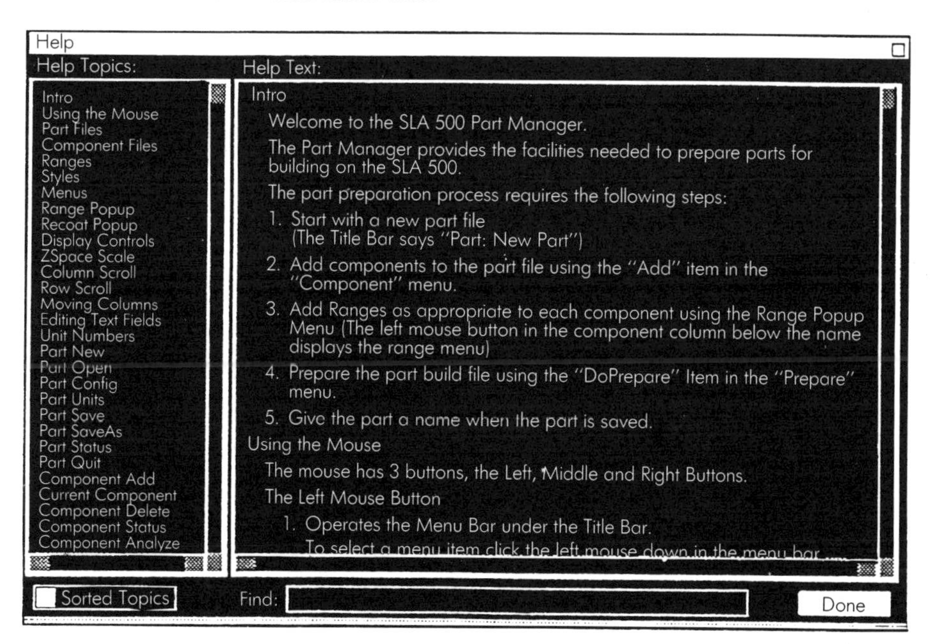

Figure 5-17. SLA-500 on-line help file.

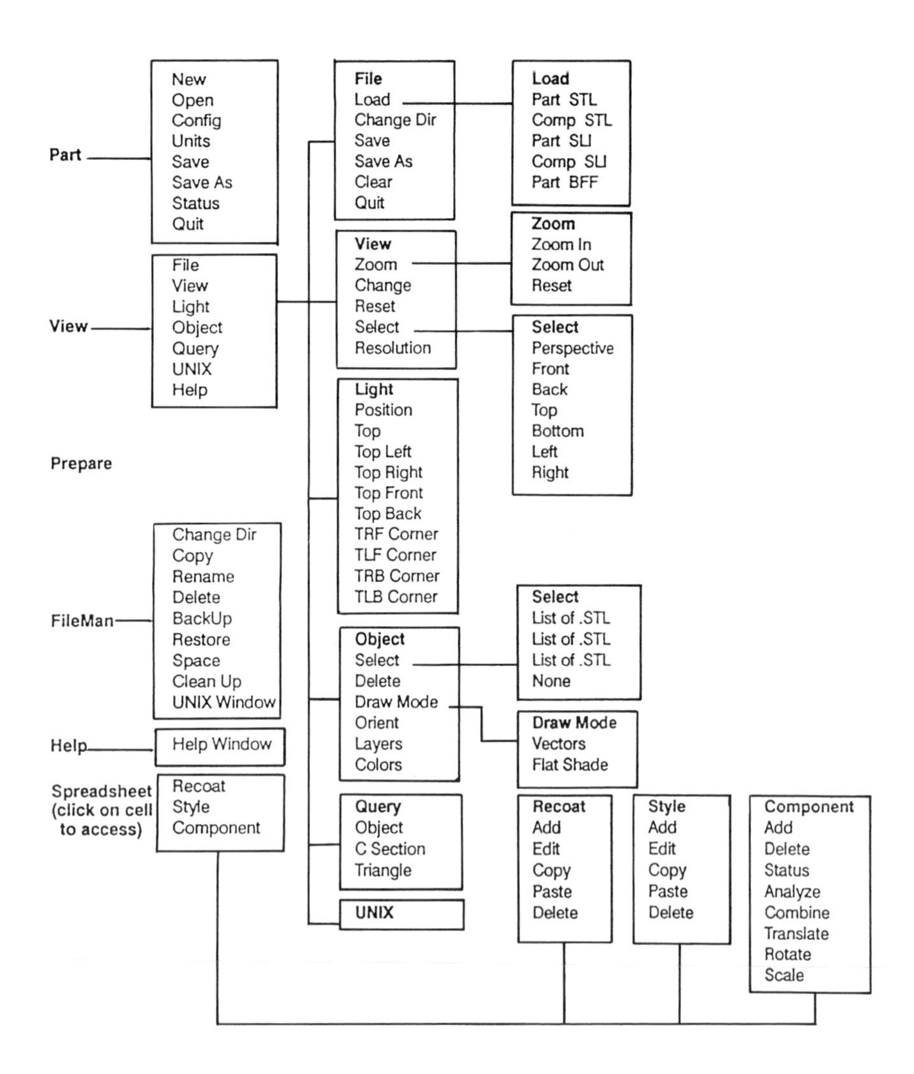

Figure 5-18. SLA-500 workstation menu tree, release 1.40.

Production and Calibration Issues

During the assembly and installation of SLA machines, the leveling systems and mirroring systems require calibration which is controlled by software programs. Once the calibrations are completed, the machines seldom require any adjustments until a laser needs to be replaced or some other component requires maintenance.

To calibrate the scanning system, a "calibration plate" is placed over the vat. On the SLA-500, the plate has 6400 tiny holes, each on a quarter inch square grid pattern. The holes lie above an array of UV sensors, and are leveled to match the height of the resin surface, as indicated in *Figure 5-19*. The mirror scans each point of the grid and stores the position of each hole. Interpolation of the mirror to vat coordinate systems occurs in the geometric correction algorithms.

Without such calibration there is no assurance that the laser is really pointed at the correct location. Small pointing errors ultimately manifest themselves as part inaccuracies. To verify that the calibration process was successful, a grid is drawn on UV-sensitive paper, which is secured tightly over the calibration plate. This Geogrid "part" is measured on the paper to ensure the final mirror positioning data and software are accurately drawing on the resin surface.

Likewise, all RP&M systems contain subsystems which require particular software communication protocols. For example, on the SLA models, the

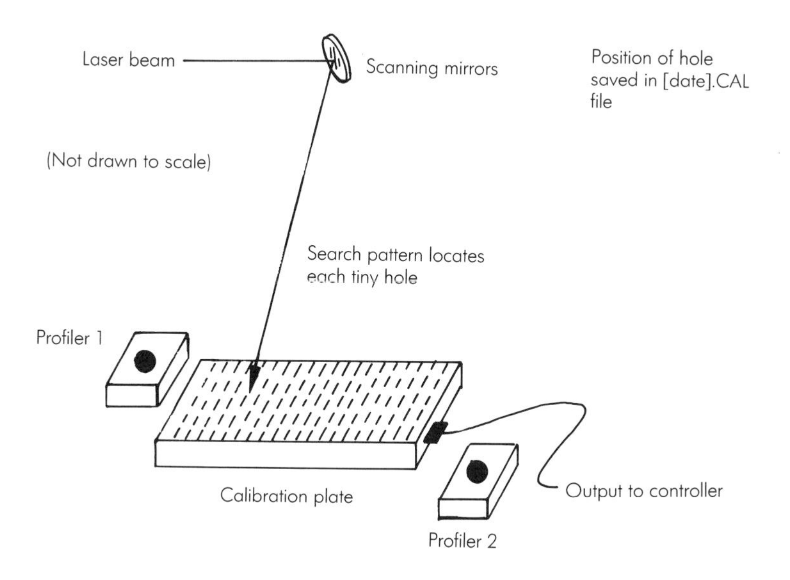

Figure 5-19. Calibration of mirroring system.

movement and location detection of the elevator and other components is accomplished by commands sent and received through a microprocessor-based indexer/encoder which is placed in the BuildStation PC Controller. The indexer/encoder has a unique software communication protocol.

Thus the software architecture is required to include libraries to communicate with whatever subsystems are contained in the hardware design. If subsystems are changed, new software must be written to ''talk'' to the new component. In durability tests on the SLA production floor, a command is sent to the elevator controller to move all the devices. The test is run overnight or longer to ensure that the SLA is functioning properly.

All of the manufacturing and field service (MFS) specific software is contained on the MFS set of disks. These disks can be released independent of SLA or other software packages. Additionally, diagnostic disk(s) generate necessary data to reset machine or resin-specific parameters without requiring a field service visit. Diagnostic disks include part files and procedures for WINDOWPANE™, CHRISTMAS-TREE™, and Reverse WINDOWPANE™ diagnostic parts (see Chapters 10 and 11).

Third Party Software

With the size of the RP&M market growing weekly, no single company is capable of meeting all of its customer demands for additional or improved software. Therefore the market for third party RP&M software becomes more attractive. These ''optional'' software packages differ from the ''core'' RP&M software packages because they are *not required* to build parts on the machine. Several developers are already selling packages, and many RP&M companies are supporting these efforts.

Third party software falls into six general categories:

1. ''Core'' software developed by a consulting firm with the RP&M company purchasing the source code. This category can prove problematic, especially if a minor bug in the software prevents the user from building parts entirely.
2. ''Optional'' software developed by a consulting firm with the RP&M company purchasing the source code. The problems are less critical because part building will never be halted.
3. Value Added Retailer (VAR) of the externally developed ''optional'' software. Adding value could include writing manuals for the software, packaging it, etc.
4. Exclusive distributorship of the externally developed software. In this category, the software is wrapped and ready to send. Royalties would be paid to the development company.

5. "Recognized" software that has been tested by the primary RP&M company. A fee may be required to offset certification process costs. The software is carried on the primary RP&M companies sales list as "available from the development company."
6. "Unrecognized software" has no approval whatsoever, and may void RP&M maintenance contracts if it causes damage to the system.

The first two entries into the third party software market are the "Layout" and "Bridgeworks" programs. "Layout" allows the user to position part outlines on a top view of the SLA build platform. Parts can be quickly translated or rotated graphically using a mouse or keyboard input. Layout does not conflict with any 3D Systems core software and simplifies the part building process. Layout is produced by VinWare Utilities of Valencia, California. VinWare Utilities also has other programs available, including a build time estimator and an RP&M data management package.

A second and more complex software package is "Bridgeworks," by Solid Concepts of Los Angeles, California. Bridgeworks automatically generates support structures for part files (STL files). The algorithms used to generate the supports utilize a number of parameters, which are well documented in the user manual. The Bridgeworks program divides support structures into five categories, as shown in *Figure 5-20*.

RP&M users without an automatic support package usually develop a library of support structures. The appropriate support model is selected from the library and then scaled in *X*, *Y*, and *Z*, translated, and/or rotated to the desired position/configuration using the View program. However, this approach does not identify *where* support structures are required. Bridgeworks' parameter set allows the user to automatically set criteria to find where support structures are required and can save significant amounts of time in the part preparation stage.

As the RP&M market grows, third party software offerings will multiply. Providing such software usually requires very little overhead. In this expansion stage, entrepreneurial software engineers will develop packages which fulfill niche and global market demands.

5.4 Comparison with Alternate Approaches

Several approaches to RP&M fabrication methods are on the market today. However, most have less than five installed customer sites. In this section, Cubital, DTM, Helisys, and Stratasys software will be briefly discussed. These four companies were selected because their machines operate using fundamentally different techniques than StereoLithography. Several other companies whose machines are very similar to 3D Systems' SLA products will not be covered because, by nature, their software is much like SLA software. For an explanation of other RP&M machines, see Chapter 16.

Gussets

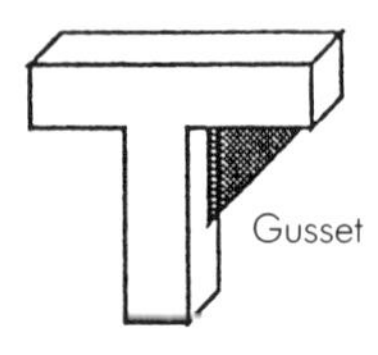

Gussets are used to support overhang areas. Gussets attach to a vertical wall near the overhang area and thus provide a surface for the overhang area to attach during building. Gussets provide the optimal support for overhang areas as minimal resources are required during building and supports are easily identified during cleanup. The gusset spacing parameter is used to control gusset spacing. The gusset length parameter is used to control the maximum gusset size.

Projected Feature Edges

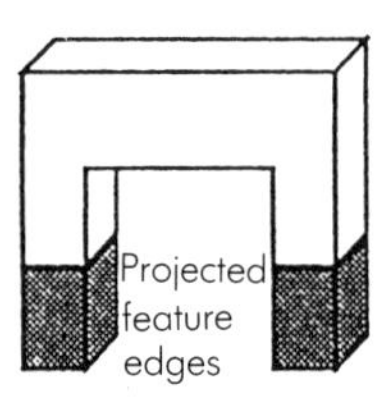

The edges of unsupported areas and overhang areas where gussets cannot reach are projected to provide support. Projected feature edges support the actual edges of the part, optimally controlling curl and providing the most accurate results. The minimum project extents parameter is used to control the projection of small feature edges.

Single Webs

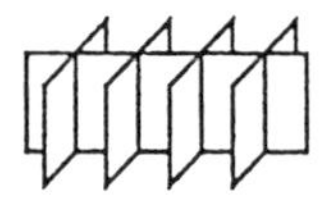

A single web support is a single wall projected from the center of a narrow feature or along a sharp edge feature. Cross members are added to provide stability. The support spacing parameter controls the spacing.

Webs

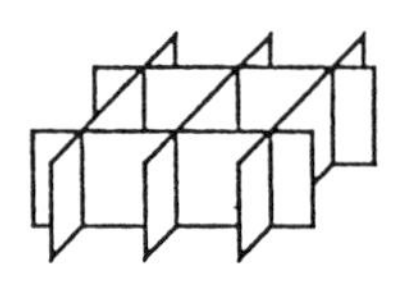

Webs are intersecting walls which provide support for the interiors of large unsupported areas. Webs also connect to the projected feature edges to provide a stable support for the area. Bridgeworks attempts to keep webs away from vertical part surfaces to preserve the finish of these surfaces. No loss of support occurs since the vertical part surfaces also provide support to the layers above. The support spacing parameter controls web spacing.

Columns

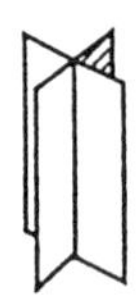

Columns are used to provide support for islands and other small unsupported areas. A column support consists of two walls which form a plus shape which is large enough to build on its own. The column width parameter controls the size of a column.

Figure 5-20. Five basic categories of support structures.[10]

Cubital

Cubital produces the Solider, physically the largest and most mechanically complex RP&M machine on the market. It is the only RP&M machine that requires a full-time operator. Cubital's software is run on a workstation referred to as the data front end (DFE).

The DFE is the interface between the designer's three-dimensional files and the model production machine (MPM). The DFE is a DecStation 5000/200 workstation made by Digital Equipment Corporation and is based on the Ultrix operating system, version 4.2. The DFE reads and checks the three-dimensional CAD files, corrects them when necessary, and prepares them for production. The input data has no geometric restrictions, and almost any CAD design can be prepared for production without special editing, such as creating support structures. The DFE contains a wide range of software tools, mostly graphically based, which are used to remove redundant information and correct flaws in the data. At this time, the current Cubital's DFE software is version 2.0.

Typical workflow is as follows:

1. Read the 3D data file into the DFE.
2. Convert the data file to the Cubital facet list (CFL) format.
3. Reduce the file size to eliminate redundant or otherwise unnecessary data.
4. Change the topology of the design as required.
5. Automatically correct deficiencies as detected throughout the file.
6. Generate a collection of parts to be made together. The collection can be nested within the available production space, using a 3D graphical editing tool.
7. Produce the collection of parts utilizing a real time slicing procedure, often referred to as "slice on the fly." However, devoting the DFE to slicing greatly reduces its ability to perform other tasks while parts are building.

Note that all steps are accomplished using graphical tools, and that one can go directly from step 2 to step 6 if no corrections or improvements to the file are required.

The accepted file formats are:

1. SDRC-IDEAS 3D CAD files.
2. CAEDS 3D CAD files.
3. Pro-Engineering 3D CAD files.
4. VDA-FS version 1.0 and 2.0 (including trim surfaces).
5. STL files, both ASCII and binary.
6. CFL files.

All of the accepted file formats are automatically converted to the CFL format. This provides an ASCII or binary representation of the polyhedral object. The CFL facets are not limited to triangles, and thus save hard disk space. The CFL file contains fields for nongeometric data, and provides:

1. A header with user-defined information.
2. Statistics on the part, such as number of points, number of facets, and number of bounding boxes.

3. No redundant point information. File size is minimized.
4. A virtually unlimited number of polygon vertices (number of edges).
5. Concave and convex polygons.
6. "Hole" data structures.

DTM

The Sinterstation 2000 operation is described in Chapter 16. The accepted data file formats include STL files and IGES surface models.

DTM's software architecture uses a 486 PC running the UNIX operating system. Different groups of building parameters are displayed on the screen depending on the material type. Recoating is accomplished by feed cylinders laying down powdered material for the subsequent layer. Parts are partially supported by the powder in the vat, and thus do not require special recoating parameters to handle "trapped volumes." The beta software has the ability to show the parts on the platform before they are constructed.

Helisys

The Helisys "Laminated Object Manufacturer" or "LOM" system is also described in Chapter 16. Helisys software emphasizes simplicity. The DOS-based software runs on Windows 3.0. The LOM-1015 uses a 386, and the LOM-2030 uses a 486. The single PC is used to drive the entire process, which includes concurrent slicing while the part is building ("slice on the fly"). A dynamic height control automatically measures the thickness of each layer of paper which is put down. Because the thickness of the paper may vary by up to a thousandth of an inch at various locations in the roll, which can be 20% of the layer thickness, this approach is required to maintain Z dimension accuracy.

The data input for Helisys is STL files only at this time. The software allows the part to be offset on the build platform, displays the part extents, and enables automatic multiple part generation. A total of 15 material specific parameters are available, including heater control parameters which activate the adhesive on the roll of paper. One particularly interesting parameter set that the software governs is "Dicing." This parameter group controls the cutting (dicing) of the extra paper which is left over from interior regions which are not a direct constituent of the part. This procedure is necessary for removal of this material after the part has been completed.

Stratasys

The Stratasys software involves Unix-based workstations which drive the 3D MODELER System. The software uses a subset of the CAMAX CAD/CAM software and also delivers a full CAD package as part of the system described in Chapter 16. The software for slicing has been integrated with the CAMAX CAD/CAM software, a menu-driven, graphical software package.

The 3D MODELER involves a NURBS-based surface modeler which:

- Can be used for stand alone prototype and design.
- Allows flexibility to make modifications to designs.
- Provides the capability to add whatever supports may be needed.

The design software capabilities allow for 3D NURBS surfaces or simple two-dimensional shapes. The system provides for the use of points, lines, arcs, and splines to construct and enhance the design model. Three different types of constructive or five different free-form surfaces are available. The software's trimmed surface features allow rapid prototype creation. Color shading and rendering software aids design verification. In addition to design capability, the software accepts a variety of CAD inputs through IGES. Either wireframe, surface, or solid CAD data from standard CAD software packages can be imported through IGES. Stratasys supports the IGES entities listed in *Figure 5-21*.

A file brought into the CAD software program in an IGES format can be edited, scaled, oriented, and even surfaced if required. Supports, if needed, may be added graphically. The 3D MODELER can import digitized data via IGES. The digitized data can be generated by probe, laser scan, sonic scan, CAT scan,

Geometric

100 Circular arc
102 Composite curve
104 Conic arc
106 Copious data
110 Line
112 Parametric spline curve
114 Parametric spline surface
116 Point
118 Ruled surface
120 Surface of revolution
122 Tabulated cylinder
124 Transformation matrix
126 Rational B-spline curve
128 Rational B-spline surface
140 Offset surface (limited support)
141 Boundary surface
142 Curve on a parametric surface (limited support)
143 Bounded surface
144 Trimmed surface

Annotation Entities

106 Cross-hatching (limited support)
202 Angular dimension
206 Diameter dimension
210 General label
108 Plane
212 General note (limited support)
214 Leader arrow
216 Linear dimension
218 Ordinate dimension
220 Point dimension
222 Radius dimension
228 General symbol

Structure Entities

308 Subfigure definition
314 Color definition (limited support)
402 Associate instance
404 Drawing (limited support)
406 Property
408 Singular subfigure instance
410 View (limited support)
412 Rectangular array subfigure instance
414 Circular array subfigure instance

Figure 5-21. IGES entities supported by CAMAX/Stratasys software.

or any other method of collecting geometric data points. This input data can be used as slices or to define surfaces. For example, information has been input directly from a CAT scan using surface point data. In addition to IGES input, the 3D MODELER can be driven directly through NC code or through the STL file format. NC code from CAM systems is supported, provided it can compute multiple continuous surfaces with an ever-increasing positive *Z* axis. The STL data can be in either ASCII or binary file formats.

A feature of this software is the easy interaction between the operator and the system through menus. Key parameters such as slice thickness, head speed, material information, and other building parameters are simply input by answering questions or filling in blanks. Automatic routines input these parameters and process the part. The software also allows multitasking so a designer can use the workstation to process, slice, and do other CAD-related tasks concurrently.

Cubital, DTM, Helisys, and Stratasys have all developed innovative and unique approaches to the RP&M field. Each company's novel machine(s) require software to enable data to be input and read, part preparation parameters to be set, and build cycles to be controlled. Perhaps each company's particular software architectures will spawn improvements to RP&M software as a whole.

5.5 Future Software Concepts

On an esoteric level, software architectures should provide a series of powerful and memorable "mental images." This allows users to feel in complete control of the operation, to clearly understand what stage of the operation they are in, and to be fully confident that completion can and will be achieved. In this final section, architectural improvements will be discussed in the "mental image" category only. The discussion of mental images begins with the well understood conceptual models of NC machines and laser printers. Next, views on how file standards might be brought into these models are discussed. The goal of a fully automated machine is covered at the end of this section.

Numerical Control Machine Model

NC machines currently exist in industrial environments in large numbers. They require a technician to monitor them in case a problem occurs, such as a milling tool crashing through a tooling plate. Special liquids such as oil and coolant are necessary to keep the machines running. NC machines are fed information via a blueprint(s) or a CAD model, which is translated to an NC language by a "dedicated" NC programmer. Many CAD packages have outputs which are already in an NC usable format. However, the NC programmer usually must alter the output to allow the part to build correctly. Often the programmer slows the feed rate of the tool at radius cuts, and does other "tricks" to make the best parts possible. Special instructions about the final part's desired characteristics may be discussed between the engineer and the NC programmer. The information conveyed from the blueprint(s) or the CAD package is used by the NC programmer and/or scheduler to decide which particular NC machine will perform the job, and if it will be run with any other work.

The paradigm between NC and RP&M machines has similarities and differences. Most RP&M machines do not require an industrial environment or an operator to monitor the machine during its build cycle; Cubital's Solider being the exception. All RP&M machines require liquids, filaments, or powders. If the machine has a large number of moving parts, it may also require oil and/or coolant. The more complex the RP&M machine, the more likely it will require a technician to solve mechanical problems.

Is the RP&M software complex enough to require a "dedicated programmer," or is the software simple, fast, and forgiving enough to allow an engineer to use the machine as a peripheral device which he or she programs? *Figure 5-22* shows the NC machine domain.

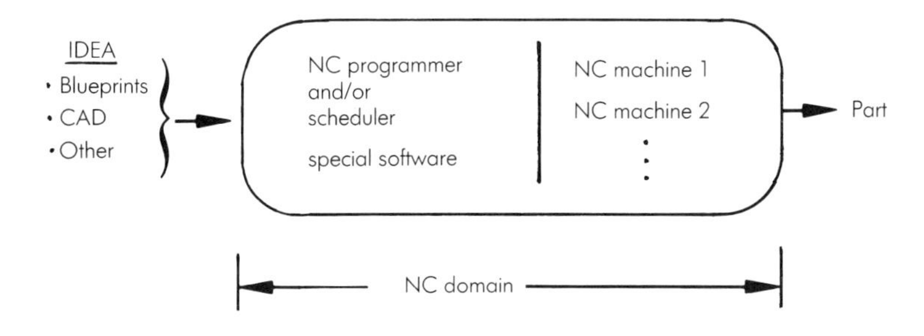

Figure 5-22. NC machine model.

Laser Printer Model

Laser printers are usually hooked into and receive files over a network. Most modern laser printers support the PostScript® format. Text, text size, fonts, and special instructions are input by various word processing software packages which output a standardized data file (PostScript®). *Figure 5-23* shows the laser printer model.

If RP&M systems are to fit into this paradigm, their files should also come over a network, standardized RP&M build files must be established, and CAD systems must be enhanced to allow simple preparation of RP&M build files. If an RP&M system can fall into this category of devices, the cost of ownership is reduced because extra people are not required to utilize and maintain the system, and integration of the RP&M system into the existing work environment is simplified.

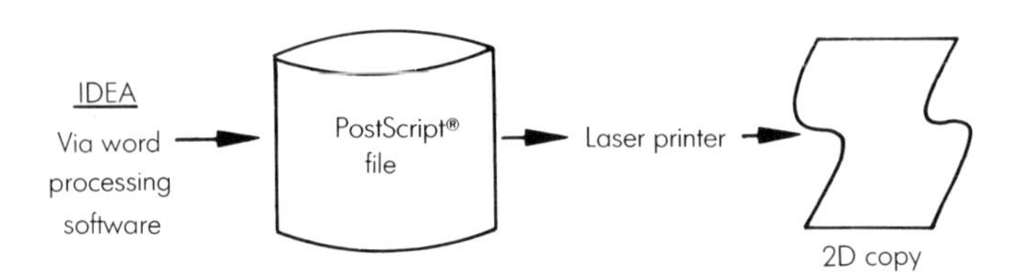

Figure 5-23. Laser printer model.

Accepting Other Data Formats/System Standards

What will be the equivalent of PostScript™ for RP&M machines? It will probably be a hybrid format based on one or more of the standardized data formats in existence today, such as IGES, NC Post, and VDA-FS. Ultimately, CAD packages should directly output buildable files, much like word processors output "ready to print" 2D image copies. As the RP&M market grows and matures, standardized operating instruction sets and protocols will almost certainly be established. Compression algorithms can and should be used to reduce the size of large files.

Also, the ability to run RP&M systems from existing PCs and workstations will further reduce the cost of ownership. To support a common look and feel, the Microsoft Windows package or X-Windows in conjunction with OSF/Motif or OpenLook tool kits can be used. RP&M systems that are internally and externally standardized make integration into existing company environments a much easier task.

Next Generation: Automation

As newer PCs and workstations deliver more computing power for reduced cost, RP&M software will probably become more graphically based with faster response times. RP&M software architectures will also generate better mental images by relying on icons and other user friendly graphic tools.

We can imagine software that works as follows:

1. The user selects a shaded three-dimensional image of a part from a group of parts on the screen.
2. A window opens up displaying the RP&M process chamber and the part automatically appears on the build platform, complete with supports, if required, and optimally oriented.
3. Additional parts are positioned in the same manner until the build platform is full.
4. No parameters need to be input to the machine since the optimal parameters are already loaded. The "GO" button is pressed, and a few hours later, the finished parts are available without requiring any manual intervention.

This RP&M software architecture would literally be easy enough for a child to use. The key words related to ease of use are "graphics" and "automation." Many other textbooks have explored the world of computer graphics, but the field of RP&M *machine automation*, with maximum accuracy for varied part geometries, is still largely uncharted. Before automation can take place, the transformation from an art (prototype SLA) to a skill (SLA-1, early SLA-250) to a science must take place. RP&M systems are currently in the metamorphosis stage, transitioning from a skill to a science.

However, claims of automation are misleading at this stage of the RP&M industry's development. The multitude of potential part geometries require the human being's innate and extraordinary pattern recognition capabilities to successfully build the most complex parts. Some day, computer programs with sophisticated pattern recognition capabilities will automatically determine optimal build techniques while the part is under construction.

In the original Turing contest, which began around 1950,[11] people in an isolated room were asked to determine if their typed communication was with a person or a computer simulating intelligence. In the spirit of the Turing contest, automated RP&M software will be competitive with skilled RP&M users only when it can build equally beautiful parts. The software architecture will be required to emulate the expert through a large decision tree of part building conventions. Perhaps eventually a neural network which actually "learns" better building techniques will be developed.

When a seasoned quality assurance inspector is unable to determine whether a skilled RP&M user or advanced software was employed to construct a complex set of parts, then the contest will be decided. The final result of that contest lies in the sophistication of the next generation of RP&M software.

ACKNOWLEDGMENTS

The author of this chapter would like to acknowledge input from the following people:

Paul Marygold, Dr. Brian Nutter, Dr. Henry Schultz, Jim Paravantes, Bart Williams, Frank Schäflein, George Teachout, Ara Bernardi, Jon Tindel, Chris Manners, Dennis Smalley, Rami Kopelman at Cubital, Donna Wyatt at DTM, Brian Hsieh at Helisys, and Cindy Wheatley at Stratasys.

REFERENCES

1. Foley, J., van Dam, A., *Fundamentals of Interactive Computer Graphics*, Addison-Wesley Publishing, Massachusetts, 1983, pp. 505-506.
2. *SLA-190/250 User Guide, Software Release 3.82*, part number 18456-M10-00, 3D Systems, Valencia, CA, September 1991.
3. Hofstadter, D.R., *Gödel, Escher, Bach: An Eternal Golden Braid*, Vintage Books, New York, 1980, pp. 276.
4. Tarnoff, H., *3D Systems Laser Controller, Version 1.21 Hardware and Software Documentation*, Tarnz Technologies (now DataPlex), Van Nuys, CA, July 1987.
5. Lesnet, N., *Software History 1988-1989*, document number 50099-C70-00, 3D Systems, Valencia, CA, January 1990.
6. *SLA-500 User Guide, Software Release 1.40 (Beta version)*, 3D Systems, Valencia, CA, November 1991.
7. Smith, P.G., and Reinertsen, D.G., *Developing Products in Half the Time*, Van Nostrand Reinhold, New York, 1991, pp. 61-79.
8. Yourdon, E., *Modern Structured Analysis*, Prentice-Hall, New Jersey, 1989.
9. Floyd, G., *Long Term Software Plan, Phase III*, 3D Systems, Valencia, CA, October 1991.
10. Bradford, R., *Bridgeworks Automated Supports for Rapid Prototyping, User Guide Version 2.0*, Solid Concepts, Los Angeles, CA, June 1991.
11. Hofstadter, D.R., *Gödel, Escher, Bach: An Eternal Golden Braid*, Vintage Books, New York, 1980, pp. 594-597.

chapter 6

CAD Processes

And as imagination bodies forth the forms of things unknown, the poets pen turns them to shapes, and gives to airy nothing a local habitation and a name.

—William Shakespeare
A Midsummer Night's Dream,
Act V, Scene 1

6.1 Introduction

Many individuals and corporations eager to purchase rapid prototyping units quickly become emersed in the various features and specifications of the RP&M systems. However, the quality and accuracy of the input data is equally important in implementing this technology. Planners must not ignore the well known basic tenet: garbage in equals garbage out. RP&M systems are highly dependent on their electronic database input. Simply stated, they are three-dimensional duplicating machines. These systems take an electronic description of a three-dimensional object and reproduce that description into a solid object. If the description is inadequate, the generated part will also be inadequate. The geometric descriptions required for current RP&M equipment are provided by computer aided design (CAD) systems.

This chapter will discuss the various procedures, requirements, and details for CAD systems intended as input to Rapid Prototyping & Manufacturing technologies. *Figure 6-1* shows a CAD system in use.

6.2 Data Requirements

RP&M software currently in wide use requires nonambiguous data descriptions of the part geometry to be generated. Nonambiguous data sets result in one

By ***David S. Reynolds****, Applications Engineer, 3D Systems, Inc., Valencia, CA.*

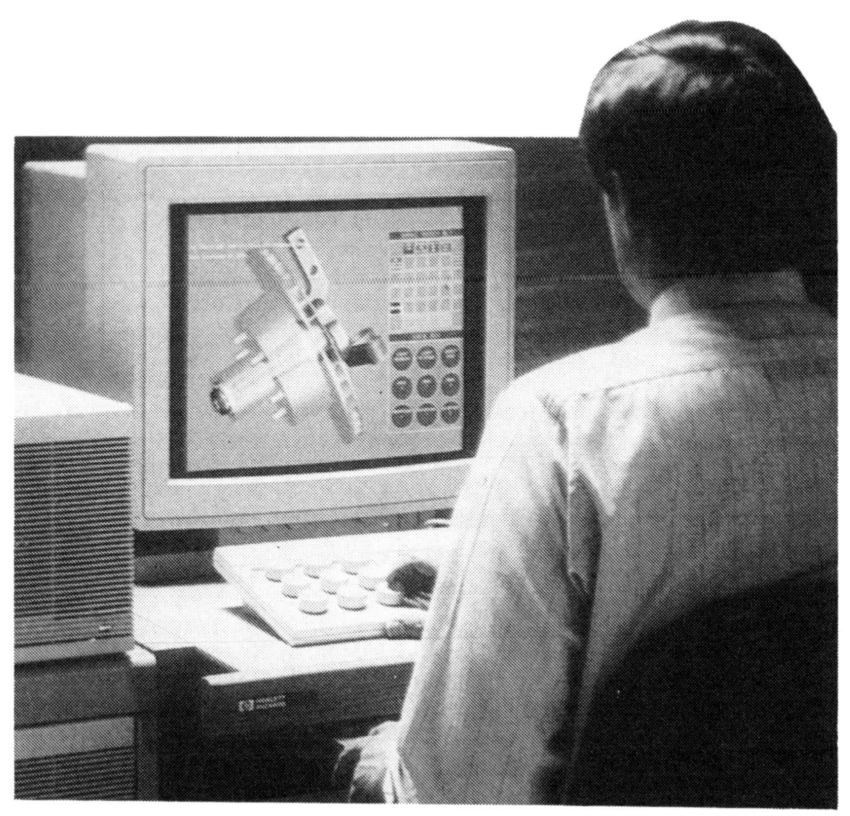

Figure 6-1. Evaluating design on a CAD/CAM system. (*Courtesy, Machine Design, June 18, 1987*)

and only one possible interpretation.[1] The model data must facilitate the generation of closed paths and differentiate between the "inside" and "outside" of the part. Imagine being able to use an electronic hot knife to cut your CAD design at any given horizontal cross-section. The knife would traverse the boundary of the object face, cut into the mass beyond the face, and across any internal feature boundaries until it passed through to the opposite side of the object. The resulting cross-section would be one or more closed paths and a complete understanding of the solid areas enveloped. *Figure 6-2* shows, as a simple example, slices through a toroid, otherwise known as a donut.

Cross-hatching algorithms create vectors that solidify the areas between the part boundaries or borders. Problems occur when the part geometry is not completely closed, because either adjacent surface vertices do not connect or whole surfaces are missing. The RP&M software must decide how to close gaps. Otherwise, the system would leave an opening that could negatively effect part building. Gaps in the borders cause the hatching vectors to be incomplete or escape outside the part.

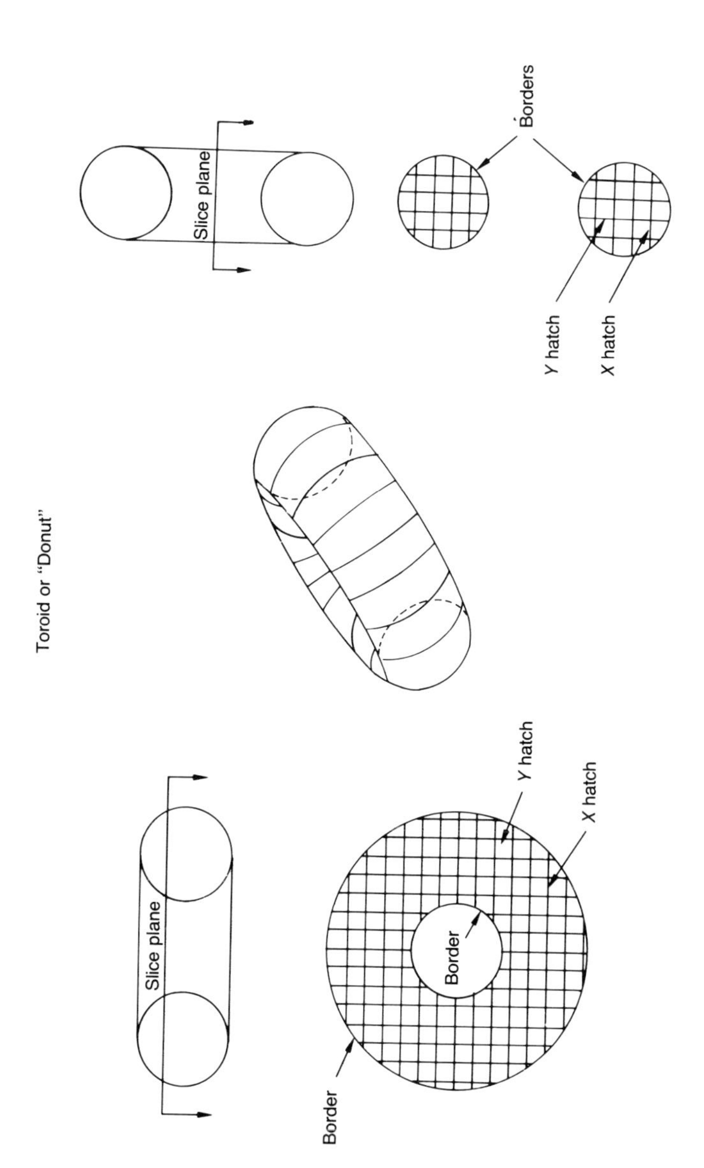

Figure 6-2. Examples of slices through a closed model.

Closed boundaries alone do not describe solid objects. The boundary data must also convey the orientation of the solid areas. Surface normal information pertaining to the object's boundary is used to indicate the orientation of the object's mass, as shown in *Figure 6-3*. If this information is incorrect, walls with zero thickness, or twisted surfaces with conflicting, impossible orientations called Möbius strips can occur. [2]

6.3 Solid Modeling

"A solid model can be defined as a geometric representation of a bounded volume. This volume is represented graphically, via curves and surfaces, as well as nongraphically through a topological tree structure which provides a logical relationship that is inherent only with solid models."[3] The topological data defines and maintains the connective relationships between the various faces and surfaces of the geometry. A requirement of solid modeling CAD systems is the generation of an unambiguous description of the geometry being modeled. Each face of the object, including its normal orientation with respect to the object's mass, are maintained. The normal is defined such that it points away from the mass. If the object's closed nature is violated, for example, due to an improper Boolean operation, the system informs the user with numerous error messages. Solid models, by definition, satisfy the requirements for RP&M input data.

Most existing CAD systems employ a hybrid array of mathematical concepts toward solid modeling. Specifically:

1. Constructive solid geometry (CSG) modelers use simple geometric shapes such as cubes, cones, and spheres as basic building blocks. The blocks are combined using Boolean operators such as union, intersection, and difference. The hierarchical representation is maintained in what is called the CSG tree.[4]
2. Boundary representation (BREP) systems are polynomial or NURBS (nonuniform rational B-splines) based. NURBS are a precise mathematical description of geometry that allows for easy manipulation of two-dimensional, three-dimensional, and surface entities, from simple to complex.[5]
3. Faceted approximation is a technique of dividing the solid object into planar surfaces. The planes may be divided into triangles or quad facets. Increasing the density of facets gives a closer approximation to a nonlinear surface. A two-dimensional example is approximating a circle with straight line segments.[6]

Systems such as SDRC's I-DEAS represent objects using a NURBS-based definition, while creating a faceted approximation. "The faceted approximation improves performance when displaying or modifying objects."[7]

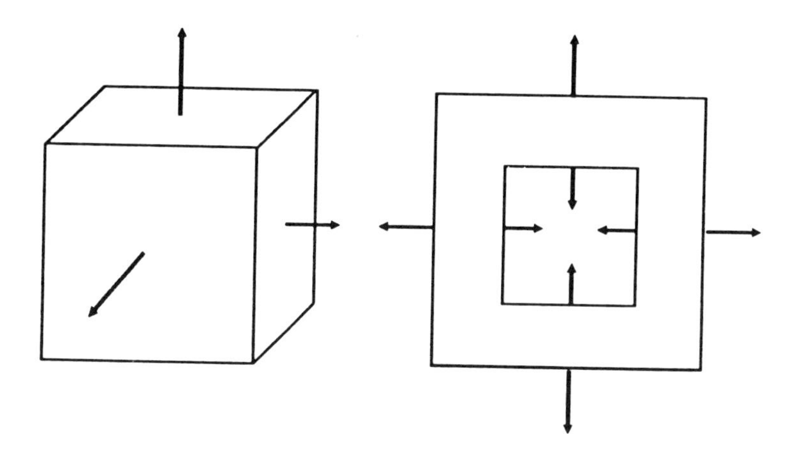

Figure 6-3. Surface normals.

The advent of RP&M during the late 1980s came at a time of increased use of solid modeling CAD/CAE/CAM systems in the U.S. and Europe. Indeed, many users have stated that RP technology has helped justify the move to solid modeling CAD systems. The aerospace and automotive industries have been on the forefront of the proliferation of solid modeling. Some systems in wide use include:

- Dassault Systems CATIA
- SDRC I-DEAS
- GM Unigraphics II
- Parametric Technologies Pro/ENGINEER
- Alias Research Inc., Alias Designer

6.4 3D Surface Modelers

Surface modelers are often used to describe aerodynamic and aesthetic shapes. The automotive industry is a large user of such systems, and auto companies have even developed their own in-house code as in the case of Ford Motor Co. and its PDGS system. A major functional difference between solid-based CAD systems and surface modelers is the absence of topological data connecting the surfaces. A surface model also lacks the capability to describe the interior of the part.

Typically, the user develops two-dimensional wireframe profiles that are revolved, extruded, swept, or blended together to form the desired surface. The surface entities can be simple, such as planes and tabulated cylinders. Complex entities require definition by mathematical interpolation schemes such as bicubic splines. Other schemes utilize polynomial expressions such as Bezier curves.[8] Most advanced systems utilize NURBS-defined entities. NURBS definition bestows optimum control over the surface shape, while allowing editing to be easily accomplished.

The individual surfaces are assembled to form the desired design. Although the surfaces themselves may be a mathematically precise description of a boundary, that alone will not render a tangible three-dimensional solution. The existence of unclosed surfaces or Möbius strips are possible since nonambiguity is not a requirement for a surface model.

The surfaces generated by 3D surface modelers are of zero thickness. Often, complex surfaces are generated by taking a specific curved line and rotating it about a given axis. This is known as a surface of revolution. To create a design for RP&M, the user must ''sew'' the adjacent surfaces to enclose the volume, and identify which side holds the mass of the part. Also, any ends of surfaces that would extend within the part must be ''trimmed.'' Although surface modelers require some additional work by the operator, they have the ability to describe complex surfaces where many solids-based systems have trouble. An additional feature is the ability to assign a finite thickness to an otherwise zero thickness surface. This technique is known as ''shelling,'' and is described further in section 6.6. Most advanced CAD/CAM/CAE systems (Catia, I-DEAS, Pro/ENGINEER, CV CADDS 5, etc.) strive for the best of both worlds: combining solid and surface generation into one homogeneous system.

6.5 CAD System Data Representation

Representation methods used to describe CAD geometries vary from one system to the next. A standard interface was needed to convey geometric descriptions from various CAD packages to rapid prototyping systems. Available standards (IGES) are not always capable of fully defining unambiguous solid geometries. RP&M systems that advertise IGES import capability may fail to mention the additional work required to generate solid prototype parts. Therefore, 3D Systems Inc. developed the STL interface specification. This has become the de-facto standard input for all RP&M systems. The STL interface is based on a polyhedral representation found in many solid modelers. Precise surfaces (NURBS-based, or primitive CSG objects) are approximated, via planes and straight lines, into a tesselated (triangulated) facet format. The level of accuracy on nonplanar surfaces is controlled by the number of facets used to represent that surface. Planar surfaces are exact since the facets already lie on the precise surface.

CAD Parameters

Since RP&M systems directly recreate the resolution given to them, the CAD data should be adjusted to reflect the required part resolution. The method for increasing the resolution of the CAD geometry varies from system to system. One method involves stating a maximum absolute deviation from the true surface. The tolerance is expressed as the distance of the maximum chord height from the true curve, as shown in *Figure 6-4*. SDRC I-DEAS Solid Modeling™ also allows the user to specify the tolerance as a percentage of chord length. Small values generate more facets to make the model more precise, but with the penalty of increasing the STL file size. Larger STL files result in increased slice time and slice file size, but have a negligible effect on increased build time. The user is prompted for facet resolution values when creating objects, or they can be refined just prior to STL file generation.

Pro/ENGINEER allows the user to specify the degree of resolution on curved surfaces by entering a quality value in its interface. The values range from 1 to 10, with 10 creating the highest resolution. Again, higher resolution values generate larger STL file sizes. Using a triangle density higher than eight seems to have an inconsequential impact on the final part. A system's internal accuracy also affects STL accuracy. SDRC I-DEAS calls this the point coincidence value. This value is often found in the system's default start-up file. Some systems allow you to change the model's accuracy interactively. With Pro/ENGINEER, one accuracy value applies to each feature within the part. Therefore, geometries with relatively large and small radii will benefit from using a tighter accuracy value. Pro/ENGINEER will regenerate all previous features using the new accuracy value.

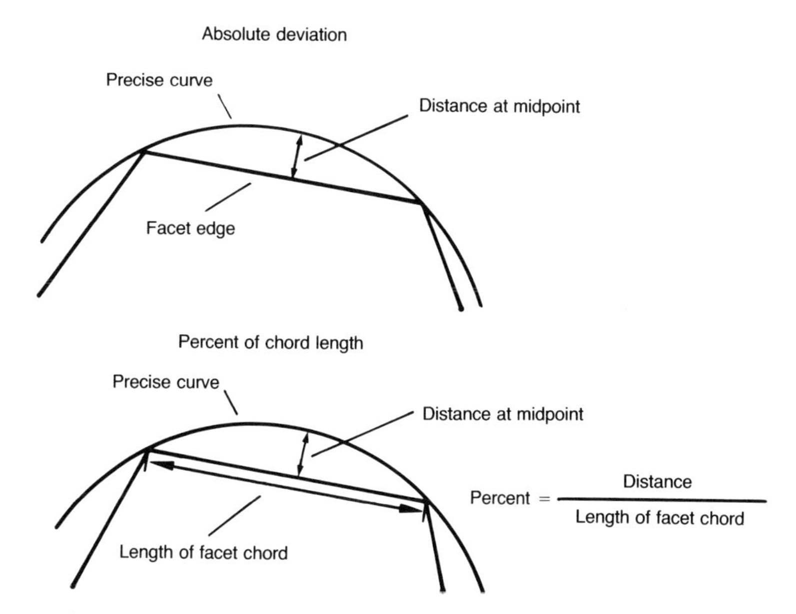

Figure 6-4. SDRC I-DEAS facet resolution methods.

Scaling the part geometry to account for shrinkages can be accomplished in the RP software. Large scaling, for parts whose applications are visualization aids, should be done inside the CAD system. This helps maintain maximum accuracy of the facets. RP&M users in the electronic connector industry often use 2X or even 5X scaled models to help visualize small features in their design reviews. Shrinkage factors for casting applications can also be incorporated during the CAD modeling phase, although these shrinkage factors are available among the RP&M software parameters.

The standard units supported by SL systems are inches and millimeters. CAD designs created in other units must be scaled to inch or millimeter values prior to slicing. This can be accomplished for SL in the 3DView or Part Manager program. Scaling can also be accomplished during the Slice program. Downstream RP&M software can then automatically set its units by examining the magnitude of the values in the slice files.

The minimum feature size that can be built is dependent upon the RP&M system used. For laser-based systems, the absolute minimum wall thickness would be the width of a single cured line. This is related to the laser beam diameter and the cure depth. The width of a single cured line can vary from 0.007″ to 0.012″ (0.178 to 0.305 mm). For a photomask system such as Cubital, the minimum feature size would be related to the resolution of the mask (200 or 300 dots per inch or 0.005″ or 0.003″). If the feature is not perfectly parallel or perpendicular to the mask's raster lines, the result is an aliased or stair-stepped definition in the slice plane, typical of laser printers.[9] Although the RP&M system may be able to draw a feature this small, the fragile nature of such a thin section may be a limiting factor. An exception is the presence of proportionally larger walls supporting thin features, as in the case of narrow cooling passages within a turbine blade.

6.6 Part Orientation

A few guidelines exist regarding part orientation in SL systems. The part geometry must reside in the positive x,y,z octant, meaning all CAD values are positive numbers. This is because the CAD real numbers will be converted to positive integers during the slice process, so any distinction between positive and negative CAD values would be lost. A positive orientation is also a convenience to the operator since he or she can reposition the STL files before slicing. Translation is accomplished using utilities found in 3DView for the SLA-250. For SLA-500 users, Part Manager software will translate STL files. Keeping the parts close to the x and y origin minimizes the extent of the part and allows the use of a maximum slice resolution factor. Supports need to occupy at least the first 0.25″ or 6.35 mm (for the SLA-250) to 0.35″ or 8.9 mm (for the SLA-500) of z space. This assures that the sweeping of the recoater blade can start safely above the metal platform on which the parts are built. Starting the first layers of the part well above the platform also improves uniformity of layer thickness, regardless of any warpage in the platform. These nonuniformities are accommodated by the supports themselves.

The RP&M operator's choice of part orientation within the building chamber will have an impact on build time, part resolution, and surface finish. Obviously, minimizing the height of the geometry (in the z axis) will reduce the number of layers required, thereby decreasing build time. Depending on the part's eventual application, the RP&M operator may sacrifice a minimum build time for the benefit of increased part resolution. Higher resolution of curved surfaces are obtained by orienting them in the horizontal plane normal to the laser beam. Sloping surfaces that proceed along the slice axis will be layered and have a distinct "stair-step" appearance, as shown in *Figure 6-5*. The height of each step is the layer thickness used in that portion of the part. StereoLithography systems allow parts to be built using multiple layer thicknesses within a building cycle. Flat up-facing regions will have the best surface finish on SL parts due to the presence of fill vectors and lack of support structures. Sections with stair-steps and intersecting supports can later be hand finished to improve their surface quality.

Liquid-based RP&M systems must also be concerned with trapped volumes. Trapped volumes are defined as spaces that hold liquid separate from the liquid in the vat. These regions may require special recoating parameters that slow the build rate. A change in orientation can often eliminate trapped volumes. For example, a cup right side up contains a trapped volume, but when turned upside down, it does not. If simple reorientation is impractical, the CAD designer can build drain holes. The holes can be plugged later, during the postprocessing stage. The diameter of a single drain hole should be approximately 10% of the

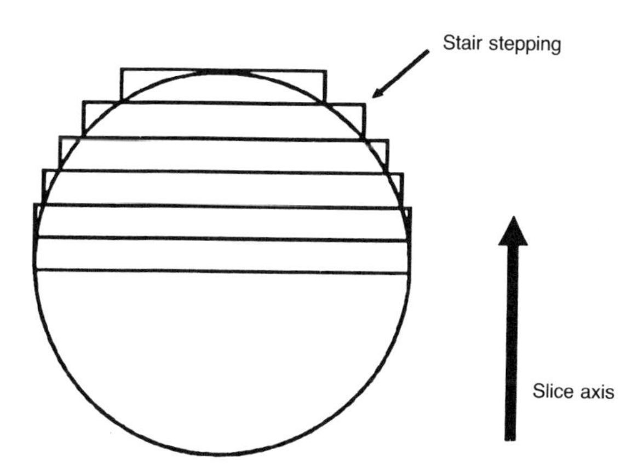

Figure 6-5. Stair stepping of curved surfaces along the slice plane.

largest diameter that would fit in the trapped cross-section on any layer. Multiple drain holes may be required on very large trapped regions. Totally enclosed volumes must have drain holes to allow the uncured resin to escape. A system using solid material like Helisys must be stopped during the process to remove the material within the enclosed volume since there is no way to remove it later.

A part's size may also determine its orientation during building. While SL involves a cubic building volume, other systems, such as DTM's SLS, utilize a cylindrical volume. An advantage of a rectangular work area is its corner-to-corner space, as shown in *Figure 6-6*. Parts may be sectioned and then assembled in the postprocessing phase. Attachment features, such as dowel pin holes, registration pins, tongue and groove features, and bolt holes can help the builder accurately assemble sections. The most common method uses standard metal dowel pins on mating surfaces. Two pins per adjoining surface will sufficiently constrain sections for proper alignment.[10] The sections can then be attached using the photopolymers themselves or any plastics compatible adhesive.

If a part is very thick and is strictly intended for use as a concept model, hollowing or shelling out the design may be advantageous. This will decrease the build time and conserve valuable resin, thereby lowering part cost. Do not forget to include drain holes to allow any enclosed resin to flow out. Pro/ENGINEER has a very useful feature called "Shell." This feature removes a surface or surfaces from the solid and then hollows out the inside, leaving a shell of specified thickness. The thickness can be consistent, or each surface can have a different offset.[11] Split patterns used for sand casting that are very massive could also be shelled and back-filled with a suitable material such as epoxy.

Inclusion of metal (or other material) inserts are possible in RP&M systems. StereoLithography allows the operator to pause the process at a specified layer to place an insert that will be enclosed at the end of the build. A requirement is that no portion of the insert protrude above the last formed layer. An insert protrusion could block subsequent curing by obscuring the laser beam, and might also result in a collision with the recoater blade.

6.7 Supports

Supports in RP&M systems are analogous to work holding devices for machining. They are created in the CAD system and output as files separate from those for the part. Additionally, third party software has been introduced that automatically provides support structures. The supports may be generated by the engineer or CAD designer, or by the RP&M operator. Many CAD systems provide an assembly mode that facilitates support creation while viewing and or referencing the part geometry. Supports hold the part in place while it is being generated. For SL systems, supports are required for every part. Some competing technologies have, in the past, advertised that supports are not a

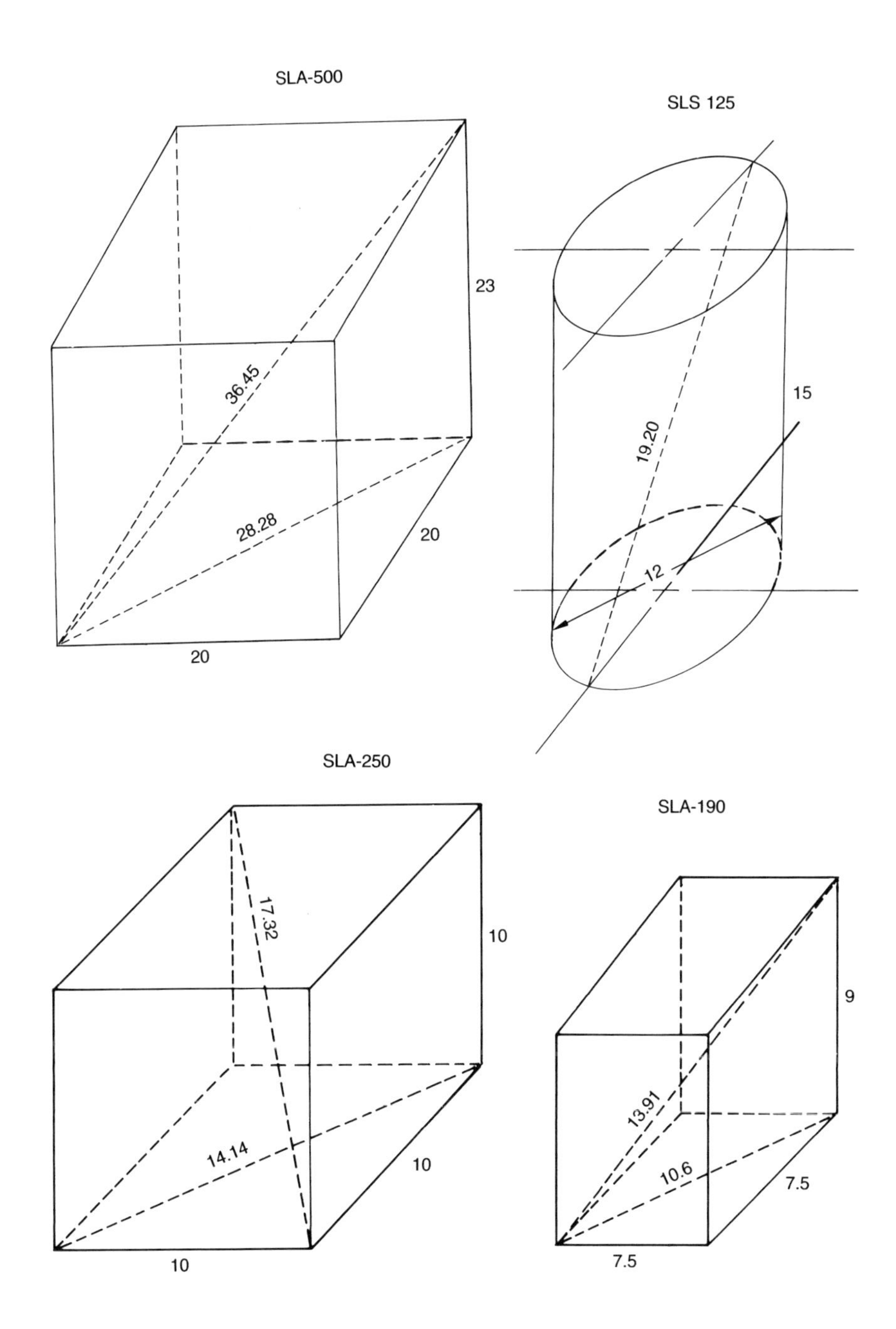

Figure 6-6. Various RP&M device work space.

requirement for their systems. But as more difficult geometries and tolerances are attempted, many have accepted their necessity. According to K. Nutt, "Recent experimentation with base structures and anchor supports has helped control warpage that may occur during part fabrication using polycarbonate and wax powders."[12] Nutt subsequently hypothesizes that future materials will eliminate the need for supporting structures. All materials, with the exception of water, shrink when they undergo a phase change from liquid to solid. Similarly, the sintering process requires densification.[13] Increased density, without any change of mass, requires a loss of volume, thereby producing shrinkage. Unconstrained shrinkage results in curl, which leads to part inaccuracy. Thus, the necessity of support structures is evident.

SL is an additive process using a layer by layer approach occurring on the liquid resin surface. To properly constrain a given layer, it must be attached to the previous layer. The very first layer is attached to the platform and is always a support. Areas that overhang the prior layer must be supported to prevent their deformation, known as curl. Curl occurs when shrinkage of layer $n+1$ pulls on layer n, similar to the bimetallic strip effect, as discussed in Chapter 10. Supports consist of combinations of thin webs, usually a single cured line width (0.007″ to 0.012″ thick or 0.18 mm to 0.3 mm thick). All support types need to overlap into the succeeding part layers by at least .02″ (0.5 mm) or two to three layer thicknesses. This intersection will ensure that the supports physically connect to the part feature.

Every SL part must have a base support to bridge the first layers of the transition from the platform to the part. The base support provides a collision avoidance buffer in which no recoater sweeping can occur. Additionally, it allows for the safe removal of the part from its platform. The base support should follow the periphery of the part's bottom layer, including its corners. This will restrict the tendency to curl as successive layers are built. Base supports can be designed as solid objects. Appropriate slicing parameters will transform the solid support into an "egg crate" pattern during build. Offsetting the outline or border of the support inward by 0.01″ (0.25 mm) is strongly recommended to prevent the part edges from being broken during support removal. Software linewidth compensation, found in the Slice program, can offset the borders of the support by adding 0.01″ (0.25 mm) to the normal linewidth compensation value. Also, projecting this retracted border down to the platform, and then using a large (0.25″ or 6.35 mm) hatch spacing will generate an excellent egg crate support system. The spacing of the egg crate pattern is produced by using 0.25″ (6.35 mm) as an x and y hatch spacing in the Slice software. Thus, the CAD user need not create this pattern vector by vector. The transformation of a solid base support into an egg crate support is shown in *Figure 6-7*.

"Islands" are layers of the part geometry that would otherwise be unconnected to any other section of the part. Islands must be anchored to the platform or to the part itself. They can be connected to the prior layers of the part,

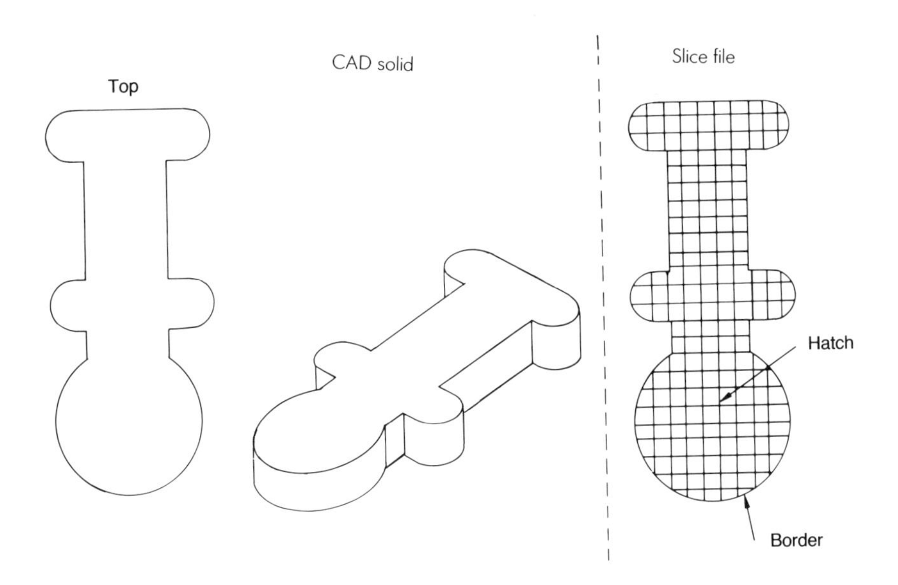

Figure 6-7. Transformation of solid model to hatched base support.

providing the structure is rigid. This method may be preferred if it solidifies less material than a connection down to the platform, thus reducing build time. This is especially true for small islands many inches above the platform.

Cantilevered sections must also be supported. An overhang that extends beyond 0.05″ (1.27 mm) will exhibit curl when unconstrained. Triangular shaped supports called gussets are strong, easily removed, and are efficient with respect to STL file size. Gussets should extend into the walls by approximately 0.015″ (0.38 mm) to ensure firm attachment. Continue the gusset into the cantilevered area by two or three layer thicknesses or 0.03″ (0.76 mm).

Roman architecture has shown that arches are self-supporting. In SL, arches or convex surfaces support themselves because the overhang between succeeding layers is very small. But if flat, down-facing areas longer than 0.05″ (1.27 mm) exist, they must be supported. Generally, a feature that slopes between 0 degrees and 30° from horizontal should be considered a flat, down-facing area. A "ceiling" for example, must be held in place or it would sag. *Figure 6-8* shows a variety of support types.

6.8 RP&M Input File

The de-facto standard input format is known as STL. It is based on a normalized, tessellated data format similar to a finite element mesh representation. Solid modeling systems generally produce files that correctly follow the

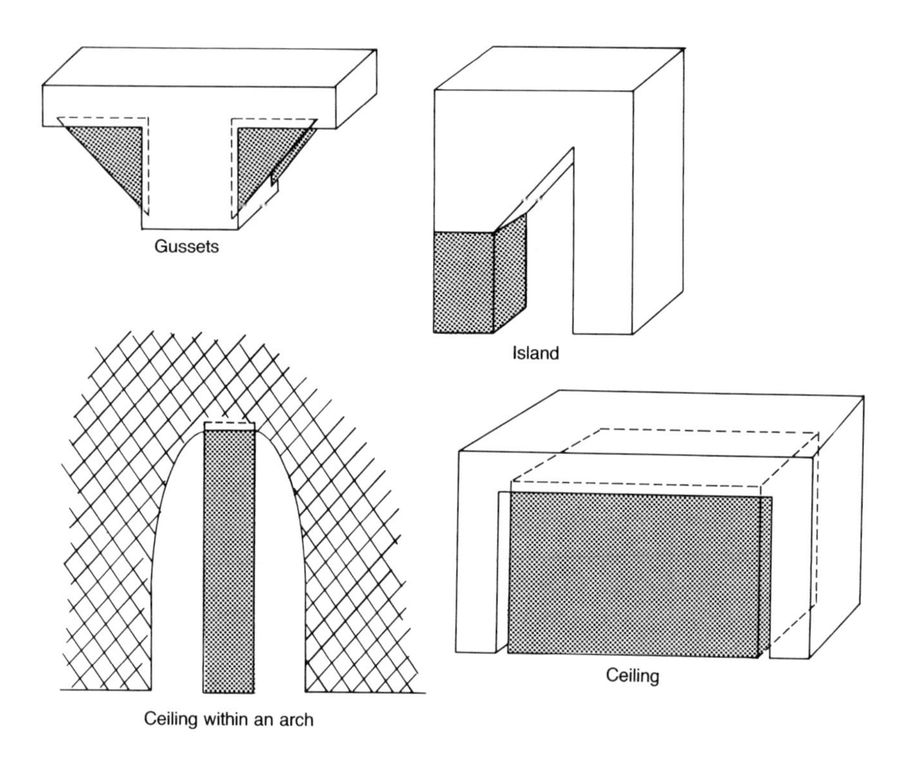

Figure 6-8. Various support types (no base supports are shown).

STL specification. This is so because they inherently satisfy the vertex to vertex rule, among others contained within the STL structure.

Each CAD system requires its own translator to convert the internal geometric database into the standard STL format. The interface software is provided by either the CAD vendors or third party software developers. Some of the larger institutions have written in-house translators in an effort to improve or customize the process.

The process and method of performing the translation varies. Some translators are activated within the interactive CAD software, while others are executed as a stand-alone program. Pro/ENGINEER is relatively straightforward. As described in Reference 14, the user activates the SLA translator under the INTERFACE main menu, EXPORT submenu. The on-screen model is active if in "Part" mode. If in "Assembly" mode, the user specifies an object on the screen with his mouse, and then chooses a quality value from one to 10. The quality value affects the facet density of curved surfaces. Generally, a value of eight gives good results according to most users. Next, one is prompted for the output type, either ASCII or binary. Always choose binary, as the STL files are much more compact and will view and slice faster. The operator now has a choice of either creating a new origin, identifying an existing origin, or accepting the default origin. Remember that orienting the parts in positive space is convenient for the SLA operator. The interface will prompt you if any values exist in negative CAD space, although the prompt often seems to be displayed no matter where the part resides.

The translation will abort if an "invalid geometry" is discovered. Try a lower quality value or change the system's default accuracy value (usually .00012″ or 0.003 mm). Changing the accuracy will regenerate the entire part and may provide a clue as to the feature causing the translator to abort. The user then enters the output file name, which is given the file extension "STL." Supports must be selected as separate output files from the part files.

SDRC's I-DEAS Rapid Prototyping™ software procedure has been simplified at the request of users. Rapid prototyping is now a separate task in I-DEAS Solid Modeling™ software. First, the user must triangulate the object's faceted approximation. Like Pro/ENGINEER, invalid facets may be found, requiring the user to locate them and fix the model. Next, under the "Create_Prototype" submenu, the user is presented with a list of SLA types, namely SLA-1, SLA-190, SLA-250, and SLA-500. The device list contains the characteristics of each machine, such as default file extension, units, and maximum object extent.[15] This list may be customized to reflect any changes the user requires, or other machines and their maximum part sizes. Next, the user can select the object to prototype. One may screen-pick the object if it is on the "workbench," use a secondary menu to select it by its label, or use a secondary menu to recall a stored object. The software will then check the object's facets, size, and position. At this point, the user may increase the facet density of the object's curved surfaces by decreasing the "absolute deviation" or "percent of chord length" values. Finally, the user specifies an output file name.

CADKEY users translate their two-dimensional and three-dimensional entities through a CADL program into the CADKEY solids program. The STL translation occurs outside of interactive CADKEY in batch mode. Every CAD system has its own STL translation procedure, and the details of any system under consideration should be investigated.

6.9 Managing CAD in an RP&M Environment

When an RP&M system is installed, engineers and designers often ask how to get their CAD designs made into physical parts. Do they simply tell the RP&M operators the name of their CAD files and assume that the parts will be made? Or, must each designer know every detail of the system and be trained along with the RP&M operators? The answer lies between these two extremes.

The process of going from CAD design to the correct STL file has been discussed in the previous sections, but the question remains: How does a company create the data necessary for the RP&M process? Ideally, the CAD designer is the best person to create the support data. He is aware of the intended use and the important features of the prototype. Also, designers are more numerous and generally more familiar with CAD than RP&M operators. If the RP operator is required to generate supports, a bottleneck may be created during periods of high demand. Utilizing support generation software such as Bridgeworks (see Chapter 5) automates the process and reduces the possibility of bottlenecks.

Initially, a new installation will be slated to build only a small number of parts. Typically, the CAD to STL interfaces are not fully installed on site, the operators are not familiar with their use, or the message gets delayed that RP&M technology is in-house and functional. At this point, the RP&M operators are recent graduates of the training class and are eager to start building parts to demonstrate what they have learned. One of the trained operators, who is also knowledgeable about the CAD system, should be assigned to investigate the specific details of the CAD interface. Each STL interface works a little differently from the others. This person should be able to log onto the CAD station and create the support files. Ideally, installation problems will have been solved before the RP&M system is installed.

If the CAD system and the slice computer are not networked, magnetic media must be used. This type of networking is whimsically known as ''sneakernet'' because a person must shuttle back and forth between the two systems. The 3D Systems slice computer is based on the Silicon Graphics Personal IRIS line of UNIX workstations. The SGI can read 1/4″ cartridge tapes with data generally written by two UNIX utilities, ''tar'' and ''cpio.'' Tar (tape archive) is the most commonly used method. Cpio (copy file archives in and out) is not well understood and is extremely cumbersome. (Note: 1/4″ cartridge tapes written on HP and IBM Unix workstations are not compatible with the SGI Personal IRIS workstation.)

Another slice processor in use is a standard DOS-based personal computer that can read 5.25″ and or 3.5″ DOS disks. A DOS floppy with data copied onto it is certainly a standard in today's business environment and the lowest common denominator for data transfer. The CAD system may indeed be based on PCs as its hardware platform, or may have a PC running DOS connected to its existing network. A limitation of using floppies is file size capacity. STL files, even when stored in binary format, can easily account for over 10 Mbytes, resulting in eight to 10 floppies being loaded into each end of the transfer.

The most efficient method of file transfer is a network connection between the CAD system and the slice computer. The physical connection used is Ethernet™, which typically uses coax cables and utilizes TCP/IP (Transmission Control Protocol/Internet Protocol) as the network transport software protocol.

File transfers can by accomplished by FTP (file transfer protocol) or the more robust NFS (networked file transfer). FTP allows the sender to "put" files across the network while the receiver can "get" files onto his local disk. NFS will allow the remote node (in this example, the slice commuter) to be identified as an apparent local disk drive.

6.10 Developing A Primer

Once the procedures regarding RP&M data creation and transfer are clearly understood, an installation-specific primer should be created that will act as a reference for design engineers who wish to utilize the technology. The primer should include design rules for RP&M, support generation, tips on maximizing facet resolution and CAD to STL file translation. The primer should also include a form with space to declare the intended application for the RP&M parts. The form would be submitted to the RP operator and updated with the most recent and accurate values of linewidth compensation and shrinkage factor(s) as obtained from the CHRISTMAS-TREE™ diagnostic part described in detail in Chapter 10. The only items the RP operator should have to choose are the resin type, the part orientation, and the layer thickness, as discussed.

REFERENCES

1. Heller, T.B., *Rapid Modeling What's the Goal?* Second International Conference on Rapid Prototyping, University of Dayton, Dayton, Ohio, Conference Proceedings, pp. 246-248, June 23-26, 1991.
2. *SLA-190/250 User Guide, Software Release 3.82*, 3D Systems, Inc., September 1991, Chapter 8, p. 8-7.
3. Donahue, R.J. and Turner, R.S., *CAD Modeling and Alternative Methods of Information Transfer for Rapid Prototyping Systems*, Second International Conference on Rapid Prototyping, University of Dayton, Conference Proceedings, pp. 221-235, June 23-26, 1991.
4. Silicon Graphics, Inc., *CAD/CAM/CAE Technology Review*, 1990, Glossary, p. 6-3.
5. Silicon Graphics, Inc., *CAD/CAM/CAE Technology Review*, 1990, Glossary, p. 6-6.
6. Silicon Graphics, Inc., *CAD/CAM/CAE Technology Review*, 1990, Glossary, p. 6-4.
7. *I-DEAS Solid Modeling Users Guide*, Structural Dynamics Research Corp., Version 5, Chapter 20, p. 20-4.
8. Mills, R., *Mechanical Design*, "Computer-Aided Engineering," Penton Publishing, Cleveland, Ohio, December 1991, pp. 14-22.

9. Cohen, A.L., *Technology Focus–Solid Ground Curing*, ''Rapid Prototyping Report,'' Summer 1991, pp. 3-8.
10. Oberg, E., et al, *Machinery's Handbook*, 23rd Edition, Industrial Press, New York, 1990, pp. 1437-1453.
11. *Pro/ENGINEER User Guide*, Parametric Technologies Corp., Version 8, Chapter 6, p. 6-47.
12. Nutt, K., *The Selective Laser Sintering Process*, ''Photonics Spectra,'' Laurin Publishing, Pittsfield, Massachusetts, September, 1991, pp. 102-104.
13. Barlow, J.W., Sun, M. and Beaman, J.J., *Technical Analysis of Selective Laser Sintering*, The International Conference on Advances in Polymer Processing, New Orleans, Conference Proceedings, April 2-4, 1991.
14. *Pro/ENGINEER Interface Guide*, Parametric Technologies Corp., Version 8, Chapter 6, pp. 6-1 to 6-2.
15. *I-DEAS Solid Modeling Users Guide*, Structural Dynamics Research Corp., Chapter 69, pp. 69-1 to 69-4.

chapter 7

Introduction to Part Building

Simplify, Simplify, Simplify.

—Henry David Thoreau
Walden (1854)

7.1 Initial Considerations

"What is the end use of this prototype?" This should be the first question asked as the RP&M project begins. The answer should be determined even before the supports are generated.

If a concept model is the requirement, it might be wise to build a half scale prototype. Building an SL part at half scale requires 87% less resin to be cured, half as many layers to be sliced, recoated, and scanned, and the support requirements are also often reduced. This further increases the speed of rapid prototyping as well as reducing the cost. Building a part at half scale may ease the justification for generating multiple copies. Four half-scale models can be produced faster than one full-scale part, and this allows *visualization* and *verification* from more individuals in less time.

When the actual part is very small, a concept model of greater scale may actually prove more valuable to the engineering, design, marketing, manufacturing, or procurement groups. Features that might be overlooked entirely at normal scale become obvious when the model has been enlarged. However, the support requirements may be increased for a part built at a scale greater than one to one. For example, a 0.12″ (3 mm) cantilever beam feature can often be built

---

By ***Todd J. Mueller***, Applications Engineer, 3D Systems, Inc., Valencia, CA.

without any support. Increase the scale to two, and the feature becomes a 6 mm (0.24″) overhang that must be supported.

If the RP&M prototype is to be used as a pattern to generate tooling, then the gating and/or parting lines can be added to the CAD model and incorporated directly into the SL part. Even if the gating or parting lines aren't built into the CAD model, the prototype itself will simplify the task of the mold maker in visualizing the parting lines, etc. Surface finish is often a top priority when generating patterns. Thinner slice layers will provide improved finish on all surfaces, even vertical walls. But very thin layers increase build time. This will be discussed in greater detail in section 7.2.

Selecting the Resin

Next, the material needs to be selected for the build. Not all resin types are currently available or appropriate for different RP&M systems. Does the prototype require ultimate tensile strength or is a tough material better suited? Some resins have greater temperature resistance, while others may allow extremely thin layers to be attained. These characteristics are explained in other chapters, but must be considered before starting the part building cycle.

Selecting the System

With various rapid prototyping equipment having different volumetric capability, the part should be matched to the most efficient machine when possible. Parts that will fit inside a 250 mm cube (including supports) can be built in one piece on an SLA-250. Larger parts can be built in sections and then bonded together using the same resin to yield a part of greater size. The CAD model must be sectioned and individual STL files created for this operation. It is important to note that STL files cannot be sectioned after they are created. For ease of assembly, the CAD models of these individual sections can have holes added for dowel pins.

The SLA-500 can build parts that fit inside a 500 × 500 × 575 mm volume, and this system has several hundred milliwatts of laser power available, while the SLA-250 is generally limited to a maximum of about 30 mW. The SLA-500 can also build a greater number of smaller parts at the same time. Alternate system approaches to RP&M should also be evaluated. These are discussed in Chapter 16.

Verifying Part Files

The most basic elements required for building good StereoLithography parts are valid STL files for both the object and the supports. Software currently exists to verify the integrity and/or provide limited correction to the STL files. The use of 3D Verify or Analyze software is the next step.

While these programs may consume from a few seconds to several hours, depending on file integrity and size, this time is often recouped during Slice processing. Corrupt STL files generate error messages and generally slow the entire process. Furthermore, the complete build cycle as well as valuable time and resin may be lost if the STL files are not initially corrected. Besides, corrupt files will have to be corrected eventually if the part is ever to be built.

Support STL Files

If third party software is being used to generate supports for the part, the STL object file should be ''Analyzed'' before proceeding. The support generating software may not recognize a feature where the triangles do not connect correctly. As a result, this feature may not be properly supported. Disk space may also be a problem if automatically generated support files are not verified. Error message files exceeding 60 megabytes have actually been generated during the slice operation for a corrupt one-megabyte STL support file.

If the support files have been created during the CAD process, it is worth the time to check the basic requirements of support design. No unsupported cantilever beam feature should exceed 6 mm and, preferably, should be less than 3 mm (*Figure 7-1*). No unsupported span should exceed 12 mm (*Figure 7-2*). All supports should have a minimum horizontal extent of 16 mm where they intersect the platform. This helps prevent the support from falling through the 7 mm diameter holes (*Figure 7-3*). Cross bracing is also required when web style supports are used.

When the support must extend vertically more than 50 mm above the platform, before it intersects the part, the support should be longer (or wider) than 16 mm. The support may fall over due to recoating forces if it is 100 mm tall and only a simple cruciform shape 16 mm across. This is one flaw of currently available automatic support generating software. These programs do not fully recognize the effects of viscous drag and leverage. The taller the support, the greater the bending moment the recoating blade can exert on that support.

Summary Considerations

Before proceeding with the StereoLithography process, the following steps should be taken:

1. Determine the end use of the prototype.
2. Verify the part or object STL file.
3. Verify the support file.
4. Check the support file to insure that it conforms to the basic support requirements and that it properly supports all important features such as cantilevers, overhangs, corners, etc.
5. Select thc appropriate resin.
6. Select the appropriately sized building system.

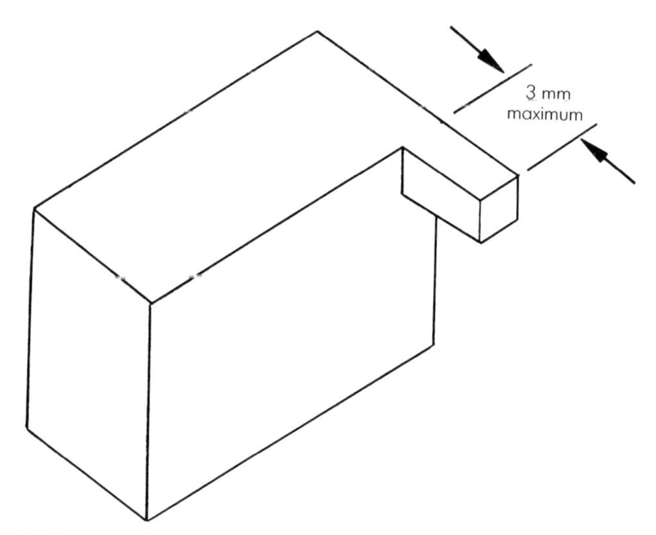

Figure 7-1. Supports for cantilever beams.

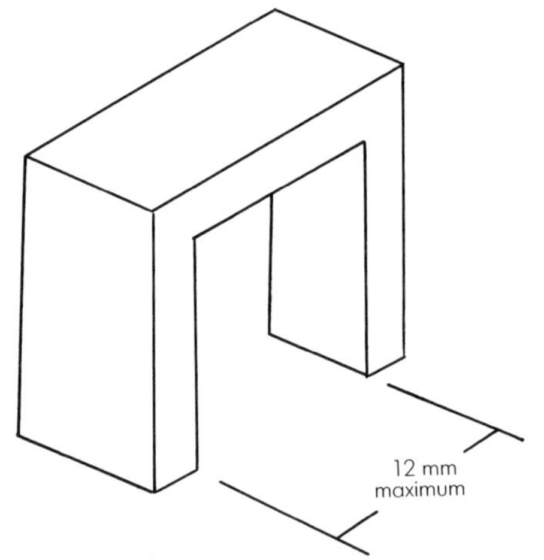

Figure 7-2. Maximum unsupported span.

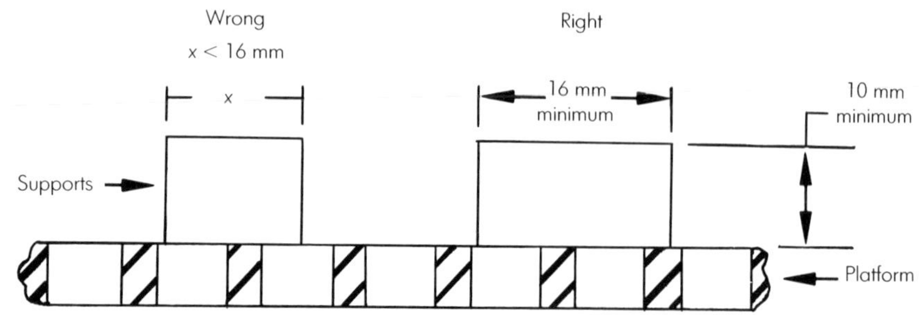

Figure 7-3. Supports spanning platform holes.

7.2 The Slice Process

Slice is the SL program used to define the geometric pattern which the laser scans to solidify the prototype. The STL file triangle information is processed into layer vector data. The slice program for StereoLithography has evolved to the point that only a few characteristics need to be fine tuned. This is true whether a very quick concept model or a functioning prototype is required. However, flexibility has been intentionally retained in the software to allow optimization of the process.

Slice Resolution and Slice Units

A slice unit is simply the value the software uses in place of CAD units. The software requires integers, therefore the CAD unit is multiplied by a value called the "slice resolution." This value normally lies between 1000 and 10,000 slice units per inch (40 to 400 slice units per millimeter). If a part exists from 2.0″ to 4.556″ and the slice resolution is 1000, then the part exits from 2000 to 4556 slice units.

The maximum allowable value of a dimension in slice units is 65,535 (2 to the 16th power, minus one). This means that if a part or support exists at a CAD value (in inches) greater than 6.5535, in any axis, then a slice resolution less than 10,000 must be used. It is important to recognize that it is the *maximum extent* (where the part exists in CAD space) that is important, not the actual length or width or height of the part. Thus, even a small part positioned far from the origin will require a smaller slice resolution.

A slice resolution of 1000 will work unless the part exists at a CAD value greater than 65.535″. If the part does exist beyond 65″ either the CAD model or STL file should be moved closer to the origin. The disadvantage of using a slice resolution of 1000 is that the software must round off CAD units to the nearest thousandth of an inch. This means the worst round off is 0.0005″ per side (for example, a part that exists from 0.9995 to 2.0004 is drawn as 1.000 to 2.000). If one side is rounded up and the other down the worst case will add or subtract less than 0.001″ to the actual part size.

If a slice resolution of 1000 is used, then all layer thickness and hatch spacing must be evenly divisible by 0.001. This means, for example, that layers can not be 0.0025″ thick but could be 0.003″ thick. Internal hatch spacing would also be affected in the same fashion. This inability to fine tune increments to the ten thousandth of an inch would consequently restrict the ability of the SLA.

Many experienced operators simply use slice resolution values of either 5000 or 1000 for most parts, when 10,000 is excessive.

The slice resolution must be the same for all files that are built together. If one part contains a CAD value of 6″ and another part exists at 25″, then the best slice resolution to use might be 1000. The SLA-500, with its spread sheet, Partman, automatically calculates the proper slice resolution for all the files intended to be built simultaneously.

One other consideration is the possibility of building a part that might be combined with a different file at a later date. A slice file generated with a resolution of 10,000 will need to be resliced if the original part is to be built later with a part that exists at a CAD value greater than 6.5535″. Depending on the file size and the slice processing computer used, this might only require a minute to complete, or in extreme cases may take longer than 24 hours!

Scale

Scale is simply the size ratio between the part to be built and the CAD model size. Using a scale of one will provide a rapid prototype of full size. Using a scale of 0.5 will build the part half scale. When making an RP&M pattern for any molding process, it is possible to accommodate the shrink factor for the final part. For example, a cast iron part might require a sand casting pattern that is 1.0104 scale.

Some SLA software is capable of scaling a file in all three axes separately. This is very helpful when neither a CAD-produced support nor third party support-generating software are available. It is then possible to use a cube STL file and generate a box support of any x, y, and z ratio.

Layer Thickness

The slice program converts three-dimensional STL files into two-dimensional cross sections. These cross sections can be "sliced" from either the CAD x, y, or z axis. The slice axis, by definition, is then perpendicular to the planes created by these cross sections. When a layer thickness is assigned during the slice process, these cross sections are derived at increments of that layer thickness. The actual part layer thickness is generated by the stepping of the elevator in the same increments. This converts the two-dimensional cross sections into the three-dimensional layers of the actual prototype.

The thickness of each slice of the part effects the surface texture, the accuracy in the z axis and the build speed. Currently, these layers can be from 0.0025″ to 0.030″ thick. Layer thickness is one of the most widely used variables in rapid prototyping. It is true that surface texture can usually be improved during postprocessing. However, when accuracy is paramount in the slice axis, a layer thickness value of 0.005″ or less should be considered.

Dimensions in the slice axis are rounded off in increments of the layer thickness, not necessarily from the bottom of the part but from the origin. If a layer thickness of 0.010″ is chosen for a part that exists in the slice axis from 0.854″ to 3.066″, it will be sliced as if it existed from 0.850″ to 3.070″. This would automatically introduce an error of 0.008″ in the slice axis before you even begin to build the part! Using 0.005″ layer thickness would slice the part from 0.855 to 3.065, resulting in an error of 0.002″. Interestingly, using a layer thickness of 0.007″ would produce a slice file from 0.854″ to 3.066″ with no round off, at least for these two vertical, albeit unusual, extents. Other slice axis features on the same part might be affected by the slice round off. The key point is to carefully assess the selection of layer thickness and not blindly use the same value for every part geometry.

Contrary to a popular misconception, thicker layers do not always result in shorter build times. *Figure 7-4* is a plot of build time versus layer thickness for various laser power levels. As discussed in Chapter 4, the laser scan velocity decreases *exponentially* with increased cure depth. Thus, thicker layers result in greatly increased laser drawing time. Depending upon laser power, resin photosensitivity, and the area being scanned, the quickest layers to build are generally between 0.005″ and 0.010″ thick. Layers 0.005″ thick scan in less time than those of 0.010″ layers but require more than twice the recoating time. The scanning time for 0.005″ would be less than half except that the hatch spacing must be decreased. The effect of decreased hatch spacing is demonstrated in *Figure 7-5*. Layers 0.015″ thick require approximately four times the scanning time but one third the recoating time of 0.005″ layers. If scanning time consumes the vast majority of the build period, it may be advisable to use a smaller layer thickness to speed up the build. *Figure 7-6* indicates the effect of part size (volume) on build time versus layer thickness. A detailed analysis, confirmed by experiment, has shown that for most SL situations, *the fastest build time is achieved near 8 mil layer thickness*. When over 200 mW of laser power are available on an SLA-500, the fastest build time will usually be

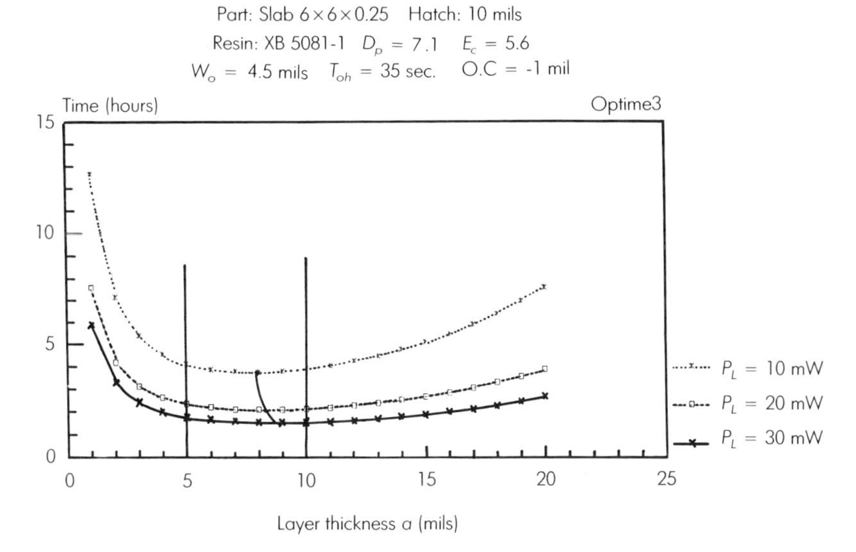

Figure 7-4. Build time versus layer thickness for various laser power levels.

achieved in the layer thickness range from 10 to 12 mils. To complete the build time/layer thickness considerations, *Figure 7-7* indicates the effect of resin type on build time.

It is possible to slice a part with variable layer thickness. The advantage of variable layer thickness is the combination of thinner layers where the surface

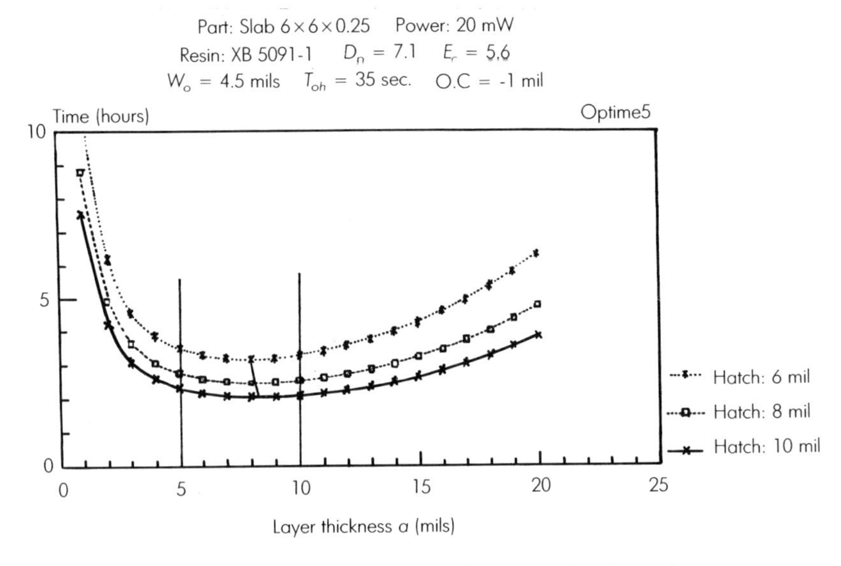

Figure 7-5. Build time versus layer thickness for various hatch spacings.

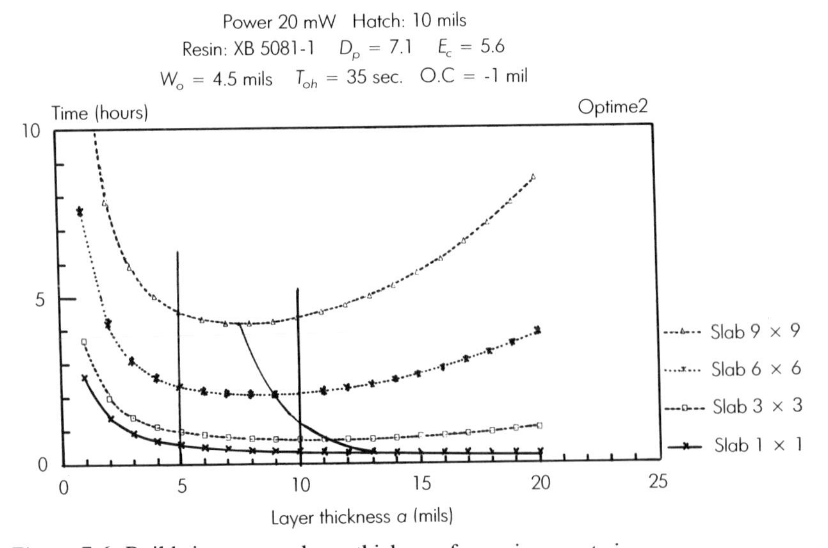

Figure 7-6. Build time versus layer thickness for various part sizes.

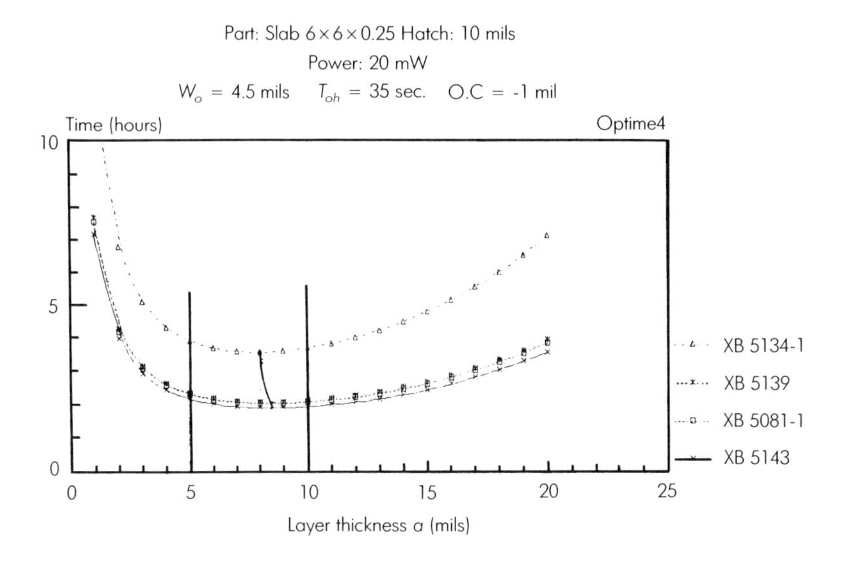

Figure 7-7. Build time versus layer thickness for various resins.

quality of a compound curved object would be improved with reduced "stair-stepping," and optimum thickness layers for faster part building where stair-stepping is not a problem. In summary, very thin layers increase total build time primarily through additional overhead, while very thick layers increase total build time through extended drawing intervals. *Figures 7-4* through *7-7* should assist the reader in determining the optimum layer thickness for their particular case.

Internal Hatch

Internal hatch is simply the method used to solidify the inside of the part, or the volume between the borders. The borders are drawn first and then the interior is scanned with any of several styles. Tri-Hatch uses a scanned line parallel to the x axis combined with lines at 60° and 120° to the x axis. This produces an internal structure of equilateral triangles. The most common spacing between these lines is 0.050″. Vertical triangular columns are thereby created which contain resin that has not been solidified. This captured resin is cured during postprocessing. Tri-Hatch was the most widely used scanning pattern until the introduction of WEAVE™.

WEAVE™ normally uses hatch scanned parallel to the x axis and the y axis. The hatch spacing is approximately 0.011″ (for 0.010″ layer thickness) for both the x and y hatch. The spacing must be reduced when building thinner layers (0.005″ layer thickness generally requires approximately 0.009″ spacing). This leaves very little resin that has not been at least partially cured. WEAVE™ and its variations are currently the best hatch parameters for the vast majority of parts. There is much more to the WEAVE™ method, and this will be discussed in detail in Chapter 8.

Skin Fills

Skin fills are closely spaced (0.003″ to 0.005″) scan lines that are drawn only on areas of a layer where there is nothing solid either immediately above or below. Skin fills are scanned after the borders and internal hatch. They were originally designed to seal off the top and bottom of the triangular columns created by Tri-Hatch. Without skin fills, Tri-Hatch parts could lose resin through gravity draining. The usefulness of skin fills has been greatly reduced by the introduction of WEAVE™. Since this technique solidifies close to 100% of the part in the vat, skin fills are no longer needed to seal the vertical columns.

This is a significant advance since skin fills can detract from part accuracy. This extra scanning exposure increases the amount of resin cured on down-facing surfaces in the z axis. Skin fills contribute an additional 0.003″ to 0.005″ depth to these down-facing surfaces. Skin fills can also induce stress in the part since the cured parallel lines overlap each other. This is precisely the type of situation that WEAVE™ avoids.

When using any variation of WEAVE™, these skin fills should either be set at zero spacing, *thereby eliminating them altogether*, or set at the widest spacing (0.005″) and scanned only on horizontal surfaces. A positive aspect of using skin fills on horizontal surfaces is the increase in the thickness of down-facing layers, enhancing their ability to resist the forces associated with recoating. Also, skinning up-facing horizontal surfaces improves their smoothness. Skin fills can be scanned parallel to either the x axis or the y axis, but only one axis should be chosen.

Minimum Surface Angle

The minimum surface angle (MSA) provides the means to determine where skin fills are scanned. It is measured in degrees from the x-y plane. If the MSA is set to zero, then only horizontal surfaces are filled. If the MSA was set at 50°, then any area that had been defined by a triangle (in the STL file) with an angle of 50° or less to the x-y plane would receive skin fills.

When using Tri-Hatch, the MSA should be minimized for the layer thickness being used. The borders overlap and seal the tops and bottoms of 0.010″ layers when the angle of the triangle is greater than about 50°. When using 0.005″ layers, the borders overlap at 40°. The thicker the layer, the larger the MSA value that is required to seal columns when using Tri-Hatch. With WEAVE™, *resin drainage is not a problem*, and the MSA may be set at zero.

Cured Linewidth Compensation

When an actinic laser beam is scanned on the photopolymer surface, the resulting solidified line has finite width. This width is dependent upon beam diameter, cure depth, and the resin penetration depth (see Chapter 4). If an STL file is sliced without taking this linewidth into consideration, part accuracy will be adversely affected. When the laser beam scans the borders, it is centered on the outer border of the part. This leaves half the beam curing resin outside the intended limits of the part (*Figure 7-8*).

The resulting part is oversized by one half the cured linewidth on all borders. If the cured linewidth is 0.010″, then a one-inch square would actually measure 1.010″ (1.0″ plus 0.005″ plus 0.005″ = 1.010″). Inside dimensions are reduced by the same amount, a one-inch diameter hole becomes 0.990″. When slicing a part, it is important to know the approximate cured linewidth in order to compensate and produce a more accurate prototype. Linewidth compensation is entered as an inch or metric dimension representing half the cured linewidth. It is important to remember that when using WEAVE™ the cured linewidth for borders is different than the cured linewidth for internal hatch since the depth of cure is different.

Slice Axis

The slice axis is the vertical axis of the rapid prototype as it is generated in the SLA, and it must be determined before the support file is generated. This is not necessarily the *z* axis of the part as it was generated in CAD. Selecting this axis efficiently can decrease build time, reduce recoating difficulties, minimize the support requirements, improve surface finish, and/or increase the number of parts that can be created during the same build cycle.

Here is a simple example: the part to be built is basically a 3″ diameter cylinder with 0.25 inch thick walls and a 0.5″ solid bottom. The entire part is 9″ tall in the *z* axis. If this part is sliced with 0.010″ layer thickness about its *z* axis, the part will contain approximately 900 layers. If it is sliced about its *y* or *x* axis, it will contain approximately 300 layers. If all other factors are equal, selecting the *x* or *y* axis as the slice axis will save the overhead time of 600 layers.

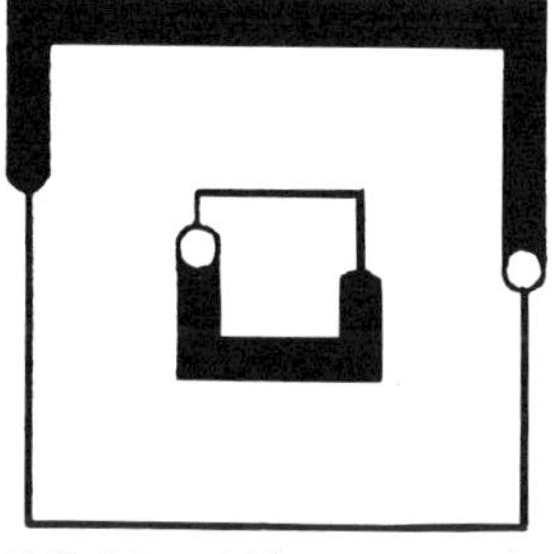

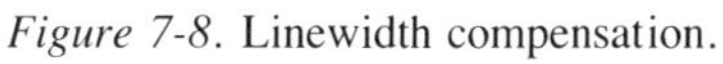

Figure 7-8. Linewidth compensation.

However, the total build time is not reduced by two thirds, since the required laser scanning time remains nearly constant.

For this example, recoating is also greatly simplified by slicing about the part's x or y axis. When sliced about the z axis, this part would generate a very serious trapped volume, and the recoating time requirement per layer could easily exceed 90 seconds, while the same part sliced about its x or y axis might only require 30 seconds per layer. This represents a time savings of (900 x 90) (300 x 30) – 72,000 seconds, or 20 hours for this example.

The amount and complexity of the supports required for this part would be minimized by slicing about its z axis. A simple base support, 3″ in diameter and 0.25″ to 0.4 ″ tall would work very well. This would also minimize both the time required for scanning the support and the time to clean off the supports during postprocessing. When this part is sliced about its x or y axis, the support requirements are greatly increased, as demonstrated in *Figure 7-9*.

If the surface finish of the cylinder walls is of paramount importance, then slicing about the parts z axis will produce the best part (provided the trapped volume can be properly treated). When this cylinder is sliced about its x or y axis, the cylinder walls exhibit the stair-stepping phenomenon caused by finite layer thickness.

It is sometimes possible to build more copies of a part or an increased number of different parts all on the same platform by carefully selecting the slice axis. To build more than three copies of the previously described cylinder, the part must be sliced about its z axis. This would permit up to nine copies to be built at the same time.

The reader should now begin to appreciate that slice axis selection is dependent upon several, sometimes conflicting, goals. Another option for the cylinder would be to rotate the STL file 180° about the x axis, turning the cylinder upside down, so the solid bottom is now on the top. This retains the greater number of layers and increases the scanning time for the supports, but would reduce recoating time to approximately 30 seconds per layer (saving roughly 15 hours), simplifies the generation and removal of the supports (*Figure 7-10*), provides the best cylinder wall surface finish, eliminates the trapped volume, and allows the maximum number of copies to be built at one time.

7.3 Build Files

Build files are the computer instructions that actually direct the RP&M system during the build cycle. Once all the necessary slice files have been generated, special programs merge these files together. Additional process information is included during this merge, although some systems require that this be added later during the prepare segment.

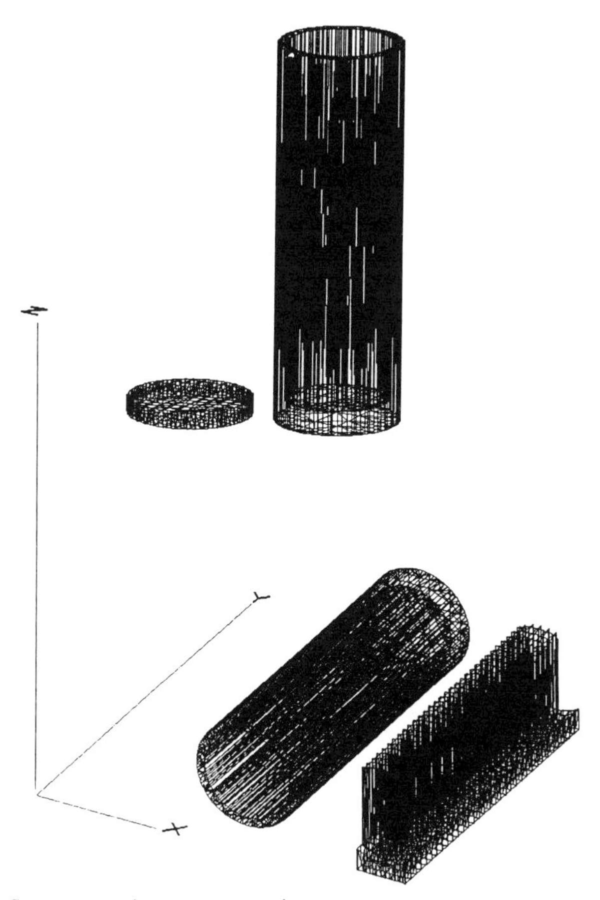

Figure 7-9. Support requirement comparison.

Types of Build Files

The slice file merge process on both the SLA-190 and SLA-250 occurs in a DOS environment on the Control Computer. This computer directs the machine during the build cycle. Four different files are generated for each individual part or set of parts to be built.

The V file contains the *vector* information. Every border, skin fill, and hatch vector has a start and end point defined by *x*, *y*, and *z* coordinates. Consequently, this file can be very large. It is the only build file that is binary, the other three files being ASCII.

The R file contains information pertaining to each *range* in the file. Ranges are groups of layers that have the same recoating parameters and use the same step increment for the elevator (see section 7.7). The R file also contains an excellent summary of the slice parameters for each separate slice file that has

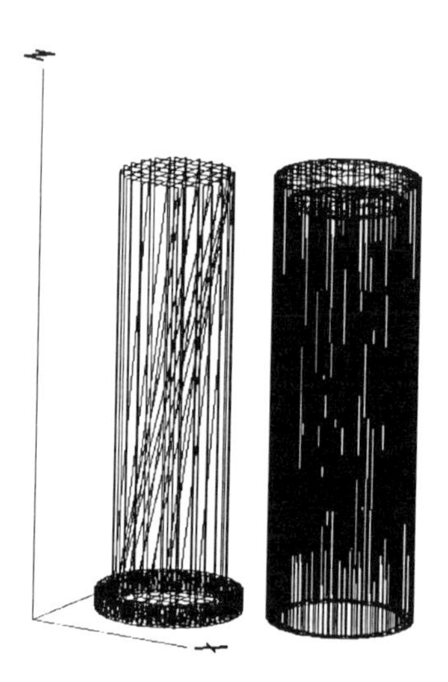

Figure 7-10. Best orientation for a successful part.

been merged into the build file. By simply viewing the R file, it is possible to determine the layer thickness, hatch spacing, the selected axis and spacing of skin fills, scale, cured linewidth compensation, etc., for each merge set being built. This information is presented in slice units along with the slice resolution. This slice information is also available in the L file.

The L file contains information regarding every *layer*. Each type of vector that occurs is listed for every layer and merge set. These vector types, with their mnemonic description, include:

LB	Layer Border	NFDB	Near Flat Down Border
LH	Layer Hatch	NFUB	Near Flat Up Border
FUB	Flat Up Border	NFUF	Near Flat Up Skin Fill
FDB	Flat Down Border	NFDF	Near Flat Down Skin Fill
FUF	Flat Up Skin Fill	NFDH	Near Flat Down Hatch
FDF	Flat Down Skin Fill		

The layer thickness is listed for each vector type. The number following the two or three letters indicates the merge set for that vector and layer thickness. For example:

> 5100, LB1, ''LT 0.010'' indicates the layer exists at 5100 slice units, the vectors are layer borders for the first merge set (the actual *x*, *y*, and *z* coordinate information is in the V file, which is a binary file), and the layer thickness is 0.010″.

The PRM, or *parameter* file, is the smallest of the build files but likely the most helpful. This build file contains the number of copies, the spacing between the copies (in CAD units), shrink factors used, and the location of each copy on the platform. These locations are presented in mirror units. Do not confuse these values with slice units.

One mirror unit equals 1/3556″ (or 0.000281″). Mirror units represent the smallest distance increment the laser beam can move at the platform. The shortest line segment that the laser beam can draw is actually two mirror units long, or 0.000562″. Since the beam is curing the polymer approximately 0.010″ in diameter, the fact that the beam cannot be placed at an infinite number of positions in the *x-y* plane has virtually no effect on the quality of the part.

SLA-500 Build Files

The SLA-500 uses a single file, the bff file, to control the machine during the build process. This file is generated by software called "Partman," and includes almost the same information that is found in the SLA 190/250 build files. The primary difference is that the bff file does not contain all the vector information of the internal hatch. Since the SLA-500 typically builds larger files, this proved to be a practical way to reduce both file size and the time required to prepare the build file.

The advantage of Partman is that all information is entered into a simple spread sheet, and the slice and merge programs are handled by a batch file. There is no need to wait for a file to complete slicing before starting merge. The one drawback to this procedure is that if an offset is required for one of the STL files, then that file must be completely resliced and a new bff created.

7.4 Merging Slice Files

Individual slice files are referred to as "merge sets" after they have been merged. Most build files contain at least two merge sets, a support, and a part. The first file entered in the merge set is considered merge set one, the second is merge set two, etc. This will allow each merge set to be individually assigned specific laser scanning parameters.

The maximum number of individual slice files that can be merged together is 11. This allows one to build five different parts with their corresponding supports all on the same platform, providing there is sufficient space. If a common support is generated that will hold all the parts being merged, then 10 different parts can be built at one time.

Merge offsets afford the operator the ability to arrange the parts on the platform in all three axes. Many parts will require offset to prevent them from being inadvertently joined because they exist in the same CAD space. This eliminates the need for psychic CAD operators who can foresee all possible combinations of parts being built together. It is even possible to merge the same file multiple times. If two of part A and four of part B need to be built together, merging one of each and then requesting four copies would provide extra parts which might not all fit on the platform. During merge, simply select part A once and part B twice. If objects A and B do not overlap, then simply apply merge offsets to the second B file. Finally, request two copies during the Prepare segment and the result will be two copies of A and four copies of B.

Another use for merge offset is "nesting" parts to increase the number of copies that can fit on the platform. The Prepare program will only allow one copy of the part in *Figure 7-11*. Up to five copies can be built together by using merge offsets as shown in *Figure 7-12*.

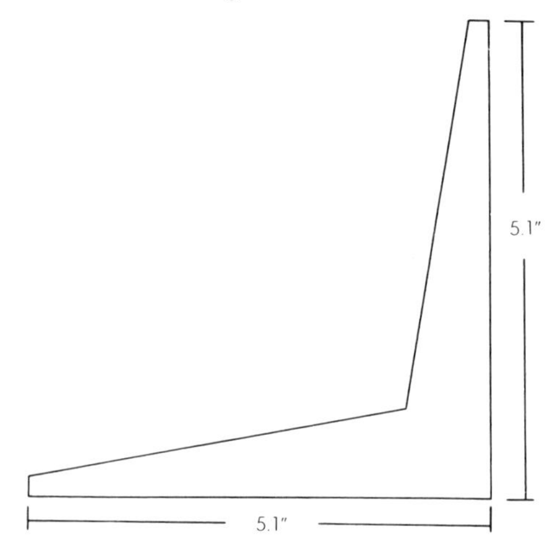

Figure 7-11. Quantity of parts allowed by the Prepare function.

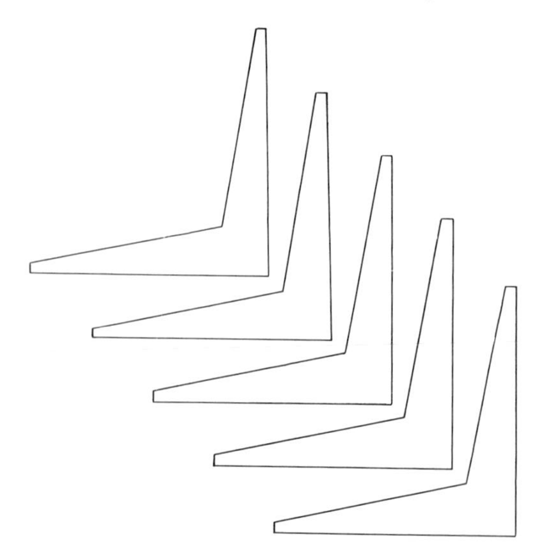

Figure 7-12. Extra parts allowed by "nesting" during the Merge function.

7.5 Customizing

When slice files are merged, there are enough default values assigned to the building parameters to build most parts. The SLA-190/250 software does provide the opportunity to customize virtually every aspect of the process. There are two separate avenues available, the Parameters Manager and the Range Manager.

The Parameters Manager

Shrinkage compensation adjusts the size of the part as the laser cures the resin. This compensation is typically required to: (1) allow for the shrinkage of the resin as it cures and (2) to intentionally make the part larger when generating a pattern for a mold cavity. Mold cavities must be made oversize if the material that is to be cast in the mold shrinks as it solidifies. This is the case for the vast majority of both plastics and metals. The x and y axes receive the same shrink factor, and the z axis can be adjusted individually. Experience shows that the resin shrinks at a slightly different percentage in the z axis than in the x-y plane. All of this adjustment is done after the STL files have been sliced and merged. Fine tuning the shrink factors is then possible without repeating the entire process. The SLA-500 software provides for individual shrink factors in the x and y axes.

Next, the number of copies and the spacing between copies is selected. Currently, the minimum spacing between copies is 0.100″. The software will permit up to 36 copies, provided enough room is available. As many as 100 copies of the same part have been built on an SLA-250 by merging four parts together and then requesting 25 copies.

7.6 Build Options

Build options add versatility to the manner in which parts are generated. There is an option to create the mirror image of a part that has already been prepared, an option to pause the build at a specific layer, etc. Most of these options can either be saved in the Parameters Manager or added at the build line when the build is started. The advantage of adding options into the Parameters Manager, as opposed to including them in the build command line, is that these commands will still be included if the build is stopped and restarted. Any option added to the build command line must be added again if a part is restarted. The following build options are available for the SLA 250/190:

/simulate allows the build to be simulated on the monitor. Each layer is "drawn" including all extra copies and all vector types. If parts accidentally intersect, don't mate as anticipated, etc., this is the least expensive time to find out. Some operators use this mode to determine their merge offsets; merging the parts in the normal positions, then running a simulated build. When interference is obvious on the monitor, simply adding approximate offsets and remerging can often resolve the problem. A simulated build will run quicker when only one copy of each part is generated on the monitor.

/start will start the build at a specified layer (in slice units) without the platform moving or checking the resin level. This allows a build to be interrupted and then resumed (for example, /start12000 for a layer at 2.4″ using a slice resolution of 5000).

/stop will cause the build to terminate at a specified layer similar to /start. The platform will rise and allow excess resin to drain from the part, and the laser will be turned off automatically. This is normally used when only a part segment is required.

/pause will cause the build to pause at a specified layer. This differs from /stop because the build is not terminated. The platform does not move, and the build can be resumed by simply pressing any key. This allows a build to be automatically suspended and then continued. Some typical examples include an electrical connector that requires metal inserts or a part geometry that may cause recoating difficulties. If a build is paused where trouble is suspected to occur, the build can be continued while the operator is present to observe.

/zoff disables the recoating functions. The elevator and the blade remain stationary while the laser continues to scan each layer. This allows several layers of support to be scanned on the platform without applying extra resin. Several passes help provide a secure bond between the supports and the platform. The build can be exited and then resumed with the /start commands minus the /zoff command. The /zoff command is also used for trouble shooting, allowing an opaque object to be placed above the resin without recoating occurring. The path of the beam can then be observed with no concern about producing polymerized resin.

/loff disables the dynamic mirrors (laser off). Not only does this prevent the laser beam from reaching the resin, but it also eliminates the time required for the computer to calculate the geometric corrections for guiding the laser. When combined with /zoff, this option operates similar to the /simulate command.

/swapxy swaps the *x* coordinate values with the *y* values. This creates a mirror image and rotates the part 90°. Only the mirror image will be built, not the regular part and the mirror image part. This is a very useful tool since only one STL, one SLI, and one set of build files are required to produce both a left and right version of the object.

/x- mirrors the part about the *x* axis. This can be combined with the /swapxy command to create a nonmirrored part that is still rotated 90°

about the z axis. This can be used to reorient a delicate part that is parallel with the blade. The force the recoater blade imposes on the part as a result of resin viscosity is reduced by rotating the part so it is essentially perpendicular to the blade.

/y- mirrors the part about the y axis.

/nodrain causes the part to remain in the vat when the build cycle is completed, as opposed to the platform and part coming out of the resin and allowing the excess resin to drain. This can be beneficial where a delicate part might distort under its own weight when left for a long period in the green state. A cured part is less than 6% more dense than the liquid resin so the resin actually helps support the part while it is submerged. Another case would be where a part will finish Saturday morning and no one will be available to postprocess it until Monday. The "nodrain" option was also intended to inhibit uncured resin from draining out through any surface imperfections when the Tri-Hatch method was used. STAR WEAVE™ eliminates this drainage. It should be noted that some resins such as XB 5134-1 exhibit a phenomenon called "swelling" (see Chapter 10), where cured polymer actually absorbs uncured resin when left in the vat for long periods. STAR-WEAVE™ reduces this problem, but the /nodrain option should be avoided when using resins that show any significant swelling.

/restart permits a part to restart at the layer it was on when the build was unavoidably interrupted, as in a power failure. It is not necessary to type in the layer number. The /restart function automatically checks the stereo.dat file for the last known elevator position and moves the platform back to this elevation. This is not a simple option to perform! When power is turned back on, the SLA-250 reinitializes the platform, sending the elevator to both limits of travel. This causes the part to be submerged in resin. When the platform is returned to the last building position, there will be excess resin on top of the part since the blade will not sweep. The SLA automatically waits 90 seconds before laser curing the layer, but this often is not enough time for the surface imperfections to relax. The leveling time may be paused at this point to allow adequate resin relaxation, or it may be necessary to exit build and use the blade to remove some of the excess resin. The layer number should be noted and then used with the /start command to continue the build.

/nosweep keeps the blade from sweeping even when the recoating parameters include sweeps. This option is normally used by field service.

There are four single key stroke build options that can only be used during part building, simulated or real, and *solely while the leveling time is counting down*.

P will pause the leveling operation when it reaches 0. This is used to allow extra leveling to occur. This command is also useful if the chamber door is going to be opened for any reason between layers. This prevents having the laser try to scan the beginning of a layer while someone is just about to close the door after inspecting the part. The laser override safety switch will automatically block the beam and the layer will not be cured while the door is open. The SLA does not "know" this has happened and will never return to the area that was not cured. This will create an imperfection in the part and could eventually cause delamination leading to part failure.

C cancels any remaining leveling time. This is primarily used while viewing a simulated build.

S affords the opportunity to skip up to a higher layer. This is used while viewing a simulated build but can also be helpful during an actual build. If a part and support file are scaled at greater than 1:1, the supports will be taller than needed. By skipping some of these support layers, it is possible to save time. It is not possible to skip to a lower layer.

X exits the build. Be careful when cancelling leveling time because the "x" and "c" buttons are next to each other on the keyboard, and more than one build has been accidentally exited.

7.7 Preparing Ranges

Merging SLI files automatically creates ranges that contain the same increment of elevator stepping. Extra ranges can be added to provide different recoating and overcure parameters. These are added in the Range Manager utility for the SLA-250/190 and in the Partman spread sheet for the SLA-500.

Range Manager

The default settings in Range Manager can properly build many parts that use web style supports and do not contain trapped volumes. It is very important to review the part and add all the ranges that might be needed before customizing existing ranges with new overcures or recoating values. *Adding a range will automatically convert all existing ranges back to the default values*. This can be annoying if five ranges and all the corresponding recoating and overcure fine tuning have already been added. When in doubt, add an extra range. If the range is not required, simply leave it untouched; the range will use the recoating and overcure values of the preceding range. Some operators add several extra ranges for complicated trapped volumes that allow them to adjust recoating values during the build. If a build is observed to be having recoating difficulties, the operation can be stopped during the leveling time, the range(s) edited, and the build can be started again at the next layer. When interrupting a build, it is important to remember that the layer indicated during the z wait (leveling) time is the number, in slice units, of the layer that has *just been completed*. If the /start command is used with the layer indicated on the screen, then the same layer is scanned twice and the part becomes taller by that layer thickness. The simplest method to obtain the correct layer is to use /start and add one slice unit to the layer that was indicated. The system automatically increments to the next layer without requiring the exact slice unit value.

Recoating

While the default values will build a great many parts, these selections are not always the most efficient. There are also some parts that will simply not build properly with the default values. To fine tune the parameters, it helps to know the default condition as well as the available range of recoating values.

Number of sweeps, *NS*, uses a default of one, but the range is one to seven sweeps. *It is very important to use an odd number of sweeps* (1, 3, 5, or 7) to avoid ''Leading Edge Bulge.'' With an odd number of sweeps, these effects do not accumulate, whereas with an even number of sweeps, accumulation of recoating nonuniformities can lead to layer delamination and/or blade collision.

Sweep Period, *Pn*, uses a default of five seconds. This means the recoater blade takes five seconds to traverse the vat. The range extends from five to 20 seconds. The vast majority of parts will use five-second sweeps, and it is rare to use values longer than 10 seconds. When using multiple sweeps, a period should be assigned to each sweep. If N3 is used, then sweep period information might be, for example, P1 5 P2 6 P3 10. Notice that there must be a space between the sweep number and the seconds indicated. If P15 is attempted, the computer considers this to be the fifteenth sweep and rejects the input.

Z wait or leveling time, *ZW*, has a range of 0 to 999 seconds. The default is 30 seconds. This is longer than required for most areas that do not have trapped volumes, but it is better to err on the long side of *ZW* than to provoke delamination resulting from layer nonuniformities. A 20-second *ZW* is certainly safe for 0.010″ layers without trapped volumes.

Elevator Velocity, *ZV*, uses a default of 0.2 (turns of the elevator screw per second). The range is from 0.051 to 0.50. This is a very important and under-utilized option. When using resins that have lower green strengths, it is necessary to slow the elevator, using values from 0.06 to 0.15. It also helps to slow the elevator for major down-facing surfaces, particularly when using thin layers (0.0025″ to 0.006″). Avoid using velocities above 0.3, unless the part and its supports are extremely robust. The small saving in time may result in diminished part quality.

Elevator Acceleration, ZA, also uses a default of 0.2 (turns per second^2) and the range is 0.051 to 0.050. Using the same value as *ZV* is a good rule of thumb.

Post Dip Delay, *PD*, has a default value of four seconds and a range of 1 to 99 seconds. This is a good place to save three seconds per layer on most parts. *PD* is the amount of time that the elevator waits at the bottom of the travel when the part has been submerged to apply more resin. If a 10″ × 10″ flat part is being created, it will require more than one second to allow the resin to flow over the edges of the part and reach the middle. As much as 10 seconds could be required depending on the resin used. But, since the platform dips 8 mm (0.31″) into the resin, most parts require only a one second *PD*. This must also have a space between the *PD* and time.

Blade Gap, G_n, allows the vertical separation between the bottom of the recoater blade and the top of the previous (cured) layer to be increased, per sweep, as a percentage of the layer thickness. This is rarely used due to improvements in the process. The one exception is increasing the value of G_n when using multiple sweeps over large part cross sections. Increasing the blade gap reduces the amount of resin the blade removes on the first sweep, lessening the viscous shear force on the part. The blade gap on the last sweep should be set to 100%. This will leave the desired layer thickness.

Trapped volumes are the major obstacle to be dealt with in determining the correct recoating parameters. The volume inside an upright coffee cup is a good example of a large trapped volume. The solid bottom of the cup will usually recoat properly with one sweep and a *ZW* of 20 seconds. But, the efficiency of the blade, with respect to removing the correct amount of resin, is reduced once the layers consist of walls that trap resin. Increasing the depth and/or width of a trapped volume requires more sweeps of the blade to achieve the correct layer dimension. The results of not compensating for a trapped volume can include layer separation and/or the blade colliding with the part. If the surface of the resin is not flat before the laser scans, this is a good indication that a trapped volume exists. Increasing the *ZW* can help if the resin level is high or low. This may not be the quickest way to level but can overcome the slight variations that occur as the trapped volume changes. *Figure 7-13* illustrates a trapped volume that is constantly changing, first becoming very wide then narrow again but with the depth constantly increasing.

There are three recoating variables that are used to deal with trapped volumes. Increasing the number of sweeps is the first recoating change. Minor trapped volumes require three sweeps. Major trapped volumes can normally employ five sweeps, although in severe cases seven sweeps are beneficial. The sweep period, P_n, also has a major effect. More resin is removed the faster the blade travels. It is possible to remove so much resin that not enough remains inside the trapped volume. Three five-second sweeps could cause this in a shallow area. Slowing the third sweep helps level the resin. A P3 8 (indicating that the third sweep takes eight seconds to traverse the vat) might resolve this problem.

The orientation of the trapped volume relative to the blade is also important. The thin wall bottle in *Figure 7-13* will benefit from having the blade parallel to the major axis of the trapped volume. The part becomes extremely difficult to build when the long dimension of the trapped volume is essentially perpendicular to the blade.

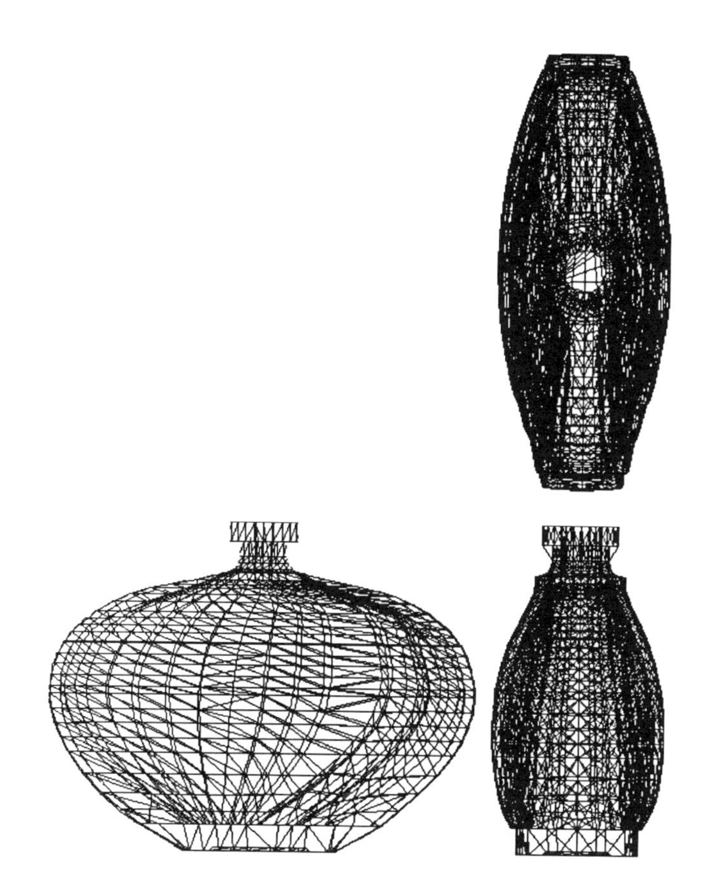

Figure 7-13. Constantly changing trapped volume.

Overcure and Absolute Cure Depth

Overcure refers to the depth that the laser cures past the layer thickness. A typical overcure for borders is 0.006″, while hatch overcure is a negative value, typically -0.001″ when using WEAVE™. Early software required the total cure depth to be entered. This required adding the overcure to the layer thickness and imputing this value for each range of each merge set. This was prone to operator error, especially when variable layer thickness was employed. Now only skin fill vectors receive absolute cure depths.

The default overcures are set for each resin type. These defaults are correct for virtually all parts with one major exception, base style supports. Since the defaults assume that the WEAVE™ style is being used, the hatch vectors of base style supports receive a negative overcure. This results in supports with layers that are only connected at the intersection of the x and y hatch, and these supports basically "wilt." This is very simple to correct. The software in Range

Manager steps the operator through the process. The overcure is primarily dependent upon the resin being used. If XB 5081-1 is used and the borders receive a 0.006″ overcure, then the hatch for the base style support merge set should also use 0.006″ overcure. Occasionally this is changed due to part geometry, with larger or taller parts having supports that receive greater overcures. Some resins, generally the more flexible types, require greater overcures for hatch supports and borders due to the lower green strength of these materials.

7.8 Summary

RP&M is becoming more efficient and less dependent on operator input. However, as with most machine tools, optimum efficiency can be influenced by the operator. Care must be exercised in selecting part orientation and resin type, and determining recoating parameters. The advanced build methods discussed in the following chapter should also significantly improve the quality and accuracy of RP&M prototypes.

chapter 8

Advanced Part Building

There are more things in heaven and earth, Horatio,
than are dreamt of in your Philosophy.

—William Shakespeare
Hamlet, Prince of Denmark
Act I, Scene V

8.1 Basic Concepts

Familiarity with various fundamental principles is important in understanding the physical basis of some of the advanced part building methods that have been recently developed. In Chapter 4 we discussed the ordering of the characteristic dimensions of SL. Specifically,

$$d_m << \lambda << S << D_p < B < R << L$$

Thus, the molecular dimensions, d_m, are the basic building blocks of SL, and are much smaller than, λ, the wavelength of the laser radiation used to initiate the resin photopolymerization. Once cross-linking of the polymer begins, "domains" with a characteristic size, S, are formed.[1] These domains are much larger than the wavelength of the laser radiation. Given sufficient actinic exposure, the adjacent domains eventually connect to form macroscopic polymerized regions.

Further, the resin penetration depth, D_p, the $1/e^2$ laser beam diameter, B, and the "zone of influence," R, are all comparable, and are about 100 times larger than the characteristic domain size, S. This is significant, since it suggests that SL technology is still about two orders of magnitude away from reaching any fundamental limits with respect to scale.

By ***Paul F. Jacobs, Ph.D.****, Director of Research and Development, 3D Systems, Inc., Valencia, CA.*

Finally, the characteristic dimension, L, of a real part is much larger than R. From d_m to L involves a span of about seven orders of magnitude in linear scale. In other words, the maximum SL part dimensions are roughly 10 million times the length of a so-called "long chain" polymer molecule.

At this point, it is also important to understand the various characteristic times involved. The shortest physically significant time scale in the SL process is the layer transit time, t_t, for a laser photon to pass through a single layer thickness. This characteristic time is defined as follows, where a is the layer thickness and c_r is the velocity of light in the resin.

$$t_t \equiv a/c_r \tag{8-1}$$

For typical values of a, in the range from 0.12 mm to 0.5 mm, the value of the characteristic layer transit time is of the order of a picosecond (1×10^{-12} second). Thus, the laser photons have completely passed through a layer in an extremely short time, and their molecular absorption as well as the generation of free radicals occurs on a similar time scale.

Now that the free radicals have been generated, the development of cross-linking (polymerization) begins to proceed at typical chemical kinetic reaction rates once oxygen inhibition has been overcome. Typically, the 1/e time constants, t_K, for chemical kinetic reactions of this type, at resin temperatures in the 25° to 35° Celsius range appropriate for SL part building, are roughly of the order of microseconds.

Next, we have the characteristic exposure time, t_e, defined in Chapter 4 as the time for the laser exposure to reach 99.99% of its full value when a Gaussian laser beam is scanned at constant velocity past a given point. Typically, t_e ranges from about 70 to 2000 microseconds.

Interestingly, as noted earlier, the photopolymer does not reach the gel state until $E = E_c$. Thus, a significant amount of exposure is required to overcome the quenching of free radicals due to oxygen in the resin. The oxygen is present in the resin as the result of diffusion from the airspace above the vat, through the resin-air interface.

Since E_c can range from about 5% to 50% of E, depending upon the desired cure depth, the time required to reach E_c will be a roughly proportional fraction of t_e. Since t_e is in the range from about 70 to 2000 microseconds, the time required for the exposure to reach E_c can range anywhere from as little as about 3 microseconds to as much as a millisecond, depending on the desired cure depth and the laser scan speed. A typical value for an SLA-250 with 9 mil cure depths would be about 200 microseconds.

Thus, we see that a considerable period of time elapses from the generation of the first free radicals (a few picoseconds) until the resin merely reaches the *gel state* (a few hundred microseconds for our example). During this time interval, continued laser exposure is generating free radicals, which are primarily being consumed by oxygen. Finally, after about 200 microseconds, the free oxygen is essentially depleted and the photopolymer reaches the gel point. Provided that the rate of actinic photon arrival, and the corresponding rate of generation of free radicals, exceeds the diffusion rate of oxygen back into the resin, the reaction process can proceed to the point of significant cross-linking.

Following this, with a considerable time lag, the photopolymer finally begins to exhibit significant and measurable shrinkage. We shall define $t_{s,o}$ as the characteristic time for the onset of measurable shrinkage, and $t_{s,c}$ as the corresponding time for the completion of measurable shrinkage. From References 8-1 and 8-2, we find for the resins currently used in SL that $t_{s,o}$ is of the order of 0.4 to one second and $t_{s,c}$ is of the order of four to 10 seconds.

Finally, the total laser drawing time for a single layer, t_d, obviously depends upon the part geometry and building technique, as well as the resin parameters D_p and E_c, the desired cure depth, C_d, and the laser parameters P_L and W_o. Nonetheless, a representative range for t_d would probably extend from about 20 to 300 seconds.

Thus, ordering the characteristic times of SL we find:

$$t_t << t_K << t_e << t_{s,o} < t_{s,c} << t_d$$

From t_t to t_d involves a span of about 14 orders of magnitude in time scale. If we define a picosecond as the basic unit of time, then inspection of the inequalities listed above reveals the following important event sequence:

1. A photon is absorbed and a free radical is generated in about one time unit.
2. Over a period of millions of time units the chemical kinetics of molecular free radical polymerization occurs.
3. After roughly a few hundred million time units, the gel point is finally reached.
4. The exposure continues for hundreds of millions, up to roughly a billion time units, with extensive cross linking occurring during this interval, leading to the formation of polymerization "domains."
5. After roughly a trillion time units, a significant fraction of the domains have interconnected and macroscopic shrinkage begins to occur.
6. Next, after roughly ten trillion time units, the shrinkage is essentially complete.
7. Finally, after tens to hundreds of trillions of time units, a single part cross-section on a single layer has been drawn.

It is clear from the relative time scales that the following physical phenomena apply:

A. The characteristic exposure time, t_e, is much larger than the characteristic time for chemical kinetics, t_K. Thus, a given location of resin continues to receive laser exposure long after polymerization has begun.

We may tend to think of the laser beam scanning over the resin surface as a rapid process; but in fact, even for the SLA-500 at the maximum possible laser beam scan speed of about 300 in./second (about 7500 mm/second) the duration of the laser exposure at a given point is still much longer than the chemical kinetic reaction time. Simply stated, "laser scanning is slow relative to chemical kinetics."

From a physical standpoint, this inequality is important. Since

$$t_K << t_e$$

any attempt to model the SL process in terms of the response to an "instantaneous" or temporal delta function in laser exposure will most likely prove unsuccessful.

The actual process is much more complicated, involving the extended exposure of an element of photopolymer for a period much longer than that required for the formation of an initial cross-linked structure. The laser exposure is still occurring while cross-linking continues. Thus, the two processes are not serial, but rather they overlap.

B. Even more important, since

$$t_{s,o} >> t_e$$

then it is also clear that shrinkage lags exposure. Physically, this inequality is particularly important since it indicates that long after a given region has been exposed it finally begins to experience significant shrinkage. This observation is absolutely crucial to the advanced build techniques described later in this chapter.

Currently, one of the significant sources of error in RP&M involves the distortion of parts due to built-in stresses. A major cause of these internal stresses are the substantial forces generated during molecular shrinkage. Unfortunately, it has not been possible to totally eliminate shrinkage in the present generation of SL resins.

However, it is possible to utilize some ingenious stratagems to significantly reduce the effects of these forces, thereby reducing distortion and improving overall part accuracy. The origins of the WEAVE™ technique are partly based on the observation that shrinkage lags exposure.

C. Finally, the inequalities

$$t_d >> t_{s,c} > t_{s,o}$$

obtain for all but the very smallest part geometries. The implication is that the laser is still drawing while some of the regions of the part have completed their shrinkage and other regions are still shrinking. This can lead not only to the development of internal stresses, but nonuniform internal stresses, or stress concentrations. Stress concentrations are inevitable with complex part geometries. However, a secondary goal of any advanced part building technique is to minimize stress concentrations as much as possible.

The development of STAR-WEAVE™ was partly based upon awareness of the inequality cited above. Specifically, the time required to draw a single cross-section of a part was often much greater than the time for the completion of shrinkage. Thus, special methods were developed to reduce internal stresses resulting from delayed shrinkage, as well as reducing nonuniform stresses. The techniques involved in STAR-WEAVE™ are effective whenever:

$$t_d >> t_{s,c}$$

Any technique based upon ''flash exposure,'' where the ''drawing'' time, or in this case the duration of the flash exposure, is less than the resin shrinkage time, will probably not benefit from some of these advanced build methods.

8.2 The Development of WEAVE™

During the first half of 1990, a research effort was initiated to develop a more accurate building method for SL. As a result of work at both 3D Systems[3] and Ciba-Geigy[4] the following information had been established:

1. The time required to draw a single laser line ranged from about one millisecond to about 500 milliseconds, depending on the length of the line, as well as the resin parameters D_p and E_c, the laser power and beam diameter, and the desired cure depth, all of which determine the laser scan velocity.
2. The time required for the laser-cured photopolymer to undergo the majority of its shrinkage within the vat was roughly three to five seconds, which is obviously greater than the time to draw a single line.
3. The relationship between the cure depth, C_d, and the maximum centerline resin exposure, E_{max}, was derived theoretically (see section 4.4, and specifically equation 4-33) and independently confirmed with extensive

experimental data.[5] The resulting functional relationship, known as the "working curve equation" and repeated here for convenience, is as follows:

$$C_d = D_p \ln [E_{max}/E_c] \tag{8-2}$$

where D_p and E_c are respectively the resin penetration depth and critical exposure parameters defined earlier.

4. Numerous measurements with a Coordinate Measuring Machine (CMM) indicated that the dominant source of error in SL at that time was post-cure distortion.[6]
5. The observation that post-cure distortion is proportional to the fraction of liquid resin remaining within a laser-cured SL part (see Reference 6).
6. The experimental observation that curl distortion (the bending of multiple unsupported cantilever layers) is related to the amount of polymer shrinkage occurring after contact with the previous layer.[7]
7. Experimental data[8] showing that photopolymer shrinkage as a function of laser exposure is small up to E_c, and then exhibits a steep positive slope in the region of exposure values just beyond E_c. For larger exposures, the slope decreases, with the shrinkage asymptotically approaching the fully cured value for essentially "infinite" exposure.

As a result of these observations, data, and theories, 3D Systems developed a new building technique, now known as WEAVE™, during July 1990. This method was based on the following concepts:

- Building SL parts with 0.05″ (50 mil or 1.25 mm) vector separation, 60°/120°/*X* equilateral triangular hatch (Tri-Hatch) left too much residual liquid resin within the part structure. This large fraction of uncured liquid resin (roughly 50%) led to excessive post-cure distortion. To reduce this error, much more resin needed to be laser-cured within the vat. Also, a building method capable of generating smoother vertical surfaces than those resulting from Tri-Hatch was definitely desired.
- Unfortunately, experiments involving reduction of the Tri-Hatch spacing (0.025″ and 0.015″ or 0.64 mm and 0.38 mm) had shown greatly increased curl distortion.
- Thus, the intent of WEAVE™ was to reduce the fraction of residual liquid resin within the part structure, while simultaneously attempting to minimize curl distortion.
- To accomplish this, a drawing method was developed involving two sets of orthogonal vectors per layer. Specifically, a set of parallel *Y*-hatch vectors was drawn on the first pass, followed by an orthogonal set of parallel *X*-hatch vectors drawn on the second pass.

However, in a departure from previous build methods, the first set of cured *Y*-hatch lines are intentionally separated from each other, in the orthogonal or *X* direction, by about 1 mil (0.025 mm).

Furthermore, the hatch lines on the first pass are also intentionally given a cure depth about 1 mil less than the layer thickness. Hence, the first set of hatch vectors drawn on a given layer neither touch each other nor the cured layer below. The following concepts are central to WEAVE™:

1. The individual hatch vectors are drawn in a time period that is much less than the characteristic time for significant polymer shrinkage.
2. The maximum centerline exposure on the first pass of hatch vectors, $E_{max}(1)$, must significantly exceed E_c to achieve a value of cure depth on the first pass, $C_d(1)$, that is only 1 mil less than the layer thickness, a.
3. Therefore, since $E_{max}(1)$ considerably exceeds E_c, a significant fraction of the polymer shrinkage will have occurred during the drawing of the *Y*-hatch vectors on the first pass of any given layer.
4. The hatch vectors on the first pass will not yet have made contact with the previous layer. Since curl distortion is related to the extent of shrinkage after contact with the previous layer, the delayed shrinkage on the first pass should not contribute significantly to curl distortion.
5. By employing an exposure on the second pass (the *X*-hatch vectors) identical to that used on the first pass, the cure depth at the hatch vector intersections after the second pass, $C_d(2)$, is given by the equation:

$$C_d(2) = D_p \ln [2\, E_{max}(1)/E_c\,] \tag{8-3}$$
$$= D_p \ln [\,2\,] + D_p \ln [\,E_{max}(1)/E_c\,]$$

However, by definition, from equation 8-2, the cure depth $C_d(1)$, of the *Y*-hatch vectors on the first pass is just

$$C_d(1) = D_p \ln [\,E_{max}(1)/E_c\,] \tag{8-4}$$

Therefore, substituting equation 8-4 into equation 8-3, we obtain the important result:

$$C_d(2) = C_d(1) + D_p \ln [\,2\,] \tag{8-5}$$

Physically, this means that whatever the cure depth was on the first or *Y*-hatch pass, provided that the exposure is kept the same on the second or *X*-hatch pass, the additional increment of cure depth, $\Delta\, C_d$, achieved on the second pass will be just

$$\boxed{\Delta\, C_d = D_p \ln [\,2\,] = 0.6931\, D_p} \tag{8-6}$$

completely independent of the absolute level of the exposure! Considering the complexity of many of the phenomena in this field, it is nice to see such a wonderfully simple result.

6. Finally, because the cured hatch lines are very closely spaced in both X and Y, the fraction of polymer that has been solidified to the gel point, or beyond, will be very high. Conversely, the fraction of residual liquid resin remaining within the part will be very low. Since postcure distortion is proportional to the fraction of liquid resin within the laser-cured part, WEAVE™ should, therefore, significantly reduce postcure distortion.

At this point, to help clarify numerous details for the reader, let us go through the WEAVE™ procedure in a step-by-step manner. In effect, this is intended as a "prescription" for the proper building of SL parts using the WEAVE™ technique. Hopefully, the following will help readers build better parts:

A. For a given resin, **accurately determine the values of D_p and E_c**. If the resin being used is one of those listed as approved by 3D Systems, it will have gone through an extensive quality control (QC) test program. This QC program is intended to ensure that both D_p and E_c are within an acceptable and very narrow tolerance band. Here, the user may input the published values for that resin.

   However, if the resin is not an approved resin, and does not have precisely defined values of both D_p and E_c, then these parameters must be determined by the user. The details for establishing accurate values of D_p and E_c are discussed in Chapter 10, under the WINDOWPANE™ experimental test procedure, which is the method used in the current QC program.

   The term WINDOWPANE™ stems from the observation that the diagnostic test part used to determine the resin parameters D_p and E_c physically looks like a tiny set of framed windowpanes.

B. **Specify the desired slice layer thickness, *a***. Typical values for a would be in the range from 5 to 10 mils (0.125 to 0.25 mm). The thinner the layers the finer the resolution, but the more sensitive the procedure becomes to small errors in recoating. Generally, users would be well advised to begin their initial part building experience with 10 mil layers (0.25 mm). After having successfully built a number of parts at this thickness, they might then more confidently proceed to thinner layers.

   A few parts have been successfully built, under special experimental laboratory conditions, with layer thickness values as small as 2.5 mils (0.063 mm). However, at this time, the practical lower limit for layer thickness is about 4 mils (0.1 mm). This is due to the potential for accumulation of tolerances in the recoating process, the resin leveling process, the measurement of the laser power and beam profile, and practical QC tolerance limits on the resin parameters D_p and E_c. Experience has shown that for layer thickness values below about 4 mils (0.1 mm), the accumulation of these tolerances can lead to layer delamination.

C. Using the working curve equation, 8-4, the system software will **automatically calculate the value of the exposure, $E_{max}(1)$**, sufficient to generate a cure depth $C_d(1)$, on the first pass (the *Y*-hatch vectors), such that:

$$C_d(1) = a - 1 \text{ mil} \tag{8-7}$$

This is the first of two selection rules that are basic to WEAVE™. The current software automatically calculates $C_d(1)$, given the slice layer thickness, a, and the desired "overcure," which in this case would be a negative 1 mil ($-0.001''$ or -0.025 mm).

D. **Specify a hatch spacing, h_s**, defined as the distance between the centerlines of adjacent parallel hatch vectors. The value of h_s is determined from the second selection rule for WEAVE™, namely

$$h_s = L_w + 1 \text{ mil} \tag{8-8}$$

where L_w is the full width of a laser-cured line having a cure depth equal to $C_d(1)$. L_w was previously defined for a Gaussian laser in section 4.5, equation 4-40.

It is worth noting that L_w may vary slightly from machine to machine owing to finite tolerances in the focused laser spot diameter, B. At present, the user must determine the spot diameter from the beam profiler data, and then calculate L_w from equation 4-40. However, if the laser beam is not Gaussian, then L_w must be determined experimentally. This will be discussed further in Chapter 10.

It is intended that future software enhancements will automatically determine B from the beam profiler data, and then calculate the corresponding value of L_w using the appropriate values of $C_d(1)$ and D_p.

E. Having completed a first-pass cure depth $C_d(1)$, the system now proceeds to the second, orthogonal pass. The system automatically sets $E(2) = E(1)$.

Thus, both sets of hatch vectors are drawn at the same laser exposure level. Provided that the laser power has not changed in the small time interval between the first pass and the second, the draw time for the second pass should be essentially identical to that for the first pass.

Wherever the laser beam moves over unexposed resin on the second pass, such as the tiny gaps between the parallel hatch vectors of the first pass, it too will generate a cure depth equal to $C_d(1)$. However, at the intersections of the orthogonal X and Y-hatch vectors the exposure will be twice that of the first exposure. The resulting cure depth at each intersection is then given by an equivalent form of equation 8-5,

$$C_d(2) = C_d(1) + 0.6931\ D_p \qquad (8\text{-}9)$$

where ln [2] = 0.6931... Substituting for $C_d(1)$ from the undercure selection rule of equation 8-7, we obtain the following condition:

$$C_d(2) = a + [\ 0.6931\ D_p - 1\ \text{mil}\ ] \qquad (8\text{-}10)$$

A necessary condition for layer adhesion is that the quantity in the bracket must be greater than zero. In this case, the cure depth at the vector intersections will be greater than the layer thickness. Therefore, provided that 0.6931 D_p is greater than 1 mil, plus any accumulated recoating and leveling tolerances, the nth layer should indeed be attached to the (n-1)th layer at every vector intersection. In effect, this places a lower bound on the penetration depth for WEAVE™; specifically, $D_p > 1.45$ mils.

The following illustrates the WEAVE™ technique with a specific numerical example:

- Resin: XB 5081-1 $D_p = 7.1$ mils $E_c = 5.6$ mJ/cm^2
- Layer thickness $a = 10$ mils
- Cure depth on first pass $C_d(1) = a - 1$ mil $= 9$ mils
- Cured Linewidth $L_w = 9$ mils
- WEAVE™ hatch spacing $h_s = L_w + 1$ mil $= 10$ mils

Using the working curve equation 8-4 to determine the maximum centerline exposure required to achieve a cure depth of 9 mils on the first pass, we find

$$E_{max}(1) = 5.6 \exp\,[9.0/7.1] = 19.89\ \text{mJ/cm}^2$$

Note that this value is determined automatically by the software. Furthermore, the software calculates the correct laser scan velocity as a function of laser power and spot size to generate the required exposure.

Since $C_d(1) = 9$ mils, while the layer thickness $a = 10$ mils, this value of $E_{max}(1)$ automatically satisfies the first selection rule of WEAVE™, as indicated in equation 8-7.

The system then proceeds to draw parallel Y-hatch vectors with centerline-to-centerline separations equal to $h_s = 10$ mils. Since the cured linewidth in this case is $L_w = 9$ mils, this leaves an intentional gap of about 1 mil between the parallel Y-vectors, thereby satisfying the second selection rule of WEAVE™.

Figure 8-1 is a schematic representation of a series of polymerized parallel Y-hatch lines, as seen from above. Note that this is essentially a magnified view of what the surface of a part would look like after the first pass on a given layer.

Figure 8-2 is a schematic of a magnified cross-sectional view of a small number of the cured lines within the nth layer.

After the first pass of Y-hatch vectors, the cross-section exhibits a series of parallel cured Y-hatch lines. As discussed in section 4.3, for the case of a Gaussian laser and a resin obeying the Beer-Lambert law, these lines would actually be parabolic cylinders with their centerlines running parallel to the Y axis. Since each has a cured linewidth of 9 mils, and is separated from the centerline of its nearest neighbors by 10 mils, this leaves a 1 mil gap between the lines at the resin surface.

Additionally, since $L_w = 9$ mils, then the choice of $h_s = 10$ mils automatically satisfies the second selection rule of WEAVE™. Also, it follows that on the first pass alone, 90% of the resin on the top surface of the nth layer has already been polymerized at or beyond the gel point. Hence, after the first pass of Y-hatch vectors has been completed, only 10% of the part cross-section at the resin surface remains in the liquid phase.

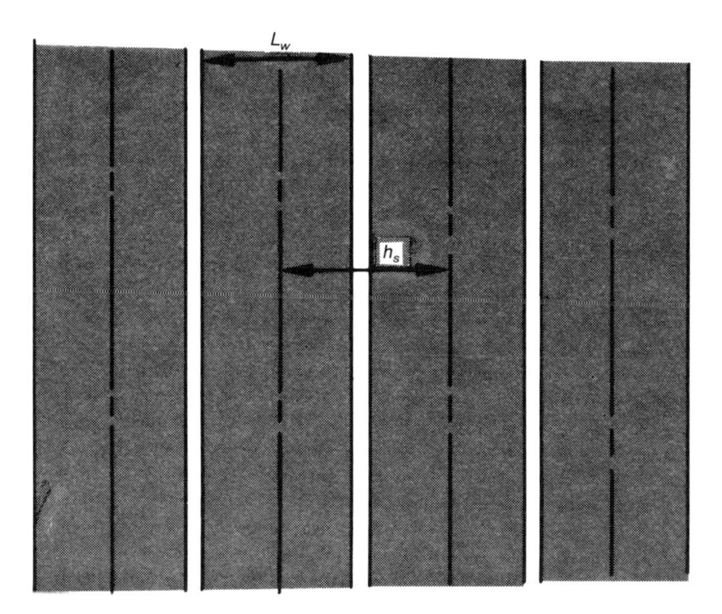

Figure 8-1. Polymerized parallel Y-hatch lines.

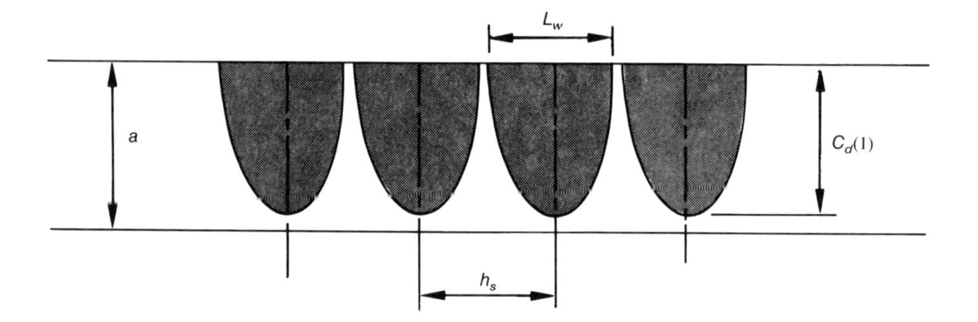

Figure 8-2. Cross-sectional view of cured lines.

The system now proceeds to the second pass. Here it draws parallel *X*-hatch vectors orthogonal to, but otherwise identical to, the earlier *Y*-hatch vectors. Thus the exposure, cured linewidth, and vector centerline-to-centerline hatch spacing parameters are identical on the second pass to the respective values used on the first pass.

However, at the vector intersections the additional increment of cure depth due to the intentional "double exposure" is simply (0.6931)(7.1) = 4.92 mils. With $C_d(1)$ set at 9 mils, then $C_d(2) = 9.0 + 4.92 = 13.92$ mils. Hence, the final result is about 4 mils of "overcure," or the amount by which the final cure depth exceeds the layer thickness.

Since the hatch vectors are separated by ten mils in both *X* and *Y*, there are 100 *X*-hatch vectors per inch intersecting 100 *Y*-hatch vectors per inch for a total of 10,000 intersections per square inch. Each intersection is attached to the layer below with about 4 mils of overcure. These numerous overcure zones act like tiny "rivets," thereby securing the *n*th layer to the (*n*-1)th layer and preventing delamination.

All these closely spaced parallel lines intersecting a second orthogonal set of equally closely spaced parallel lines cause the cured surface to resemble a tightly woven piece of material. The cloth-like appearance of the surface on each cured cross-section accounts for the origin of the name WEAVE™. The title is, however, not intended to suggest that the individual vectors pass over and under one another.

Finally, after two passes of orthogonal hatch vectors, a magnified schematic of the surface of the *n*th layer, as seen from above, appears similar to *Figure 8-3*.

It is clear by inspection of this figure that if $h_s = 10$ mils and $L_w = 9$ mils, then the top surface of the *n*th layer is 99% solidified at or beyond the gel point. Determining the volumetric percentage of the *n*th layer that has been cured at or beyond the gel point is a more complicated problem. If we assume that the individual cured lines are parabolic cylinders and the line intersections are

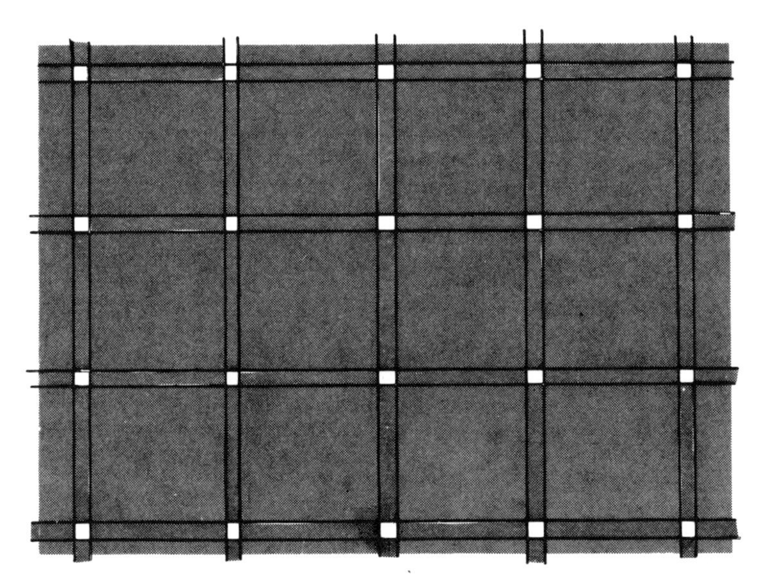

Figure 8-3. View of the surface after two passes of orthogonal hatch vectors.

truncated paraboloids of revolution (truncated due to the overcure resulting from the double exposure), then calculations based on volumetric integration indicate that approximately 96% of the resin is solidified when building with WEAVE™.

As a consequence, the residual liquid resin within a laser-cured part built with WEAVE™, in the manner described above, amounts to only about 4% of the part volume. As noted earlier, post-cure distortion is proportional to the fraction of liquid resin remaining in a laser-cured part. It is this characteristic that accounts for the dramatically reduced postcure distortion measured on diagnostic parts built with WEAVE™ relative to those built with 50 mil Tri-Hatch. Data from these measurements are presented in Chapter 10.

It is true that WEAVE™ significantly improves part building accuracy, as is evident from numerous test results presented in Chapters 10 and 11. However, the following comments are also appropriate regarding potential difficulties that may be encountered when using WEAVE™:

1. As noted above, success with WEAVE™ depends upon accurate control of cure depth. This in turn requires accurate control of the two resin parameters D_p and E_c. Without precise resin quality control, batch-to-batch variations can cause errors in cure depth. These errors can lead to problems with layer delamination if $C_d(2)$ is too small, or excessive part distortion if $C_d(1)$ is too large.

   This effect can be particularly pernicious when the system is frequently used to build many moderate to large parts. In this situation, the user is required to add resin rather often. The net result can be a bewildering set of concentrations of resins from various batches. Thus, the "pedigree" of what is actually in the vat can quite rapidly become very uncertain. The only practical solution to this problem is to ensure that careful QC procedures are employed and that the acceptable tolerances on D_p and E_c are maintained at as small a level as is both functionally and financially prudent.

2. Furthermore, to achieve the correct cure depth, the correct exposure must be delivered, which is proportional to the laser power and inversely proportional to the product of the beam radius and the laser scan velocity. Obviously, errors in determining any or all of the parameters P_L, W_o, and V_s will also lead to errors in C_d. This is the reason that it is particularly important to ensure that the system is carefully calibrated on a regular basis. As in many situations, deferred maintenance eventually proves costly.
3. Also, to properly assign a hatch spacing, h_s, the cured linewidth L_w must be accurately determined. Since the linewidth depends upon the focused laser spot diameter B, any significant variation in spot diameter or substantial beam eccentricity can result in values of L_w that are outside the specified range. These problems may occur when a Helium-Cadmium laser undergoes "mode-hopping" near the end of its tube life. The second selection rule of WEAVE™ requires that the hatch spacing exceed the cured linewidth by about 1 mil.

   Thus, if L_w is too small and h_s is not appropriately set, the resulting parts will probably be too soft. Conversely, if L_w is too large, the hatch vectors drawn on the first pass of a given layer may touch one another, resulting in excessive distortion. This can be somewhat of a "lions on the left, tigers on the right" problem. The clues to watch for are either soft parts or excessive distortion. We shall discuss this issue further in Chapter 10, with a description of some diagnostic techniques to help keep the user on safe ground.
4. WEAVE™ also requires accurate control of layer thickness. If the recoating system is not accurately calibrated, the resin layer thickness may be too large or too small. If too large, $C_d(2)$ may be less than the actual resin layer thickness, as opposed to the intended slice layer thickness, a. In this case, layer delamination may occur. Conversely, if the resin layer thickness is too small, then $C_d(1)$ may be greater than this value, the hatch vectors on the first pass may adhere to the lower layer, and excessive curl distortion may occur.

WEAVE™ does require exacting specifications on the system hardware, software, and resins. Nonetheless, it is worth noting that thousands of significantly more accurate parts have been successfully built with this technique.

Finally, in addition to enhanced accuracy, WEAVE™ also increases productivity. Most parts actually build more rapidly with WEAVE™ than with 50 mil Tri-Hatch. Although WEAVE™ involves greater total hatch vector length, the system draws at much higher speeds since the cure depths are smaller. As we shall see, the increase in laser drawing speed more than offsets the increased hatch vector lengths.

Rearranging equation 8-4, we obtain after simple algebra:

$$E_{max}(1) = E_c \exp\left[C_d(1)/D_p\right] \tag{8-11}$$

Thus, it is evident by inspection of equation 8-11 that **reduced cure depths require exponentially reduced exposure levels**.

Further, as has been shown in Chapter 4, equation 4-32, E_{max} is inversely proportional to the scan velocity, V_s. Thus, reducing the cure depth greatly reduces E_{max}, which results in a significant increase in laser scan velocity.

The laser drawing time is essentially determined by the total hatch vector length divided by the laser scan velocity. Here we are neglecting servo/mirror finite acceleration and deceleration effects, and assuming that the laser drawing time is dominated by hatching, with borders and fills constituting a small fraction of the total scan time. Furthermore, we also assume that the time required to accomplish all other aspects of the problem, such as recoating, leveling, analysis, etc., remains constant.

As an example, let us consider 50 mil, 60°/120°/*X* Tri-Hatch, with $a = 10$ mils and $C_d = 16$ mils (6 mils overcure typical of Tri-Hatch). Using resin XB 5143, with $D_p = 5.6$ mils and $E_c = 4.3$ mJ/cm^2, we may, therefore, calculate the required value of $E_{max}(1)$ from equation 8-11.

$$E_{max}(1) = 4.3 \exp\left[16/5.6\right] = 74.9 \text{ mJ/cm}^2$$

The corresponding value for WEAVE™, where $C_d(1) = 9$ mils is,

$$E_{max}(1) = 4.3 \exp\left[9/5.6\right] = 21.4 \text{ mJ/cm}^2$$

Thus, we see that the required maximum centerline exposure for Tri-Hatch is about 3.5 times that for WEAVE™.

Let us now consider a small element of area interior to the cross-section of a part. The intent is to determine the total hatch vector length for WEAVE™ relative to that for Tri-Hatch. While one could, in principle, select any arbitrarily shaped area, it is convenient from a bookkeeping and computational standpoint to pick a parallelogram, as shown in *Figure 8-4*.

The major advantage of this shape lies in the fact that it is a perfect overlay of an integer number of equilateral triangles (four), while it also results in all the WEAVE™ *X*-hatch vectors being of constant length. Nonetheless, it is still important to note that some of these vectors are common to exterior portions of hatch and should not be "double counted."

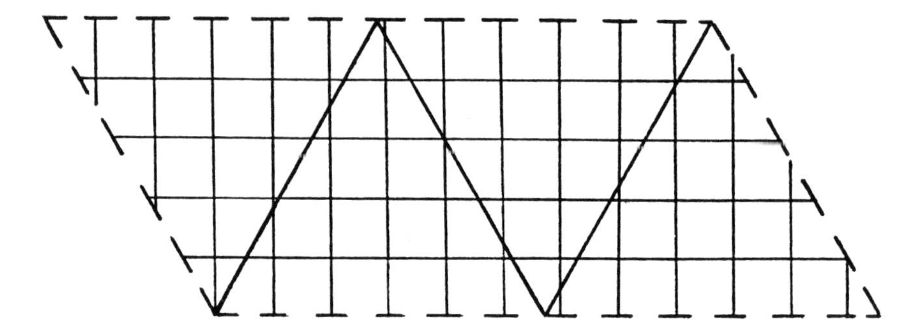

Figure 8-4. Parallelogram.

For convenience, all hatch vectors that share their function with an adjacent exterior hatch area have been drawn as dashed lines. All hatch vectors solely related to the interior parallelogram area being studied are drawn as solid lines. For example, the perimeter of the parallelogram is shown dashed. This is because in Tri-Hatch, all these vectors are common not only to the four equilateral triangles in the subject area, but to exterior triangles as well.

Similarly, the upper and lower borders of the parallelogram also represent *X*-hatch WEAVE™ vectors shared by the subject area as well as the exterior region. To properly maintain the "bookkeeping," solid hatch vector lines are considered to have their full length, while dashed hatch vectors are counted at only half their actual length.

This is done since the dashed vectors need only be drawn once, but they serve a dual function for both the interior and exterior areas. Precisely because they are accomplishing a dual function, only half of their draw time should be apportioned to the parallelogram. It is, therefore, necessary to treat such vectors as halved in length to ensure that neither build method is unfairly penalized, by double counting such vectors.

For an equilateral triangle having an altitude of 50 mils, simple trigonometry easily shows that the length of each side of the triangle is given by 50 mils/sin $60° = 57.74$ mils. For the Tri-Hatch case, there are 6 common border vectors and 3 interior vectors, for a total equivalent vector length equal to $6 \times 57.74 = 346.4$ mils.

For the WEAVE™ case, there are 4 such common lengths and 8 interior lengths in *X*-hatch vectors. Thus, the total equivalent *X*-hatch vector length amounts to $10 \times 57.7 = 577$ mils. For the *Y*-hatch vectors, the total equivalent length is $11 \times 50 = 550$ mils. Thus, the total effective length for the WEAVE™ case is 1127 mils.

Therefore, the hatch vector length ratio for WEAVE™, relative to Tri-Hatch, is given to close approximation by 1127/346.4 = 3.25. However, the laser scan velocity ratio for WEAVE™ vectors relative to those for Tri-Hatch was determined earlier to be 3.50.

Hence, the final result is that the total drawing time for WEAVE™, relative to that for Tri-Hatch, is roughly equal to the quotient of the hatch vector length ratio and the laser scan velocity ratio, or 3.25/3.50 = 0.929.

In conclusion, based on this analysis:

WEAVE™ will actually draw about 7% faster than Tri-Hatch when building with the same resin, the same laser power, and the same layer thickness value.

This result is in general agreement with experimental observations for numerous parts of different geometry which had been built in the same resin using WEAVE™ and Tri-Hatch. In all cases, WEAVE™ was 5% to 10% faster! Thus, when properly executed, WEAVE™ will not only generate more accurate parts, but will do so with enhanced productivity.

8.3 The Development of STAR-WEAVE™

With the advent of WEAVE™ in late 1990, the improvements in part accuracy were considerable when compared to parts built using the former Tri-Hatch method. Detailed diagnostic test results, involving many thousands of data points, are presented for both methods in Chapter 10. Quantitative part accuracy test measurements, using both building techniques, are presented in Chapter 11.

Nonetheless, while WEAVE™ was a major advance, there was still considerable room for further improvements in part accuracy. Specifically, the following problem areas/anomalies still existed:

1. When building large flat slabs, it was observed that most of the departure from a true plane occurred at the corners of the slab.
2. Furthermore, one of the corners would always show an error distinctly larger than the other three.
3. Additionally, it would always be the same corner that produced the largest error, independent of the resin type, the specific SLA machine used, the absolute level of the laser power, or the layer thickness.
4. Certain resins, and most especially XB 5143, would exhibit very tiny microfissures when built with WEAVE™.
5. In certain specific geometries, such as a flat plate with a hole passing through the plate, as shown schematically in *Figure 8-5*, the occurrence of a macrofissure was often observed. In extreme cases, the macrofissure would pass through the entire thickness of the plate, rendering the part structurally useless.

6. When such macrofissures did occur, they were always oriented in such a fashion as to be tangent to the diameter of the hole. This behavior is also shown schematically in *Figure 8-5*.

As recently as early 1991, explanations were not available for any of these phenomena. However, since the advent of WEAVE™, postcure distortion had gone from the dominant source of error in SL to a relatively minor role. Then, the dominant source of error became part distortion due to internal stresses developed during part building, with the most significant distortions occurring on large, nominally flat slabs.

The first new concept directed at hopefully eliminating, or at least reducing these problems, was STaggered hatch. The reason for capitalizing the first two letters of the word "staggered" will be evident later. *Figure 8-6* is a schematic diagram of the difference between conventional hatch and STaggered hatch.

With conventional hatch, the X vectors on the nth layer are positioned directly above the X vectors on the $(n\text{-}1)$th layer. The Y vectors are also positioned directly above one another. The resulting photomodulus contours are such that the maximum values of polymer modulus are located directly down the centerline of each vector, with the weaker regions lying between the vectors.

As a result, the regions between the vectors represent potential weak spots where internal stresses, developed during part building, could lead to tiny cracks or "microfissures." Thus, regular WEAVE™ was analogous to building a brick wall with all the bricks located directly on top of one another, and the vertical mortar lines also directly on top of one another, extending the full height of the wall in a straight line.

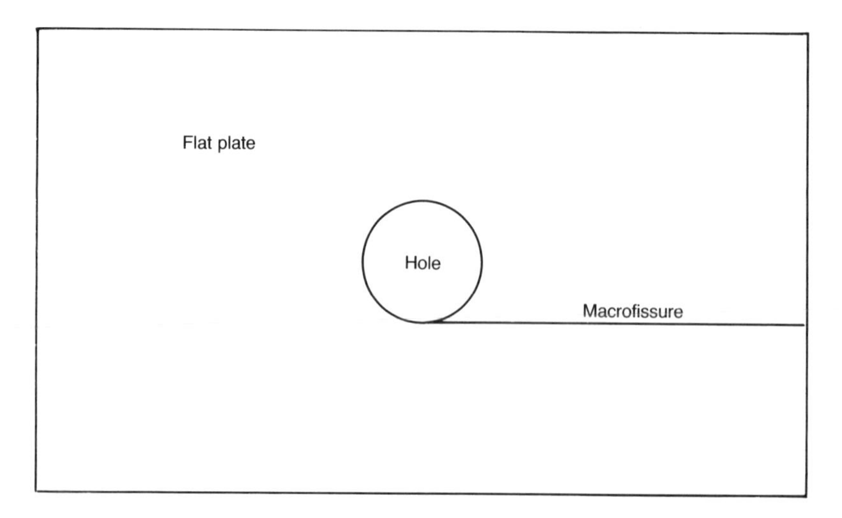

Figure 8-5. Macrofissure.

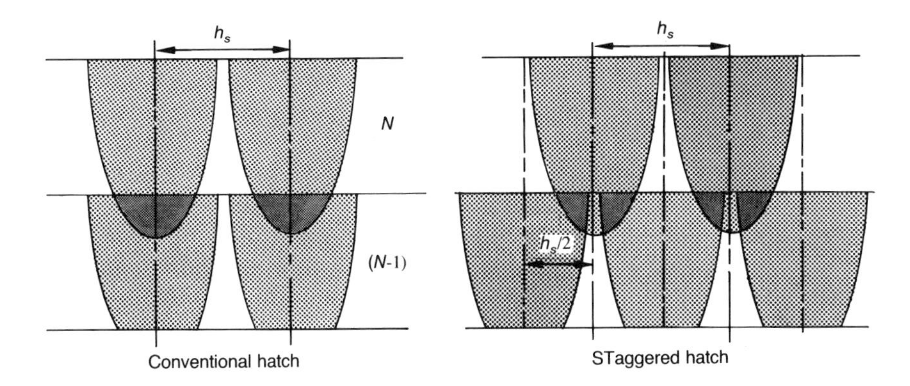

Figure 8-6. Variation between hatch types.

Obviously, simple experience shows that real brick walls are not built in this manner. Rather, the bricks are "staggered" to avoid weak spots and a tendency to develop long cracks directly down a given mortar line. This simple insight led to the concept of STaggered hatch. Since resin XB 5143 had shown a significant tendency towards the formation of microfissures, as previously noted in item number 4, use of STaggered hatch was attempted on this resin.

Special custom software was written to accomplish STaggered hatch. Parts were then built using XB 5143. The result was dramatic! As soon as the hatch vectors on the *n*th layer were offset by exactly half the regular hatch spacing ($h_s/2$), relative to those on the (n-1)th layer, all evidence of microfissures vanished. Specifically, as shown in *Figure 8-6*, the key to STaggered hatch is to intentionally offset the *X* and *Y*-hatch vectors on successive layers to reduce stress concentrations along the relatively weaker regions between vectors.

The next area of concern was the reduction of stress concentration effects as previously described in items 1 through 3. Specifically, not only was it important to improve the absolute flatness of large slabs, but it was also highly desirable that such parts be more uniform, without any corner exhibiting significantly more distortion than the others.

Initially, the nature of the support structures were studied in some detail. After building numerous slab 6×6 diagnostic parts, with different support spacings, geometries, and overcure levels, the departure from a planar structure was found to be influenced very little by the supports. Of course, this assumes that the supports are properly positioned, and do not separate from either the platform or the part.

Since the support structure was clearly not the answer, a different approach was taken. *Figure 8-7* shows a schematic top view of one of the process diagnostic parts, known as Twin Vertical Walls, which is used as a measure of postcure distortion. Careful observation of this part during building led to the recognition that an improved drawing sequence was definitely possible.

With regular WEAVE™, the short vectors, shown in *Figure 8-7*, would be drawn first, and always in the same order (from left to right, and propagating from the front of the vat toward the back).

Next, this sequence would always be followed by the long vectors also being drawn in their own identical sequence (from front to back, and propagating from left to right). This is also shown in *Figure 8-7*.

As a result, a given layer had been partially polymerized while the *X*-hatch vectors were drawn. When the first of the *Y*-hatch vectors was drawn, on the left long side, this region would attach to the previous layer. As the additional *Y*-vectors were drawn, a propagating "wave" of curing, attachment to the lower layer, and subsequent shrinkage would proceed from the left toward the right.

Since the last portion to shrink was the right long side, the net result was to generate a curvature that was concave as viewed from the right. *Figure 8-7* also shows one such thin vertical wall in its final distorted form. The extent of the distortion has been greatly exaggerated for the purposes of visual representation.

It seemed reasonable to conclude that the resulting bending moment might therefore be dependent upon the drawing sequence. Thus, a given drawing sequence might result in an internal bending moment of such sense as to produce a concave curvature to the right. Conversely, by reversing the drawing sequence on the next layer, the moment generated therein should be of opposite sense. This would tend to produce a concave curvature to the left.

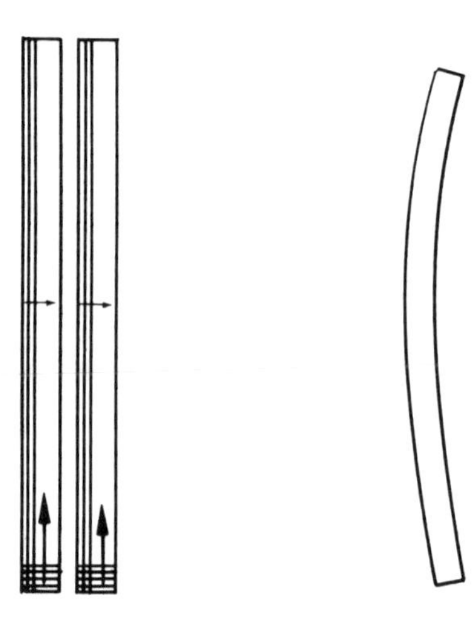

Figure 8-7. Twin vertical walls.

Over many layers, there should be a tendency for these internally generated bending moments to at least partially cancel one another. The final result would then be reduced part distortion, a greatly reduced tendency for stress concentrations, and improved overall part accuracy. Clearly, all these effects are potentially beneficial.

Again, special custom software was written to accomplish "alternate sequencing." The term alternate sequencing is used to designate the fact that the drawing sequence is alternated from layer to layer.

Furthermore, WEAVE™ involves two different types of vectors (*X* and *Y*), with four possible scan propagation directions (to the right, to the left, to the front, and to the back).

Simple analysis will show that there are eight possible combinations of drawing sequences on any given layer. Thus, the implementation employed in software release 3.82 involves so-called "eight fold" alternate sequencing. With this software, the *X* and *Y* vectors will alternate in the order of the drawing sequence. Thus, for example, the *X* vectors might be drawn first on even numbered layers, while the *Y* vectors would then be drawn first on odd numbered layers.

Furthermore, the direction of vector propagation is also alternated. Thus, for example, on the *n*th layer the *X* vectors may be drawn first, propagating from the front of the vat towards the back of the vat. However, on the $(n+1)$th layer, the *X* vectors would then be drawn second (after the *Y* vectors), and they might then propagate from the back of the vat toward the front of the vat.

A careful observer will be able to detect the fact that with the new building method, any consecutive eight layers will each be drawn in a different sequence. The entire procedure will then repeat identically from the ninth through sixteenth layers, etc. The alternate sequencing process will then continue throughout all the layers of the part.

As soon as experimental tests were run on the diagnostic test part slab 6×6, the parts appeared clearly more planar, and showed much less stress concentration. Specifically, the tendency for the same corner to exhibit the worst distortion was essentially eliminated! Overall part accuracy was improved, and the formerly puzzling tendency for one corner always to be significantly more distorted than the others, and for this corner to be the same one in different machines using different resins, was essentially eliminated, and its cause understood.

Finally, as described earlier in items 5 and 6, parts built with the tough resin XB 5143 using regular WEAVE™ occasionally showed major macrofissures. Further, these features were almost always oriented tangent to the diameter of any interior hole. *Figure 8-8* is a schematic drawing of such a part, showing where the macrofissure would typically occur. In some cases, these macrofissures would threaten the structural integrity of the entire part.

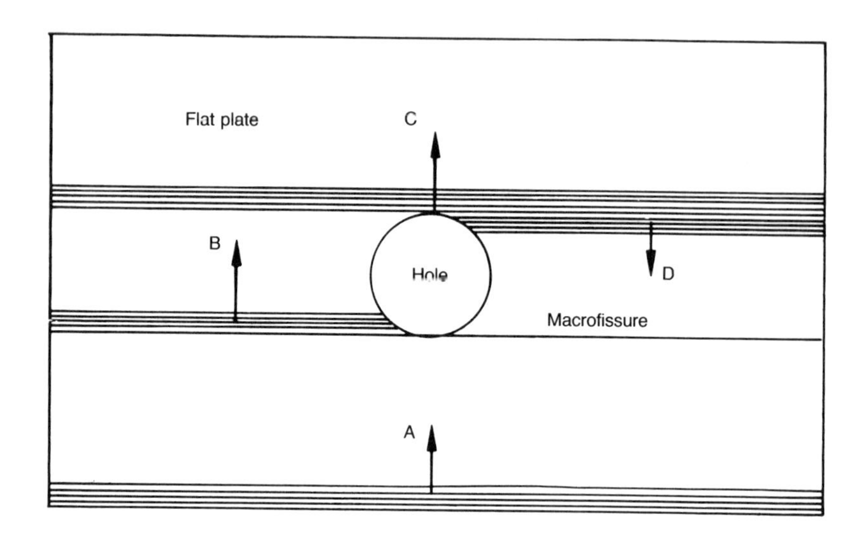

Figure 8-8. Schematic drawing of a typical macrofissure.

Again, as soon as alternate sequencing was evaluated on such a part, the entire macrofissure problem disappeared! In retrospect, the cause of the problem is clear, as is the reason why alternate sequencing eliminated macrofissures.

Inspection of *Figure 8-8* shows that with regular WEAVE™ the *X* vectors would always be drawn from left to right, propagating from the front of the vat toward the back of the vat (A).

When they arrived at the front edge of the hole, the drawing sequence would "divide," in the sense that the *X* vectors would propagate toward the back of the vat, but only on the left side of the hole (B).

Upon reaching the rear tangent point of the hole, the *X* vectors would resume drawing the full width of the part, but still propagating toward the back of the vat (C).

Only when this operation had been completed would the drawing sequence return to complete the missing hatch area (D). Also, this area, on "the return swing," would then be hatched with *X* vectors propagating from the back of the vat toward the front.

Unfortunately, during the time interval from the drawing of the *X* vector tangent to the front of the hole, until the drawing of its closest neighbor at the very end of the *X*-hatching sequence, the cured resin was shrinking. Since, for example, the portion of the part from the front to the tangent line might be 3″ or 3000 mils, then even with only about 0.6% linear shrinkage when building with WEAVE™, the total shrinkage in this case would be 18 mils. Thus, the tangent line drawn just prior to "dividing" the *X*-hatch vectors may have moved about 9 mils toward the front of the vat by the time its "nearest neighbor" would be drawn at the end of the *X*-hatch sequence.

The net result was, therefore, very much like ''continental drift.'' By the time the vector sequence returned to complete the last of the ''missing area,'' the final X-hatch line was actually positioned with a gap of about 9 mils relative to its intended ''nearest neighbor''. This is essentially equivalent to missing an entire X-hatch vector.

By itself, this probably would not be catastrophic, although the gap might show up in the finished part as a blemish. What was significant, however, was the fact that this gap would occur in exactly the same location layer after layer. The net result was a major weak zone, or so-called ''dislocation line,'' ultimately leading to a ''macrofissure.'' In certain cases, these macrofissures were so deep that the part would actually cleave under its own weight in the green state

When this part was built using alternate sequencing, macrofissures were eliminated. The reason is now clear. Alternate sequencing varies both the order of the X and Y vectors, as well as their propagation directions. Thus, the tangent vector on a given layer would cause a single dislocation line originating at a specific point on the circumference of the hole.

However, on the next layer, the alternate sequencing algorithm would result in a dislocation line along a different tangent to the hole. This tangent would not only be orthogonal to the dislocation line on the previous layer, but it would also be tangent to the hole at a different point. The new point would be located one quarter of the circumference of the hole away from the corresponding point on the previous layer.

Therefore, dislocation lines on adjacent layers never overlap when built with alternate sequencing. Furthermore, in general, they would never even ''cross'' one another, notwithstanding that they would be separated vertically by a layer thickness. As a consequence, the dislocation lines are always singular and isolated, with the result that they never develop into macrofissures. The net effect is the preservation of part integrity in the green state.

Finally, in drawing slab 6x6 with WEAVE™, it also became clear that hatch vector shrinkage must be responsible for a considerable fraction of the internal stresses leading to part distortion. This was confirmed by drawing some test parts without any up-facing skins. These parts still exhibited considerable distortion, although somewhat less than those drawn with up-facing skins.

From Newton's Third Law, shrinkage clearly cannot generate an action on a given cured border vector without a reaction. Specifically, to have an effect on part accuracy, the shrinkage must exert a reaction force on the borders. Regular WEAVE™ attaches all hatch vectors to the borders, as did Tri-Hatch before.

Therefore, when the hatch vectors on the first pass undergo shrinkage, they are attached to the borders on both ends. The net effect is that the shrinkage forces occurring within these hatch vectors cause reaction forces on the borders. This distortion of the borders leads to a reduction in part accuracy.

One of the basic concepts of WEAVE™ was to accomplish as much shrinkage as possible on the first pass of hatch vectors, without attaching to the layer below. In this way, the maximum extent of shrinkage would be occurring on the first pass without leading to distortion through contact with the lower layer. However, what was perhaps less apparent was the fact that, provided the vectors were attached at both borders, they could still generate considerable internal forces.

The situation was recognized as analogous to tightening the string on a bow. As the bowstring continues to be tightened, corresponding to the shrinkage of a hatch vector, the bow will bend. Here, the bow is directly analogous to the previous layer, with the ends of the bow comparable to the borders. Despite the fact that the bowstring is only attached to the bow at the ends, increased bowstring tension will nonetheless result in increased bending of the entire bow.

Clearly, it is not appropriate to eliminate either the hatch vectors or the borders. However, it is possible to avoid securing one end of each hatch vector to one of the borders. In this way, the first pass hatch vectors would be free to shrink without causing any reactive forces to be exerted on the borders. This, in effect, would be like cutting the bowstring very near one end of the bow. Without attachment at both ends of a bow, a bowstring has no effect whatsoever.

Thus, the concept of retracted hatch was developed. Here, each hatch vector, whether *X* or *Y*, is attached at one and only one border. The other end of the hatch vector is retracted a tiny distance, typically about 10 mils (0.25 mm) from the adjacent border. *Figure 8-9* schematically shows the nature of retracted hatch.

Note that the hatch vectors are not only retracted from the borders, but that the retraction sequence itself is alternated. Thus, a given *X*-hatch vector (shown darker for emphasis) may attach to the border on one end, while its nearest neighbors will not be attached at that end. Conversely, at the opposite border that *X*-hatch vector will not be attached, while its nearest neighbors will be attached.

Once more, special custom software was generated to allow for the implementation of retracted hatch. After this had been accomplished, diagnostic test parts were built with the new feature. The immediate result was a further level

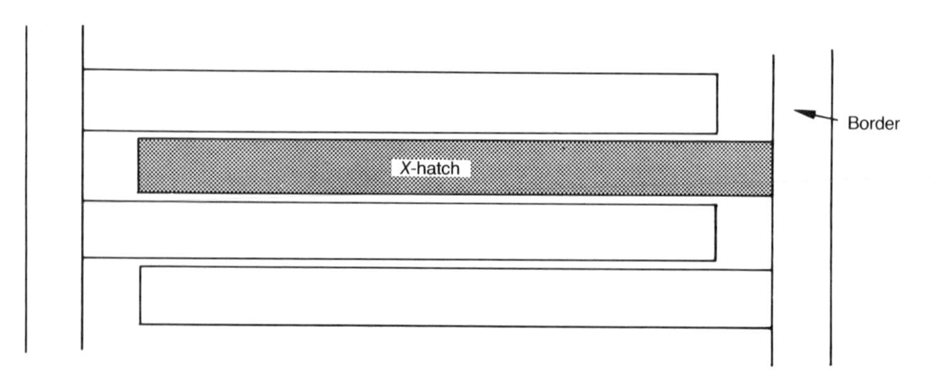

Figure 8-9. Retracted hatch.

of improvement in part accuracy. As a consequence, the decision was made to perform a full battery of tests with various combinations of all three of the new features:

- STaggered hatch.
- Alternate sequencing.
- Retracted hatch.

The test results are presented in Chapters 10 and 11. Since the results were beneficial in all resins tested, and especially significant in XB 5143, the development of new software was undertaken to enable the use of these techniques. This release, known as 3.82, was made available to users in October, 1991.

The new build style, which was a derivative of WEAVE™, and which included all three of the new concepts described above, is known by the acronym STAR-WEAVE™, where the "ST" refers to STaggered hatch. The "A" stems from Alternate sequencing and the "R" derives from Retracted hatch.

8.4 Future Advances

At this point it is not possible to precisely specify what future advanced part building techniques will evolve. However, we can say a few things in general.

Over the past two years, SL technology has advanced rapidly. Part accuracy results, described in detail in Chapter 11, show considerable progress from Tri-Hatch to WEAVE™ to STAR-WEAVE™. At this time, 3D Systems is already working on yet another, even more advanced building method intended to further extend the improvements that have already been achieved with STAR-WEAVE™. While the initial results definitely look promising, it is still far too early to draw any significant conclusions.

Also, the reader should realize that testing a new building technique is a very time-consuming process. First, a detailed set of diagnostic tests must be completed for each of the approved resins. This data must show sufficient progress to justify all the labor associated with developing, testing, debugging, internal alpha testing, external beta testing, and finally releasing updated software to support the new building method.

Furthermore, at no point do we want to go backwards. Thus, not only are significant improvements necessary in some key areas, but other characteristics must not be adversely affected. In short, hitting a home run on each trip to the plate is not easy, and grand slam home runs are very rare indeed.

Nonetheless, it is reasonable to at least list some of the future goals of SL with respect to advanced part building methods. Each of the goals listed below is already the subject of R&D efforts:

1. Further reductions of internal stresses.
2. Additional reductions of stress concentrations.
3. Continued improvement in overall part accuracy.
4. Additional reductions in part building time, with a corresponding improvement in productivity.
5. New methods to improve the surface finish of parts as they come directly from the vat.
6. Developing building techniques capable of exploiting the optimum performance characteristics of new resins having improved physical properties.
7. Perhaps most important of all, gaining a deeper and more fundamental understanding of the basic physical phenomena. It has been this kind of improved understanding that led to the current developments, and will provide the impetus for further advances.

REFERENCES

1. Kloosterboer, J.G., *Network Formation by Crosslinking Photopolymerization and Its Applications in Electronics*, published in *Advances in Polymer Science*, Volume 84, *Polymers in Electronics*, pp. 1-61, 1988.
2. Hofmann, M., *Real Time Shrinkage Experiment*, personal communication, May 21, 1990.
3. Leyden, R.N. and Pang, T., *Measurements of Resin Shrinkage with Ultraviolet Exposure*, 3D Systems internal report, July 3, 1990.
4. Bernhard, P., Hofmann, M. and Hunziker, M., *Advanced Testing of StereoLithography Resins*, Second International Conference on Rapid Prototyping, University of Dayton, Conference Proceedings, pp. 86-89, June 23-26, 1991.
5. Nguyen, H., *Techniques for the Measurement of Resin Working Curves*, 3D Systems Technical Report No. 56, April 5, 1990.
6. Richter, J. and Jacobs, P.F., *The Present State of Accuracy in StereoLithography*, Second International Conference on Rapid Prototyping, University of Dayton, Conference Proceedings, pp. 269-294, June 23-26, 1991.
7. *Ibid.*
8. Bernhard, *loc. cit.*

chapter 9

Postprocessing

It ain't over 'till it's over.

—Lawrence Peter "Yogi" Berra
Catcher, New York Yankees
Circa 1965

9.1 Introduction

This chapter will discuss four of the final steps in the evolution of a StereoLithography prototype. These are as follows:

1. Part Removal.
2. Part Cleaning.
3. Postcuring.
4. Part Finishing.

These steps may seem prosaic. In fact, they are critical to the outcome of the effort. If parts are not properly removed, at a minimum, valuable resin will be wasted. In extreme cases, prototypes containing fragile geometries may be damaged.

Furthermore, improper cleaning can leave a thin film of resinous solvent solution on the part. If the solvent sufficiently dilutes the resin, without stripping it completely free from the green surface, it is possible that the photoinitiator will be ineffective at completing any further polymerization. In this situation, the residual resin film will never properly postcure, leading to perpetually "tacky" surfaces. Also, incomplete cleaning can leave uncured resin in corners, blind holes, or narrow passages. Upon postcure, this excess but relatively undiluted liquid resin may polymerize, leading to errors in the final

---

*By **Paul F. Jacobs, Ph.D.**, Director of Research and Development, 3D Systems, Inc., Valencia, CA.*

dimensions. Excessive exposure to strong solvents definitely can degrade part accuracy.

Also, incorrect postcuring can lead to substantial postcure distortion. Insufficient actinic postcure irradiance can result in tacky surfaces. Excessive actinic irradiance can also cause rapid polymerization, resulting in a substantial exothermic heat release. In extreme cases, the resulting high internal resin temperatures can actually generate sufficient thermal stresses to fracture the part.

Finally, lack of patience can destroy fragile parts during finishing operations. Support removal, as well as sanding and glass bead blasting must be done carefully to avoid damage, especially to delicate geometries. Once you have come this far, you really don't want to strike out in the ninth inning with the bases loaded.

9.2 Part Removal

As discussed in Chapter 2, photopolymer resins, and especially acrylate based resins, can cause skin irritation. For this reason, appropriate rubber or Neoprene gloves should be worn whenever handling parts or platforms that are wet with liquid resin. To minimize waste, heavy duty gloves can be reused many times. In the long run, this is actually less expensive than using disposable gloves. Further, heavy duty reusable gloves are actually safer, since they are generally less permeable than the disposable variety, and are much less prone to tearing and leaking.

Unfortunately, neither the U.S. Occupational Safety and Health Administration (OSHA) nor the American Conference of Governmental Industrial Hygienists (ACGIH) have established occupational exposure limits for the acrylate resins currently used in StereoLithography. According to General Aniline and Film (GAF), the data accumulated to date suggests that safe breathing levels should be below 10 parts per million (ppm). The odor thresholds for these materials are about 0.05 to 0.1 ppm. Thus, the user is readily alerted to their presence.

However, the equilibrium concentrations of the more volatile resin components within the SLA are much higher than those in the room. Therefore, it is prudent to open the SLA door/window for at least one minute prior to reaching into the working space. Better still, install an exhaust fan with an in-line charcoal filter. This fan would run for about one minute after the part has been completed but prior to part removal. In this way, the concentration of potentially hazardous volatile components is further reduced.

At this point, depending upon the object geometry, the platform/part can be either removed or tilted on edge to drain excess liquid resin back into the vat. Commonly, the perforated platform can be tipped on its edge and balanced on the platform support arms. In this way, the part can be positioned at nearly 90° to the original build axis. This facilitates excellent resin drainage from trapped volumes.

After a few minutes of draining, both the platform and the attached part are ready for removal from the SLA. At 3D Systems, the recommended method involves placing the platform on a shallow-rimmed stainless steel tray with handles at each end. In this manner, the platform/part can be moved to the cleaning station without allowing resin to drip onto the floor and without generating any hazardous waste. The liquid resin that does get on the tray can be cleaned, along with the platform/part, in the resin solvent.

An alternate approach is to use thick, highly absorbent cellulose padding, which is placed under the platform/part prior to removal from the SLA. The advantage of this approach is that it is effective. A disadvantage associated with this method is that the cellulose padding material, after absorbing liquid resin, must be disposed of as hazardous waste.

In principle, one could cure the resinous padding material either in sunlight or in a postcuring apparatus (PCA), and simply dispose of it in the trash. This sounds like a good plan. History indicates that most users opt for tossing the resinous padding into the hazardous waste containers. Thus, the trays really do reduce disposal problems.

9.3 Part Cleaning

The task of part cleaning has traditionally been one of the least beloved in the entire RP&M sequence. Most people view it very much like washing dishes. Everyone enjoys eating Thanksgiving dinner. Few relish the cleanup. However, part cleaning is a necessary component of the entire process. Only recently has it become clear that proper cleaning is quite critical to the successful generation of accurate parts.

For about three years, from late 1987 to late 1990, the dominant cleaning solvent for SL parts at 3D Systems was alcohol. Either methyl, isopropyl, or denatured alcohol were commonly employed. Alcohol was also the solvent of choice at the great majority of SL user installations.

However, various experimental tests[1] performed at 3D Systems indicated that exposure to alcohol could cause nontrivial swelling distortion. More extensive tests were performed with a number of different solvents. *Figure 9-1* shows the results of these tests upon a diagnostic part built with resin XB 5081-1, using WEAVE™.

From this figure, it is clear that alcohol, Therm MS (a solvent used at various European SL sites), and propylene carbonate, all show significant part distortion due to swelling (in excess of 25 mils on a 2.000″ or 50.80 mm dimension within 24 hours and as much as 2 mils in one hour).

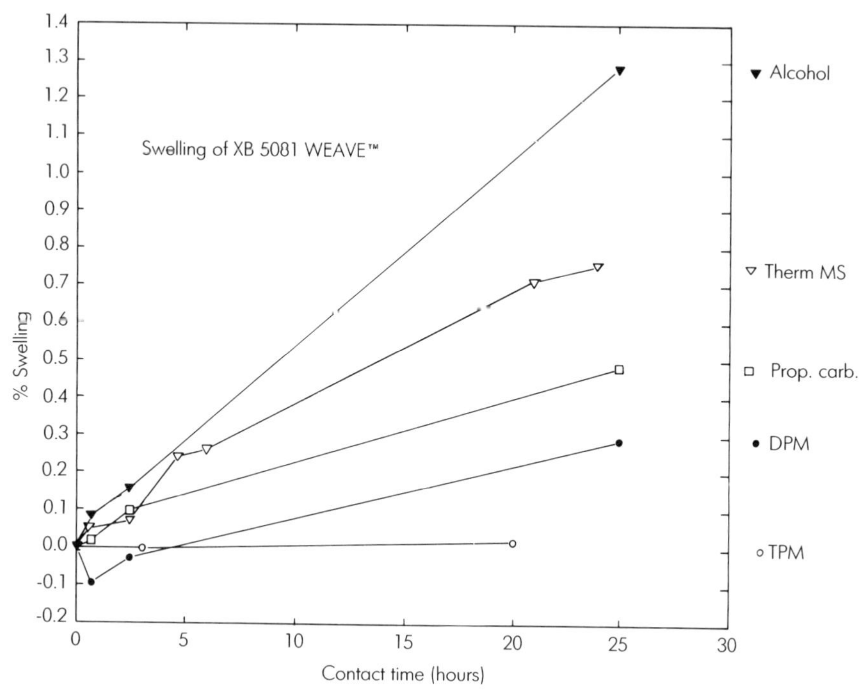

Figure 9-1. Swelling of XB 5081-1 versus contact time for various solvents.

During 1989 through 1991, SL technology achieved substantial progress in improved part accuracy (as discussed in Chapter 11). The near-term goal of SL is to achieve 5 mil tolerances on dimensions of a part as large as the platform. This requires limiting the root-mean-square (RMS) errors to less than 0.2 mils per inch, or 0.02%. Clearly, at these exacting levels, one cannot permit solvent induced swelling errors that are many times larger.

Figure 9-1 presents experimental data for a number of different solvents. While some solvents were very effective at stripping one or two resins, they were significantly less effective with others. Ideally, a given solvent should work well with all SL resins. Otherwise, different solvents would be required for different resins. This would add both cost and complexity for the users, and was considered undesirable.

Figure 9-2 shows similar test results for the same diagnostic part built in resin XB 5143, also using WEAVE™. The data clearly shows that alcohol can lead to significant part distortion. Errors as great as 24 mils were measured on the same 2.000″ diagnostic dimension when immersed in alcohol for 2.5 hours. Immersion in alcohol for only 30 minutes resulted in a swelling error of about 7.4 mils, or 0.37%, which is greater than 18 times the entire RMS error budget!

Sufficiently detailed data is not available to be certain that swelling errors increase linearly with alcohol contact time for very brief immersion intervals. Nonetheless, interpolating this data strongly suggests that swelling distortions

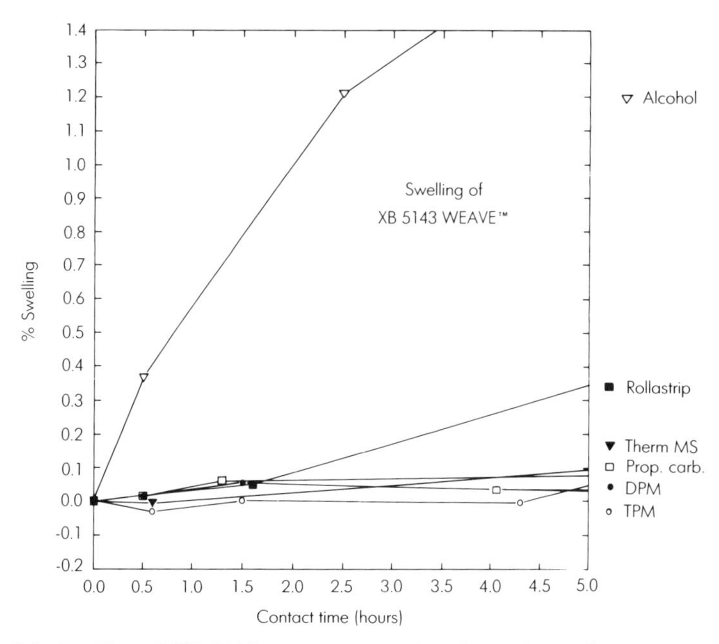

Figure 9-2. Swelling of XB 5143 versus contact time for various solvents.

comparable to the entire RMS error budget for accurate part building can occur with alcohol within about one minute!

These experimental test results also showed that Di-Propylene Glycol Monomethyl Ether (known commercially as DPM) as well as Tri-Propylene Glycol Monomethyl Ether (known commercially as TPM), were both very effective at stripping these resins. While the results of *Figure 9-2* suggest that they are about equivalent, the results of *Figure 9-1* indicate that TPM produces substantially less swelling distortion with XB 5081-1 than does DPM. In both cases alcohol caused significant swelling distortion.

Therefore, in the interest of part accuracy, it is clear that alcohol should NOT be used as a resin solvent. We are well aware that many SL installations initially used alcohol, including 3D Systems. Also, users may have grown accustomed to working with this solvent. Nonetheless, it is important to point out that the data of *Figure 9-2* indicates that even a brief exposure of a green SL part to alcohol can result in a significant degradation of part accuracy.

For those users interested in achieving maximum RP&M part accuracy, alcohol as a part cleaning solvent is not recommended. As to DPM and TPM, they are both effective resin solvents for all of the approved SL resins. As previously noted, DPM does cause more swelling distortion than TPM in resin XB 5081-1. Also, TPM has a much higher flash point (93°C, or about 200°F) than DPM (57°C, or about 135°F). A test conducted outdoors at 21°C (70°F) indicated that a lighted match is actually extinguished in TPM.

Overall, TPM is much safer than either alcohol or DPM. First, there is no explosion hazard with TPM. Second, TPM is much less volatile than either alcohol or DPM, and has little odor. Third, TPM is very effective with all approved and recognized SL resins, when used properly. Fourth, TPM does not require expensive ultrasonic cleaning equipment to function efficiently. Fifth, TPM is reasonably priced, between $7 and $8 per gallon. And sixth, TPM remains fully operational up to about 12% resin loading.

Beyond 12% resin concentration, the solvent action of TPM begins to be somewhat less effective. Above about 15% resin concentration, the solvent should be considered to be "resin saturated," and no longer useful for part cleaning. At this point, new TPM is required.

However, it is worth noting that at 12% loading, the monetary value of the resin in the solvent is roughly six times the cost of the TPM. Therefore, it is good practice to replace the used TPM when it reaches about 12% resin concentration. Any small savings that may be realized by using the solvent beyond this point will almost certainly increase the part cleaning times substantially, and may also leave the part surfaces permanently "tacky." Since the prototypes are far more valuable than spent TPM, this would be false economy.

Fortunately, there is a significant difference between the specific gravity of the liquid SL resins (1.14 $\pm$0.01) and that of pure TPM (0.96). An appropriately sensitive hydrometer can readily detect if the resin has reached the 12% concentration level. When the specific gravity of the mixture reaches 0.98, it is time to replace the TPM. Should the specific gravity reach 0.99, the solvent will no longer function as an effective cleaning agent.

The used TPM and resin mixture must be treated as a hazardous waste and should be handled in the same manner as other spent resin solvents. Since both TPM and SL resins are hydrocarbons, the most effective means of disposal is by burning at an authorized waste treatment facility. With proper incineration, the final products should be primarily carbon dioxide, water vapor, and energy. In a modern cogeneration plant, this energy can then generate electrical power.

Tests of various methods of employing TPM as a resin stripper have resulted in the following.

First, moderate but continuous motion (5 to 10 cm/sec) of the part relative to the TPM is by far the most effective method of part cleaning with this solvent. Apparently, the TPM significantly reduces the local resin viscosity at their mutual interface. If no relative motion occurs, molecular diffusion at the interface will cause the local resin concentration to gradually become so high that effective stripping action becomes greatly retarded.

However, if the part is simply moved through the solvent, an action similar to forced convection will cause the outermost resin layer to slough off into the solvent. The relative motion will then expose lower resin concentration solvent to a fresh film of resin. This fresh solvent reduces the viscosity of the residual resin at the new interface, and so on. This process continues as long as relative motion exists between the solvent and the part, and a film of liquid resin remains on the part surfaces.

At some point, depending upon the freshness of the TPM, the extent and duration of relative motion, and the geometric complexity of the part, the resin will have effectively been cleaned from all of the surfaces. For simple flat plate parts (Slab 6*6), in fresh TPM, with continuous relative motion between the part and the solvent, thorough cleaning takes less than five minutes.

However, for complex parts with numerous holes having depth to diameter ratios greater than 4:1, the cleaning period will increase. When using TPM that is already at the 12% resin concentration level, while still employing good relative motion, the required cleaning time may extend to about 15 minutes for a simple flat surface and 40 to 50 minutes for parts with complex geometries.

Without relative motion (simply immersing the part in TPM), complex parts may take many hours to be cleaned. For parts with holes having depth to diameter ratios greater than 4:1, the liquid resin at the base of these holes may never be truly cleaned using immersion only. Hence, it should be clear that relative motion between the platform/part and the TPM is essential.

Secondly, ultrasonic equipment is not very effective. Certainly, it is far less effective at resin stripping than simple relative motion. Parts that were fully cleaned within 10 minutes using recommended levels of relative motion required about 40 minutes to reach the same state when employing ultrasonics with no relative motion. Evidently, the density difference between the resin and the TPM is too small (compared to metallic objects in an aqueous solution) for the ultrasonic energy to provide good cleaning action.

For users who have already purchased an ultrasonics system, do not despair. A practical expedient involves installing a simple sparging apparatus in the form of a pump and some submerged tubing with numerous small holes. One inch (2.54 cm) outside diameter tubing with a single 0.125″ (3.18 mm) diameter hole every inch or so will work very well.

In this implementation, the ultrasonic transducers can be turned off, as they are of marginal value. The TPM is recirculated by pumping it through the tubes, creating numerous small jets. The overall effect is very much like a whirlpool bath. Here, relative motion between the solvent and the resin film is the result of the part remaining at rest while the solvent moves past. Nonetheless, if you have not yet purchased any resin stripping equipment, a more effective and lower cost option is described in the following paragraphs.

Third, considering the results of the first two statements, 3D Systems investigated what might be termed a "semiautomatic" approach to a Resin Stripping Apparatus, or RSA. The "automatic" in this context is intended to imply a cleaning operation which does not require any manual intervention. The "semi" implies that some modest amount of manual action may be needed (putting the platform/part into the unit, turning the RSA on, turning the RSA off when the cleaning operation is finished, and subsequently removing the platform/part from the RSA).

The most effective resin stripping method discovered to date involves pneumatically generated relative motion of the part and platform in a solvent reservoir. An excellent example is a unit manufactured by Ramco Equipment Corporation.[2] This is a commercial product originally designed for degreasing automotive components. It is very rugged and has a long useful life. Some units have been operational for as long as 10 years.

Also, this unit, in an RSA configuration, is reasonably priced (about $6000), and is constructed of stainless steel, so corrosion is not a problem. Furthermore, the Ramco MK 36 model can accommodate both SLA-250 and SLA-500 platforms. This model holds 90 gallons of TPM, costing about $700. A single tank of TPM will remain effective until about 11 gallons of resin are dissolved therein.

For those sites involving a single SLA-250, this quantity of TPM should be effective for about two years, provided reasonable precleaning steps are taken as discussed earlier. At 3D Systems, there are 2 SLA-1 units, 5 SLA-190 units, 12 SLA-250 units, and 5 SLA-500 units. This is probably the equivalent of at least 25 SLA-250 systems. Further, these units collectively tend to be building parts with about a 60% average system utilization.

Thus, the net result, in terms of resin loading rate, is roughly equal to about 15 SLA-250 systems operating 24 hours per day, seven days a week. A single 90-gallon tank of fresh TPM in the 3D Systems RSA typically lasts about seven weeks. The reader can approximate the effective lifetime of a 90-gallon quantity of TPM for their site by dividing 100 weeks by the number of full-time equivalent SLA-250 systems. Within this spirit of approximation, an SLA-500 should be treated, from a part cleaning standpoint, to be roughly equivalent to two SLA-250 systems. *Figure 9-3* shows a cleaning unit.

Once the part has been thoroughly cleaned of excess resin, it can then be removed from the TPM. Since the immersion period will have been for less than one hour in almost all cases, negligible swelling should occur. After the platform and part (which ideally should still be attached) have been removed from the TPM, they should be allowed to drip back into the RSA for at least one minute. This step avoids spilling TPM on the floor and minimizes the amount of solvent that is ultimately transferred into the rinse station.

Since TPM is extremely nonvolatile at room temperature, spills *must* be immediately wiped up with a wet cloth. This is very effective since TPM is especially soluble in water. If not cleaned, the TPM will remain on the floor and will present a safety hazard since it is quite slippery.

Next, both the platform and the part need to be rinsed with ordinary tap water, as the TPM leaves a shiny film on the part. If not removed, this TPM film could impede surface postcure. As previously noted, TPM is very soluble in water and will dissolve rapidly. Obviously, the rinse water will eventually accumulate a quantity of TPM as well as a smaller amount of resin.

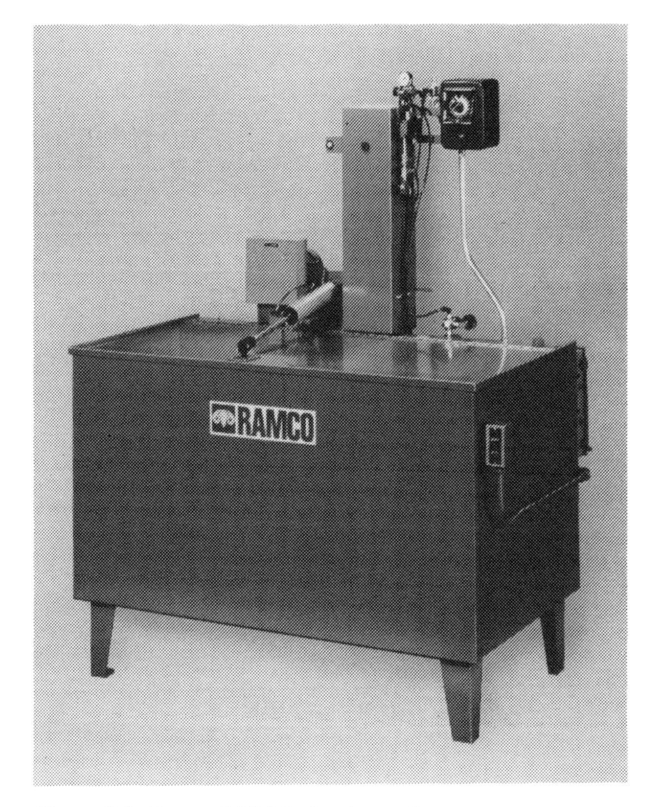

Figure 9-3. Ramco MK 36 cleaning unit.

Fortunately, TPM is not considered to be especially hazardous. A Material Safety Data Sheet, or MSDS, for TPM is shown in *Figure 9-4*. Thus, disposal of small concentrations of TPM in the rinse water is not, in itself, a problem. Unfortunately, the 12% to 15% resin loading in the TPM, despite considerable dilution, still results in a finite concentration of resin in the rinse water. This resinous water can present a potential disposal problem. The approach developed by 3D Systems for rinsing TPM from both the part and the platform is shown in *Figure 9-5*.

Tap water is sprayed onto the part and platform with a flexible hose connected to an adjustable "pistol grip" nozzle. This simple arrangement is easily operated with one gloved hand, while the platform and the attached part are held with the other gloved hand. The spray is contained by a large (roughly 30″ by 24″ by 24″) Nalgene tank.

The rinse water, along with the diluted TPM and resin, then flows by gravity into a Nalgene settling tank. The settling tank can be somewhat smaller than the rinse tank. Since resin is not soluble in water, and is also more dense than water, it will tend to slowly settle to the bottom of the tank. When the liquid in this tank

MATERIAL SAFETY DATA SHEET

ARCOSOLV® TPM

MSDS No. HCR0014 1S
Rev. Date 11/02/90

ARCO CHEMICAL COMPANY
3801 WEST CHESTER PIKE
NEWTOWN SQUARE, PA. 19073

IMPORTANT: Read this MSDS befor handling and disposing of this product and pass this information on to employees, customers, and users of this product
This product is covered by the OSHA Hazard Communication Rule and this document has been prepared in accord with the MSDS requirements of the n

I. General — SEE SUPPLEMENT BEGINNING ON PAGE 6

Trade Name	ARCOSOLV TPM	Telephone Numbers	EMERGENCY 800/424-9300 CHEMTREC; 215/353-8300 ARCO CHEM; CUSTOMER SERVICE 800/321-7000 INFO ONLY
Other Names	PROPANOL, (2(2-METHOXYMETHYLETHOXY)METHYLETHOXY)-; TRIPROPYLENE GLYCOL METHYL ETHER; TPGME		
Chemical Family	ALIPHATIC GLYCOL ETHER	DOT Hazardous Materials Proper Shipping Name	NOT REGULATED
Generic Name	TRIPROPYLENE GLYCOL METHYL ETHER	DOT Hazard Class	NOT REGULATED
CAS No.	SEE SECTION IX	Company ID No.	E000141900
UN/NA ID No.	N/AP		

II. CAUTION Summary of Hazards

PHYSICAL HAZARDS: SLIGHTLY COMBUSTIBLE LIQUID

ACUTE HEALTH EFFECTS: (SHORT-TERM)
SLIGHT EYE IRRITANT
SLIGHT SKIN ABSORPTION HAZARD
SLIGHT SKIN IRRITANT
SLIGHT INGESTION HAZARD
NO INHALATION HAZARD IDENTIFIED FROM DATA FOUND

CHRONIC HEALTH EFFECTS: (LONG-TERM)
REPEATED, PROLONGED, EXCESSIVE EXPOSURE MAY RESULT IN KIDNEY DAMAGE.

SEE SUPPLEMENT, PAGE 6, FOR NPCA HMIS RATINGS.

III. Fire and Explosion

Flash Point (Method)	Autoignition Temperature (Method)	Flammable Limits (% Vol. in Air)
GT 200° F (PMCC)	N/AP	Lower N/AP Upper N/AP

Fire and Explosion Hazards: HEAT FROM FIRE CAN GENERATE FLAMMABLE VAPOR. WHEN MIXED WITH AIR AND EXPOSED TO IGNITION SOURCE, VAPORS CAN BURN IN OPEN OR EXPLODE IF CONFINED. VAPORS MAY BE HEAVIER THAN AIR. MAY TRAVEL LONG DISTANCES ALONG GROUND BEFORE IGNITING/FLASHING BACK TO VAPOR SOURCE. FINE SPRAYS/MISTS MAY BE COMBUSTIBLE AT TEMPERATURES BELOW NORMAL FLASH POINT.

Extinguishing Media: DRY CHEMICAL; CO2; WATERSPRAY; FOAM FOR ALCOHOLS; WATER FOG

Special Firefighting Procedures: DO NOT ENTER FIRE AREA WITHOUT PROPER PROTECTION. SEE "DECOMPOSITION PRODUCTS POSSIBLE." FIGHT FIRE FROM SAFE DISTANCE/PROTECTED LOCATION. HEAT MAY BUILD PRESSURE/RUPTURE CLOSED CONTAINERS, SPREADING FIRE, INCREASING RISK OF BURNS/INJURIES. USE WATER SPRAY/FOG FOR COOLING. AVOID FROTHING/STEAM EXPLOSION. BURNING LIQUID MAY FLOAT ON WATER. ALTHOUGH SOLUBLE, MAY NOT BE PRACTICAL TO EXTINGUISH FIRE BY WATER DILUTION. NOTIFY AUTHORITIES IF LIQUID ENTERS SEWER/PUBLIC WATERS.

SEQ: 43 ***FOR "DISCLAIMER OF LIABILITY," SEE THE STATEMENT ON PAGE 4***

Figure 9-4. Material safety data sheet for TPM solvent.

reaches a preset level, a dual position automatic float switch is activated, which delivers electrical power to the pump.

The liquid in this tank (except near the bottom), is mostly water due to gravimetric settling, as well as the initially low resin and TPM concentrations. When the pump is activated, this liquid is pumped through a large capacity swimming pool type filter. This filter traps the great majority of the residual resin.

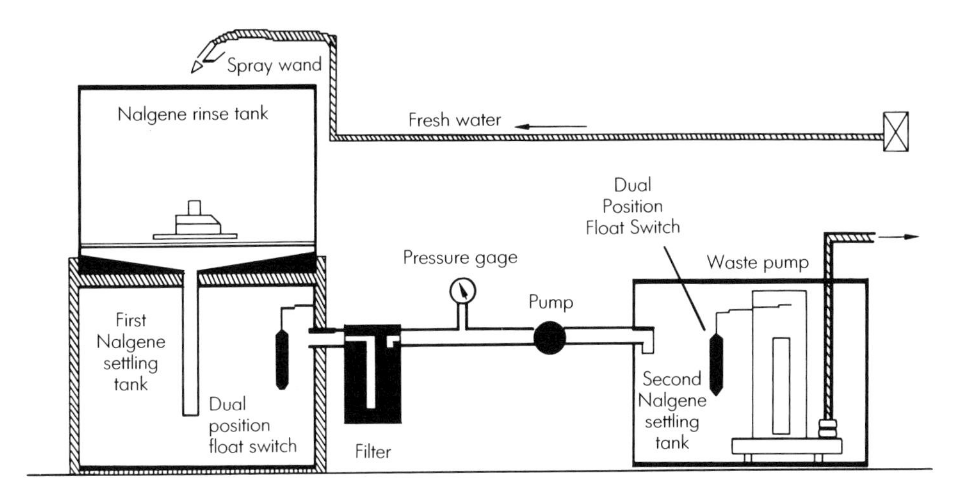

Figure 9-5. Design for a prototype rinsing station.

When the liquid level in the settling tank has been lowered to within approximately 4″ (100 mm) of the bottom, the dual position float switch automatically deactivates the pump. In this way, one is never pumping the resinous sludge near the bottom of the settling tank, and the pump only needs to operate occasionally, when the liquid level nears the top. The system should be checked often to assure proper function of the float switch and pump. Failure of either component can lead to an overflow condition and consequent floor mopping.

Eventually, the filter will become clogged with resin, and the pressure drop across the filter will increase above a preset value. When the pressure drop increases significantly, the filter element should be replaced with a new one. The old filter element may be cured in sunlight and disposed of in the trash.

Next, the liquid proceeds to a second Nalgene settling tank, identical to the first. In this way, any resin that has managed to get through the filter can again be allowed to settle. A second pump, also activated by another dual position float switch, is then connected to the sewer lines.

About every six months, depending upon usage, the settling tanks will require cleaning. The resinous sludge should be scraped into disposable aluminum foil pans and cured in sunlight. Judicious timing relative to good sunlight periods is well advised. Otherwise, this material can also be treated as hazardous waste. Since hazardous waste disposal is not inexpensive, the former approach is recommended.

More immediate, however, is the matter of rinse water disposal. Samples of the final rinse water effluent should be submitted to the local water authorities. With this precaution, authorities may determine that the resulting rinse water can be directly discharged into the sewer lines. The final determination at any given site will depend upon the effectiveness of the settling and filtration steps, the concentration of the resin in the TPM solvent, the care exercised to minimize the amount of TPM and resin transported from the cleaning station into the rinse station, and the local regulations and enforcement practices.

At the very worst, the rinse water would have to be treated as a hazardous waste. Since TPM is very soluble in water, great quantities of rinse water are not necessary and hence, even in the worst case, the disposal costs should not be onerous. On the other hand, if your site is granted a permit for waste water disposal into the existing sewer lines, the overall system becomes even simpler and more economical.

It is important to point out that this disposal information is not cause for alarm. The details were presented to help the user, or potential user, understand all the necessary requirements for a complete installation. We quite intentionally did not want to "gloss over" some of the slightly messy or inconvenient aspects of RP&M. The situation is analogous to describing the steps in changing the oil in an automobile. The job can be messy, the spent oil is considered a hazardous waste, and stuffing it under yesterday's newspaper in a trash can cannot be justified. However, while an automobile's sales brochure may never mention changing oil, failure to do so will produce poor results.

At this point, the water rinsed platform and part are removed from the rinse station and air dried. We have found that low pressure (20 to 30 psi) compressed air is very helpful for three reasons.

First, if the water droplets are simply allowed to air dry, water spots may appear on the surfaces of the part. These spots will be more likely in those geographic areas where tap water contains numerous minerals. While not critical to part accuracy, water spots may degrade aesthetic appeal.

Second, air drying can be rather slow, especially for parts containing intricate passages, blind holes, recessed corners, etc.. Compressed air sweeps the water droplets away, and greatly speeds this step. We find that even the most geometrically complex parts can be dried in about one minute using compressed air. Care should be used when directing the air stream onto those parts with especially delicate geometries.

Third, this method also provides a simple but very effective means of establishing if the platform and part are truly clean. If all surfaces are free of liquid resin, the air jet directed onto the part or platform will simply sweep water droplets along in the direction of the air flow. This will then leave the surface dry and free of any "rippled smudges." However, if any residual resin film is still present, the airjet, after moving the water droplets, will reveal shiny, sometimes rippled-looking smudges or streaks, which are simply the residual liquid resin being physically dispersed by the air stream. In this case, the platform and the attached part should be returned to the RSA, the rinse station, and again air dried until such rippled smudges are no longer evident.

While the term ''rippled smudges'' may not be perfectly clear to one who has not yet performed these steps, it does indeed become quite obvious when actually observed. In most cases a second cleaning will be sufficient. In especially stubborn cases, a Q-tip dipped in TPM and applied to the problem spot may be helpful. Even here, remember to subsequently rinse the part and dry it.

Table 9-1 provides the user with a list of commercial sources where TPM may be purchased. In a few cases, only very large quantities may be available. Most of the suppliers listed will be able to provide 55-gallon drums of TPM. Smaller quantities, such as five gallon containers of TPM, are only available from secondary suppliers. Unit costs in the smaller quantities are somewhat higher. As noted earlier, TPM does not present an explosion hazard at room temperatures. Thus, the extra 20 gallons are easily kept for later use without the need for storage in explosion-proof cabinets, as is the case with alcohol.

Table 9-1
Commercial Sources of TPM Solvent

Source	Solvent Name
ARCO Chemical Corp.	Arcosolv TPM
Dow Chemical Corp.	Dowanol TPM
Olin Chemical Corp.	TPM
Union Carbide Corp.	TPM
Van Waters & Rogers	TPM
Orange County Chemical*	TPM

* *Available in five gallon containers.*

9.4 Postcuring

Until the development of WEAVE™ in late 1990, and STAR-WEAVE™ in 1991 (as discussed in Chapter 8), postcuring was the largest source of error in StereoLithography. Detailed diagnostic measurements of postcure distortion are presented in Chapter 10. These measurements point to one aspect of the considerable progress achieved in SL part accuracy during the period from 1989 through 1991.

During the first few years of SL (1988 through 1989), postcuring was done with high-pressure, high-irradiance, UV emitting, mercury arc lamps. These lamps certainly seemed like a logical choice, since the spectral output from the mercury arcs definitely was effective at polymerizing SL resins. In fact, it was too effective! This seemingly curious statement follows from exhaustive studies of postcure distortion.[3,4,5]

It must be remembered that during this time period, most parts were being built with 60/120/X, 50 mil Tri-Hatch. As a result, about 50% of the volume of a part intended to be postcured was still in the liquid phase. The high actinic UV irradiance levels of the original mercury arc lamp PCA presented two problems.

First, a large proportion of the UV radiation from the mercury lamps was very near the wavelength band of maximum photoinitiator absorption. The result was very short postcure penetration depths, in the range of 0.005″ to 0.007″ (0.13 mm to 0.18 mm). Consequently, most of the UV radiation was absorbed in a very thin layer near the outer surfaces of the part.

Second, the mercury lamps emitted very high levels of actinic spectral irradiance. Typical values were about 20 milliwatts per square centimeter per 10 nanometer wavelength band. The reader may not think that these values are very large. However, as a point of comparison, one could achieve PCA exposure levels comparable to those produced by the SLA laser, within one second! Within just a few minutes of postcure in this PCA, the part would attain extremely large actinic exposure dosage.

SL resins are generally poor thermal conductors. Also, very large actinic exposures were being absorbed within quite thin resin layers near the surface of the part. Furthermore, photopolymerization is an exothermic reaction. The net effect was that the resin tends to polymerize rapidly, while generating considerable heat in a small volume of a material with a low value of thermal conductivity. Thus, the resin would tend to undergo rather significant heating. The resulting elevated temperatures would also further accelerate the reaction rate, which would then generate even more heat. Finally, due to the use of Tri-Hatch, considerable unreacted resin was present.

The net result was a very substantial temperature increase of the part, especially in thick regions where the surface to volume ratio was small and heat transfer to the outside was less effective. A simple adiabatic thermal analysis indicated that maximum resin temperatures as high as about 200°C, or almost 400°F were possible. These extreme temperatures could cause significant internal stresses.

For this reason, we believed it was appropriate to perform some postcure thermal experiments. *Figure 9-6* shows the test results obtained when using a mercury lamp PCA to cure resin XB 5081-1. The data are presented in the form of a plot of temperature versus time. The temperature measurements were made using a Chromel-Alumel thermocouple junction. The thermocouple was fixed near the center of a small, actinic UV transparent, Pyrex beaker. This beaker was then filled with about 200 mL of resin.

The beaker, resin, and thermocouple wires were then positioned at the center of the PCA. At a convenient moment, the mercury lamps in the PCA would then be turned on. Resin temperature measurements were taken approximately once every minute. To test the repeatability of the experimental results, the entire test sequence was performed three times. The results were repeatable within 3°C. The system was calibrated, prior to testing, and again after testing, at 0°C (32°F) and 100°C (212°F), using ice water and boiling water respectively. All calibration errors were within ±1°C.

Figure 9-6 shows the following important results:

1. The maximum resin temperatures reached values as high as 189°C (372°F) for the resin sample postcured in air. This experimentally measured peak temperature is close to the adiabatic estimate.
2. The warm-up period for the mercury lamps was about four minutes. The lamps were turned on for a total of nine minutes, corresponding to about five minutes of full irradiance. However, the resin did not reach its maximum temperature until about 30 minutes after the lamps were turned on, or some 21 minutes after the lamps had been turned off.
3. The maximum rate of change of temperature occurs in the time interval from about 20 to 25 minutes after the lamps were turned on, or about 11 to 16 minutes after the lamps had been turned off. Since there were no other external sources of heat once the lamps were extinguished, this strongly suggests that exothermic polymerization reactions continue well after the actinic UV radiation has been terminated.
4. Even when the Pyrex beaker containing the resin was almost fully immersed in many kilograms of water in a much larger transparent Pyrex beaker, the maximum resin temperature was still almost 148°C (300°F). Here, the maximum temperature occurred almost 45 minutes after the lamps were turned on, or well over a half hour after the actinic UV radiation had been terminated. Thus, even postcuring in water did not materially change the results.

What is not evident from *Figure 9-6* is that the resin showed numerous tiny stress cracks during the initial cooling phase. The cracks became much worse later in the cooling period, and the cured volume actually fractured in the air cooled case, after about two hours. Admittedly, postcuring beakers of liquid resin is not the same as postcuring actual parts. Nonetheless, it is clear that even for actual SL parts, curing with the original mercury lamp PCA units led to the generation of very high resin temperatures, considerable thermal stresses, and corresponding thermal strains. These strains manifested themselves as postcure distortion and, in extreme cases, fracture of the part.

Thus, the task was to develop a postcuring method that would accomplish all of the following:

- Find a source of radiation which, while still emitting actinic UV photons, would have a spectral distribution resulting in greater postcure penetration depths. In this way, one could achieve much more uniform overall postcure.
- Find a source of radiation with significantly lower spectral irradiance to promote slower rates of polymerization, correspondingly reduced part

temperatures, and hence reduced thermal stresses and strains.

- Develop a new PCA that would significantly reduce postcure distortion, resulting in improved part accuracy and ideally providing this enhanced capability in a cost-effective manner.

While these goals seemed ambitious, all of them were met, with varying degrees of success. After much analysis, test, and evaluation, a series of actinic fluorescent lamps manufactured by Phillips Lighting Co. was selected. *Figure 9-7* is a plot of the spectral power distribution of the specific lamp TLK 40W/05.

Note that the "K" is a designation for a 24-inch long tube, and the 40W indicates that the input electrical power for this lamp is 40 watts. The "05" is the critical designation in terms of the spectral output. *Figure 9-8* is a similar plot for the specific lamp TLK 40W/03.

These lamps have proven to be extraordinarily effective at postcuring resins optimized for use in the SLA-250, involving a helium-cadmium laser emitting at

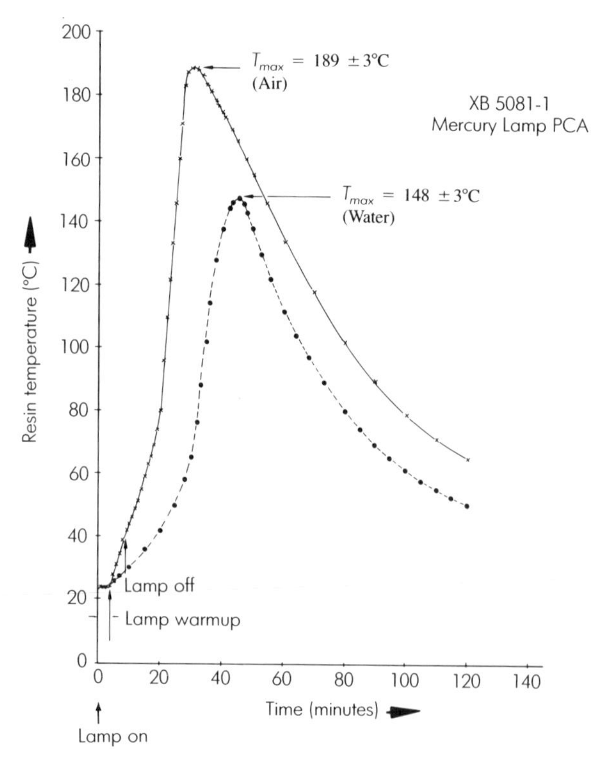

Figure 9-6. Temperature versus time for XB 5081-1 in air and water using a mercury lamp PCA.

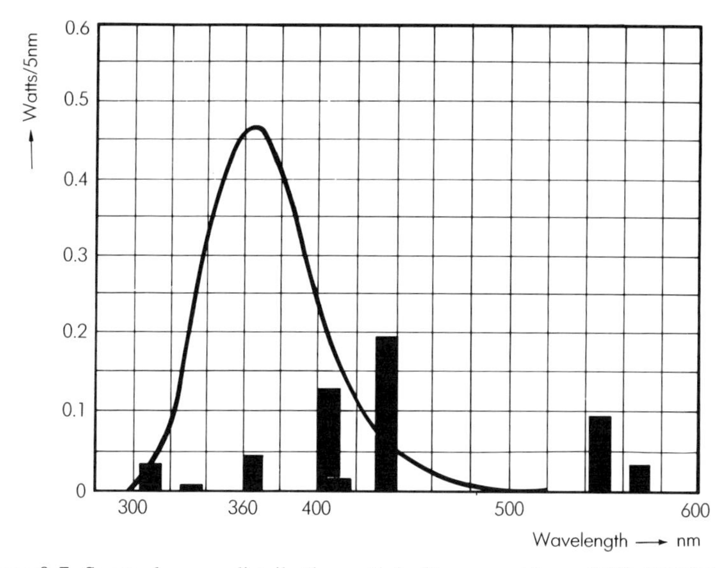

Figure 9-7. Spectral power distribution actinic fluorescent lamp TLK 40W/05.

325 nanometers, and the SLA-500, employing an argon ion laser emitting at 351 and 364 nanometers, respectively. The reasons for their effectiveness will become apparent shortly.

Next, a series of experiments were performed to determine the penetration depth, as a function of wavelength, for a number of SL resins. This was accomplished by measuring cure depth as a function of exposure, while using a continuum UV source with a known spectral irradiance distribution, in conjunction with a series of narrow bandpass filters and a calibrated radiometer. The details of the experimental technique are presented in Reference 4.

Essentially, a resin sample was exposed to UV radiation after transmission through one of 12 possible narrow bandpass filters. These filters, from Andover Corp., had their nominal center wavelengths (the wavelengths at their respective transmission peaks) at 10 nm intervals from 320 nm to 430 nm. From the transmission curves supplied for these filters, it was clear that all the actual peaks were within 2 nm of the nominal center wavelengths.

Also, these filters had a nominal Full Width @ Half Maximum (the difference in wavelength between the points where the transmission curve passes through 50% on the ascending and descending sides) wavelength bandwidth of 10 nanometers, or ±5 nm about the center wavelength. In short, the errors in filter wavelength positions were not at all significant for our problem.

The transmitted irradiance, passing through the filter, was then measured using the calibrated radiometer. Since the resin exposure is simply the incident irradiance multiplied by the exposure interval, data could be gathered for a range of exposures at each wavelength, simply by varying this interval. *Figure 9-9*

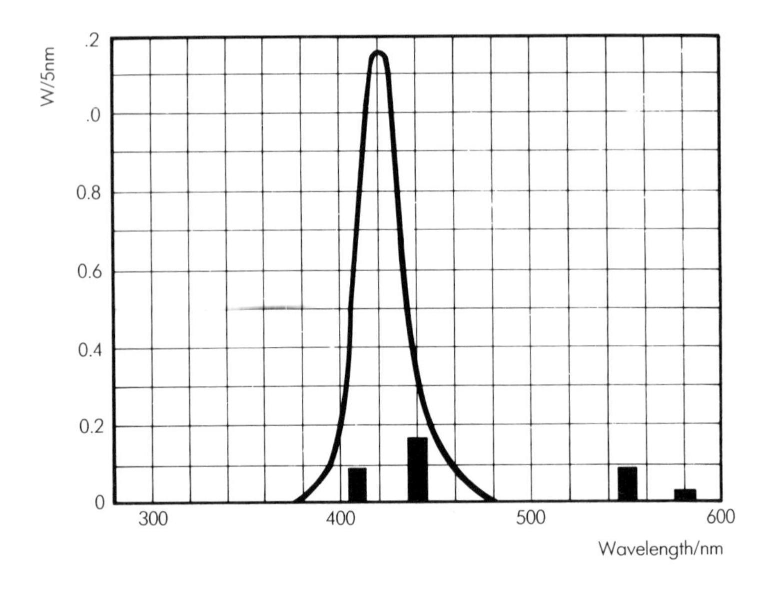

Figure 9-8. Spectral power distribution actinic fluorescent lamp TLK 40W/03.

shows a series of ''postcure working curves,'' where cure depth is plotted against the logarithm of the exposure. The data of *Figure 9-9* is for SL resin XB 5081-1. *Figure 9-10* shows similar results for SL resin XB 5139.

The values of the penetration depth, D_p, as a function of the nominal centerline cure wavelength, λ, are presented in *Figure 9-11*, which plots the resulting values for both resins. These postcure spectral penetration depths are still determined from the slope of the individual working curve.

Note that for these two SLA-250 resins, the values of D_p at the HeCd laser wavelength (325 nm) are in good agreement with the usual laser-generated values. However, with increasing wavelength, the values of $D_p(\lambda)$, for both of these resins, increase rapidly as one approaches a wavelength near 390 nm. No photopolymerization occurred beyond this wavelength for either resin. This ''limiting wavelength'' appears to involve some type of photon energy threshold for the particular photoinitiator used in these resins.

The experimental test results presented in this figure clearly indicate that to achieve the goal of longer postcure penetration depths, it is very important to use radiation having a wavelength greater than the appropriate laser wavelength, but below the limiting wavelength. Above the limiting wavelength, the photons are no longer actinic. Below the laser wavelength, the penetration depths are too short. Optimum postcure performance requires intermediate wavelengths.

Figure 9-12 is a plot of the critical energy, E_c, as a function of wavelength, for both resins. The values of E_c were determined from the exposure intercepts of the data presented in *Figures 9-9* and *9-10*, for the same two resins. *Figure 9-12* also shows an increase in $E_c(\lambda)$ from the laser wavelength toward the

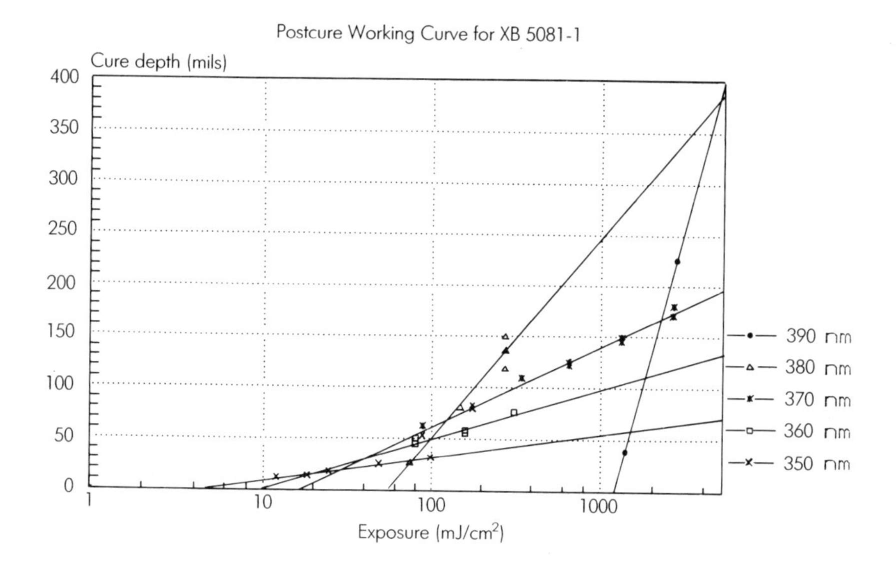

Figure 9-9. Postcure working curves at various wavelengths.

limiting wavelength, except that greater divergence of the data for these two resins is noted beyond 380 nm.

We shall now define the so-called "volumetric efficacy" of a resin by the quantity $\gamma_v(\lambda)$, where

$$\gamma_v(\lambda) \equiv D_p/E_c \tag{9-1}$$

Since D_p can be expressed in units of centimeters, and E_c in Joules per square centimeter, then the units of $\gamma_v(\lambda)$, from equation 9-1, are cubic centimeters per Joule, or cured volume per unit of absorbed actinic energy. Clearly, "volumetric efficacy" is an excellent measure of the effectiveness of the postcure process (*Figure 9-13*).

Three important results are evident from inspection of this figure.

1. Both resins exhibit peaks in $\gamma_v(\lambda)$. Thus, for each resin there is an optimum postcure wavelength which will result in the maximum volume of postcured resin for a given amount of absorbed actinic radiant energy. For XB 5081-1, the optimum postcure wavelength is near 360 nm, while for XB 5139 it is near 375nm.
2. The peaks of these $\gamma_v(\lambda)$ functions are sufficiently broad that actinic postcure radiation in the wavelength range from 360nm to 370nm will provide volumetric efficacy values in excess of 5 cm^3 per Joule for both resins.
3. In this wavelength band, the average postcure penetration depth will be almost an order of magnitude greater than the values at the HeCd laser wavelength of 325 nm.

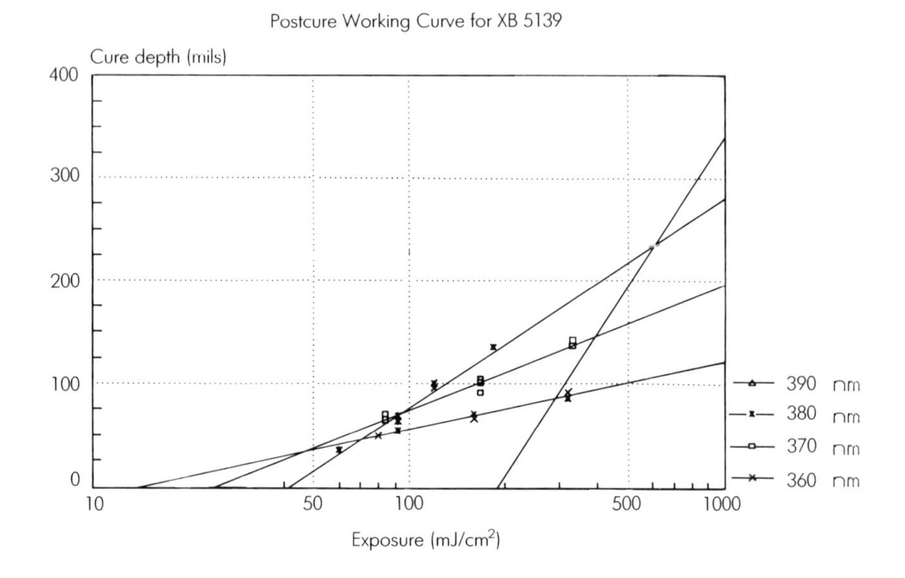

Figure 9-10. Postcure working curves at various wavelengths.

Returning to the data of *Figure 9-7*, it is now clear why the actinic fluorescent lamp, model TLK 40W/05, is so effective at postcuring resins used in the SLA-250. The spectral power distribution of this lamp peaks exactly in the optimum wavelength region. Furthermore, the absolute values of the spectral irradiance within the new PCA-250 are almost two orders of magnitude lower than those of the earlier mercury lamp PCA (about 250 microwatts/cm^2 versus about 20 milliwatts/cm^2). Therefore, returning to our earlier three goals for improved postcure:

- The TLK 40W/05 lamps certainly emit actinic radiation, they provide much greater postcure penetration depths than the high pressure mercury lamps, and, hence, they achieve much more uniform postcure.
- These lamps do indeed provide significantly lower spectral irradiance values than the earlier mercury lamps, as noted. As a result, the rate of polymerization, and the corresponding rate of exothermic energy release, are both much lower. Measurements of peak temperatures incurred when using the new PCA-250 to postcure the same two resins showed maximum values in the 40°C to 45°C (104°F to 113°F) range. These experiments were performed in the identical manner as described for the mercury lamp tests. The peak temperature values occurring in the new PCA-250 are dramatically reduced from the 189°C (372°F) values recorded with the mercury lamps, involving resins postcured in air. They are even well below the peak temperatures recorded for resins cured in water with the older PCA.

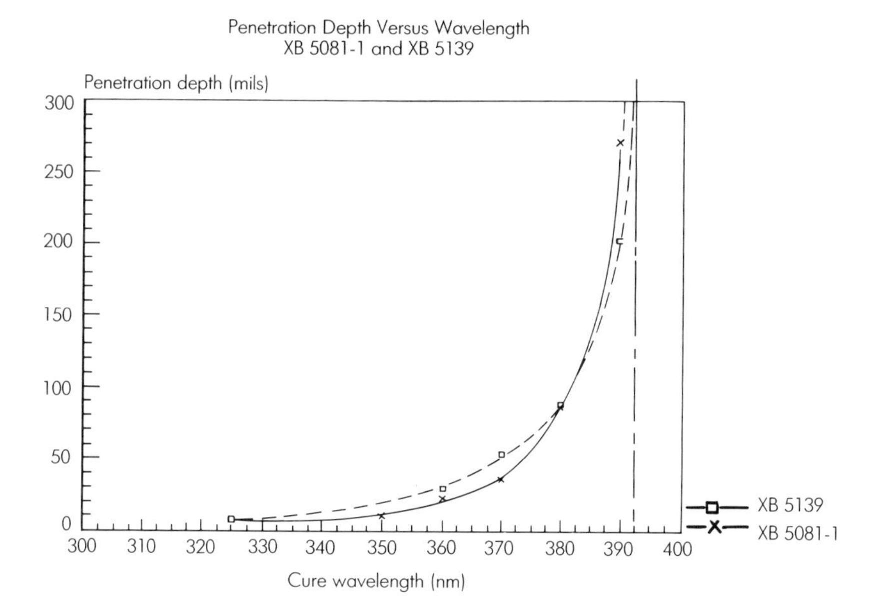

Figure 9-11. Penetration depth as a function of wavelength.

- With the greatly reduced temperatures measured when using the actinic fluorescent lamps, it follows that thermal stresses would also be correspondingly reduced. The result should be decreased postcure distortion, and improved part accuracy, relative to the earlier mercury lamp PCA.

It should be noted that the results for the SLA-500 resin, XB 5131, were quite similar to those shown in *Figures 9-9* through *9-13*. The major exception was that as a result of utilizing different photoinitiators, the curves for this resin are shifted about 50 nm towards longer wavelengths. In resin XB 5131, the volumetric efficacy reaches a maximum very near 420 nm. Thus, it is clear from *Figure 9-8* that the type TLK 40W/03 actinic fluorescent lamp is virtually ideal for this resin.

As a consequence of these results, the new PCA-250 is based upon the use of type TLK 40W/05 lamps, while the larger PCA-500 uses a mix of both type TLK 40W/03 and TLK 40W/05 lamps. This has been done to assure that the PCA-500 will be effective for parts built on either system. The converse is not necessary since a user having only an SLA-250 need not be concerned with curing SLA-500 resins. Finally, there is another reason for mixing both lamp types in the PCA-500, even if a user had only an SLA-500. The reason involves surface curing effects.

It is worth noting that the combination of the newer PCAs coupled with the advanced part building methods WEAVE™ and STAR-WEAVE™ have dramatically reduced postcure distortion. Where postcure distortion was formerly the

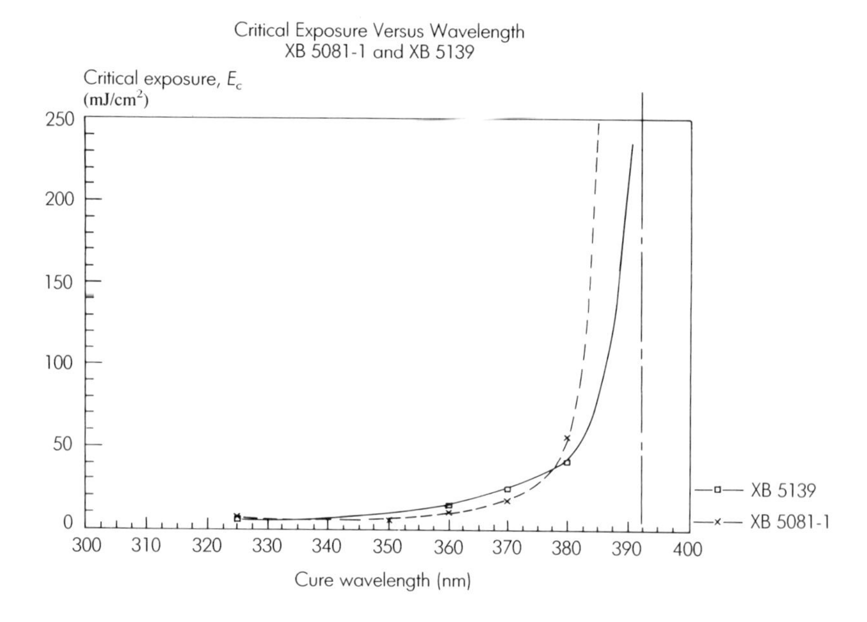

Figure 9-12. Critical exposure as a function of wavelength.

dominant source of error in SL, it has been reduced so substantially since mid-1990, as to now be among the minor sources of overall part error.

The information provided herein explains the rationale behind the design of the newer PCAs. Clearly, by reducing the incident actinic irradiance, we have made a conscious decision to tradeoff postcure interval against improved part accuracy. Of course, this begs the obvious question: "How long must a given part be postcured in the new PCAs?" An exact answer would depend upon the details of the part geometry. This is the case because the exponential absorption of radiation, in accord with the Beer-Lambert Law, will result in thicker sections taking longer to postcure than thinner sections.

Polymerization in general, and postcure in particular, as the point of this discussion, depend upon actinic exposure. However, actinic exposure is simply actinic irradiance times exposure duration. Therefore, in the middle of thicker regions, the actinic irradiance is reduced through absorption, so the exposure duration must be increased to achieve the same level of exposure. Hence the increase in postcuring time. Since an arbitrary part may have many sections of different thickness, it would seem at first that a general rule would be unlikely.

However, if we make one reasonable approximation, namely that the absorption of actinic photons in a given path length of air is completely negligible compared to their absorption in the same path length of resin, we may then generate a simple model for the maximum required postcure interval.[6]

Consider a part of arbitrary geometry, with many complex interior hollow regions. Since we have just agreed that it is a very reasonable approximation to neglect actinic photon absorption in air, then, from a postcuring standpoint, the air might just as well not be there. Therefore, one could imagine a gigantic press that exerts enormous forces upon the part, compressing it into a sphere. If the density of the resin is ρ (grams per cubic centimeter), and the mass of the part is M (grams), and the part volume is V (cubic centimeters), then these quantities are related by the simple expression:

$$M = \rho V \tag{9-2}$$

However, for a sphere of radius R,

$$V = \frac{4}{3} \pi R^3 \tag{9-3}$$

Substituting equation 9-3 into equation 9-2, after some simple algebra we obtain the result:

$$R = [3 M / 4 \pi \rho]^{1/3} \tag{9-4}$$

Experimental results obtained during a series of postcure evaluations[6] showed that the cure depth advances very much like a propagating front, moving at a very nearly constant postcure velocity, C_{pc}. Thus, the maximum possible postcure time, defined as $(t_{pc})_{max}$, is given by:

$$(t_{pc})_{max} \equiv R/C_{pc} = [3M / 4 \pi \rho]^{1/3}/C_{pc} \tag{9-5}$$

Equation 9-5 indicates that the maximum postcure interval should scale with

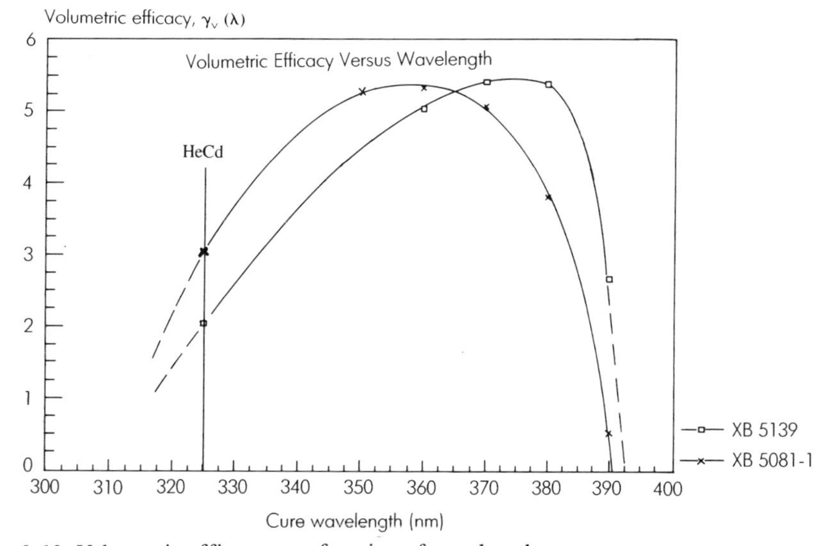

Figure 9-13. Volumetric efficacy as a function of wavelength.

the cube root of the part mass. Thus, a log-log plot of maximum postcure time versus part mass should result in a straight line with a slope equal to 1/3.

In other words, a part with a mass of one kilogram should take only 10 times, rather than 1000 times as long to postcure as a one gram part. In general, this is in agreement with the test data, subject to one additional condition. That condition involves surface phenomena, described in the following paragraph.

The previous analysis (equations 9-2 through 9-5) describes the maximum volumetric postcure interval. However, this would imply that a part of very small mass should be able to be postcured in a brief interval. For example, equation 9-5 would predict that the required postcure interval for a one gram part made from SL resin XB 5131, and postcured in a PCA-500, should require only 0.17 hours or about 10 minutes to complete its cure. Actual data shows that this is true for the interior of the part, but the exterior surfaces would still be quite tacky after only 10 minutes of postcure.

However, this simple model does not consider the effects of oxygen inhibition at the exterior surfaces of a part in contact with ambient air. Clearly, a unit surface area of infinitesimal thickness has effectively no mass at all. Nonetheless, this surface area will require a finite time to reach sufficient exposure dosage to overcome the effects of oxygen inhibition on the photoinitiator. Therefore, surface curing time must be independent of part mass. From numerous experimental results (see Reference 6), the time required to cure the exterior surfaces of a part depends primarily upon the incident actinic irradiance.

Figure 9-14 is a log-log plot of the maximum postcure time versus part mass, or a so-called "Postcure-Time Map," for resin XB 5131 in a PCA-500. The volumetric cure function is described by a straight line with a slope equal to one third as noted earlier. The surface cure function is simply described by a horizontal line, independent of part mass. The position of this horizontal line is a function of the incident actinic irradiance, as noted above.

For parts having a mass less than the "transition mass," at the intersection of the two straight lines on the Postcure-Time Map (about 80 grams for this case), the required postcure interval is determined solely by the surface cure time. Simply stated, since the part is small, internal postcure propagation will reach the center of the thickest section before the surface is fully cured. For the case of XB 5131 in a PCA-500, with an actinic irradiance level at the center equal to about 225 microwatts/cm^2, the required surface postcure interval, from *Figure 9-14*, is 0.7 hours, or about 42 minutes.

However, for parts having a mass greater than the transition mass, the maximum postcure interval is determined by the volumetric cure requirements, which increase with part mass. Here, the propagation of the cure front *may* take longer than the surface cure. The reason we say *may* is because one could have a part with a mass greater than the transition mass, which happens to consist of a number of large area thin walls. In such a case, it is possible that the interior postcure could still be completed before the surface cure.

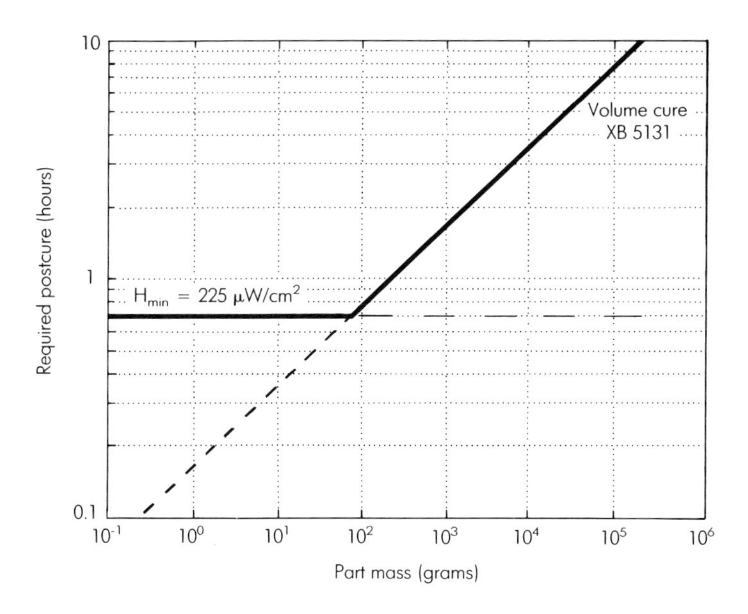

Figure 9-14. Postcure-Time Map.

Nonetheless, the Postcure-Time Map does represent an upper limit postcure interval, which is still valuable for the user. Since part accuracy does not degrade when postcure is performed for longer periods, in either the PCA-250 or the PCA-500, the only negative aspect of leaving a part in one of these PCAs "too long" is that some time will have been wasted. However, since postcure intervals are almost always shorter than the part building time, postcuring is rarely a bottleneck.

In conclusion, all that is necessary to estimate the required postcure interval for a given part is a scale to weigh the part and an appropriate Postcure-Time Map for the resin used. For example, a 1.0 kg (2.2 lbs.) part produced on an SLA-500 from resin XB 5131 and postcured in a PCA-500 will require a maximum interval of about 1.7 hours. The biggest "part" that can be built on an SLA-500 would be an enormous cube equal in size to the entire vat. The mass of this part would be about 300 kg, or roughly one third of a ton!

Clearly, this would not be a part that anyone would be likely to build, but it is instructive as an upper limit. *Figure 9-14* illustrates that the required postcure interval in this case would still be only about 10 hours.

9.5 Part Finishing

Having successfully designed, set up, built, removed, cleaned, rinsed, dried, and postcured the prototype, one is now ready for part finishing. As noted earlier, if the part is intended to serve solely as a concept model, simply removing the supports will probably suffice. With the original brittle polymers, this generally involved inserting a dull edged blade or putty knife between the part and the platform. This was appropriate since the supports would break off prior to being cut. On large sections this is not a problem. However, on fragile sections great care is necessary in order to avoid damaging the part.

Fortunately, with the newer tough resins, such as XB 5143, support removal is much less of a problem. These resins are significantly less brittle. As a result, they will nicely slice or cut rather than undergoing brittle fracture. In most cases, supports in resin XB 5143 can be easily removed using a sharp, long-bladed knife for large parts. For small parts, an X-acto knife is effective. In the case of extremely delicate part geometries, a fine, sharp scissor may also work well.

Further, these newer resins may be effectively and accurately machined. Resin XB 5143 has been successfully milled, turned, drilled, bored, and tapped. Threaded inserts, screws, dowel pins, etc., may be used with this resin. This is especially useful if one wishes to build a part that is too big for the SLA. It may now be built in two or more portions and be bolted together into its final form.

Once the supports have been removed, even a concept model will profit from some minor sanding to eliminate residual traces of the supports. For concept models, a few minutes of light sanding with either 100 or 150 grade sandpaper is appropriate. Since the polymerized resin will generate a fine white powder after sanding, rinse the part in tap water and air dry it with compressed air, as discussed previously. One need not be concerned about this powder being rinsed down the drain lines. Once postcured and essentially fully polymerized, these materials need not be treated as hazardous waste.

On the other hand, if the part is intended for additional RP&M applications, such as a pattern for soft tooling or a master for investment casting, additional finishing steps are warranted. While this is not the only effective method and the materials suggested are not the sole products that should be considered, the process description that follows has been successfully implemented on thousands of SL prototypes. Simply stated, it works very well.

First, the postcured prototype should be very lightly bead blasted with the finest grade of glass beads commercially available. During bead blasting, special attention should be paid to fine detail in the prototype geometry. Remaining too long in one spot can result in dimensional errors. The user should practice on a few sacrificial pieces made from the same resin prior to bead blasting a valuable prototype. In this manner, one can get a feel for the best way to work with that material, while discovering the consequences of errors on objects of no great value.

Second, after bead blasting, the prototype should be rinsed in tap water and air dried. This will wash away not only any powdered polymer but also residual glass beads that may be present in small holes, narrow passages, or sharp corners.

Third, the prototype should be lightly hand-sanded with 100 grade sandpaper to eliminate traces of the supports, burrs, or other tiny surface defects, including the effects of stair stepping. Upon completion of this sanding step, the part should again be rinsed in tap water and air dried.

At this point, one may start to hear inner voices whispering, "Do I really need to rinse and dry after every step?" Ignore them, because the answer is *yes*. The increase in touch labor is only minutes per step. You have saved weeks or months on your prototype to this point. *Do not* cut corners at this late stage and risk damaging the final result.

Fourth, the prototype should be very lightly and carefully hand-sanded in the following sequence: 150 grade sandpaper, rinsed in tap water and dried; 220 grade sandpaper, rinsed and dried; 320 grade sandpaper, rinsed and dried (hearing any inner voices yet?); 400 grade sandpaper, rinsed and dried; and then 600 grade sandpaper, rinsed and thoroughly dried. Basically, all these steps are used to remove the progressively finer scratches created by the previous step. While the process may sound very time-consuming, one to two hours is typical for most parts. Very large or intricate parts clearly will take longer.

Fifth, to attain a high-quality polished surface finish, use a solid "rouge-like" material on a buffing wheel. Again, after buffing, rinse in tap water and air dry. This will eliminate any residual buffing material.

Sixth, and finally, to achieve an almost mirror-like finish, a liquid polishing compound is recommended. The best results have been achieved using this material with either a very soft piece of flannel cloth or using bare fingers for just a few minutes. Essentially, this material is removing most of the microfine scratches produced by the previous buffing composition.

Once more, hopefully for the last time, the prototype should be rinsed in tap water, air dried, and then polished with a soft flannel cloth. If you have faithfully followed these steps, the prototype will "look like a million dollars." Depending upon its ultimate application, it might well be worth a fair fraction of that amount.

REFERENCES

1. Ertley, E., *Resin Stripping Development*, 3D Systems Report, April 5, 1991, pp. 1-9.
2. Patterson, R., Ramco Equipment Corporation, 32 Montgomery Street, Hillside, New Jersey.
3. Evans, B. and Larsen, T., *Optimum Post Cure Parameters for Stereo-Lithography Resin XB 5139*, 3D Systems Technical Report No. 90-74, June 29, 1990, pp. 1-18.
4. Evans, B. and Larsen, T., *Optimum Post Cure Parameters for Stereo-Lithography Resin XB 5134-1*, 3D Systems Technical Report No. 90-189, August 8, 1990, pp. 1-10.
5. Evans, B. and Larsen, T., *Optimum Post Cure Parameters for Stereo-*

Lithography Resin XB 5131, 3D Systems Technical Report No. 90-194, August 13, 1990, pp. 1-10.

6. Feldman, L. and Jacobs, P.F., *XB 5131 Post Cure Tests*, 3D Systems Technical Report No. 90-223, September 11, 1990, pp. 1-8.

chapter 10

Diagnostic Testing

Even when the external and scientific requirements for the birth of an idea have long been there, it generally needs an external stimulus to make it happen; man has, so to speak, to stumble right up against the thing before the idea comes.

—Albert Einstein
Essays in Science, 1934

10.1 Introduction

Many of the recent advances in SL technology owe their development to the initiation of an extensive diagnostic testing program at 3D Systems during 1989. When the diagnostic program first began, the tests were rather primitive and, as the quote above suggests, we were stumbling about in the dark. Once the foundations were laid for diagnostic testing, the following positive results were realized:

1. We began to establish a database for various resins using a given part building style.
2. Comparisons of resins began to be quantitative and, therefore, objective rather than subjective.
3. As we gathered data, certain trends became gradually evident. These trends pointed to improved part building methods.
4. An objective means became available to separate positive from negative results, and a criterion for determining if a new building method provided an improvement was at hand.

By ***Hop Nguyen****, Senior Research Scientist,* ***Jan Richter****, Senior Research Engineer, and* ***Paul F. Jacobs, Ph.D.****, Director of Research and Development, 3D Systems, Inc. Valencia, CA.*

5. More accurate measurement techniques were developed. This was critical to further advances in part quality. More accurate measurements uncovered even more subtle phenomena. This led to the development of additional diagnostic tests.
6. When applying these new diagnostic tests to the various resins, additional trends became apparent. This led to clearer definitions of the properties desired in SL resins.
7. These redefined resin requirements led to the development of newer resins with improved physical properties. However, with these new resins a different set of problems became evident.
8. The solutions to these problems required newer and more powerful software to enable still newer part building techniques. These new techniques then required additional diagnostic testing.
9. With each new part building technique, it became necessary to determine the optimum parameters for each resin. The criteria for comparison would then be the diagnostic test results.
10. It also became apparent that to build accurate parts, the resin parameters must be tightly controlled within narrow tolerance limits. This implied an extensive resin quality control (QC) program.
11. The diagnostic test parts, as well as the large and growing database, became the basis for much of the resin QC program.

10.2 R&D Diagnostic Tests

Introduction

The following diagnostic tests were developed by 3D Systems to characterize the resins and part building methods used in StereoLithography. Originally, we were interested in determining the extent of postcure distortion and the magnitude of the resin shrinkage. Today we are measuring postcure shrinkage, swelling, cantilever curl distortion, vertical wall postcure distortion, and horizontal slab distortion.

Each diagnostic test covers a matrix involving different resins and part building techniques. The building techniques started with the original 50 mil equilateral triangular hatch (Tri-Hatch), followed by 25 mil Tri-Hatch. When the WEAVE™ build method was introduced, we added an 11 mil hatch spacing WEAVE™ to the matrix, and when STAR-WEAVE™ was developed, an 11 mil hatch spacing STAR-WEAVE™ was also added. Resins that have been tested, approved and continue in the diagnostic test matrix are Ciba-Tool XB 5081-1, XB 5134-1, XB 5139, and XB 5143 for HeCd lasers, as well as XB 5131 for Argon lasers.

Postcure Shrinkage

Postcure shrinkage, S_{pc}, is defined as the "green" measurement, M_g, of a linear dimension minus the postcured measurement, M_{pc}, of the same dimension, divided by the "green" value. The shrinkage is given in percent (%).

$$S_{pc} = \frac{(M_g - M_{pc})}{M_g} * 100$$

Note, that S_{pc} is related to, but *not* identical to the shrinkage compensation factor that is applied in building a StereoLithographic part. That shrink factor will be covered in section 10.5.

3D Systems measures postcure shrinkage with a series of diagnostic test parts called "Nine-Box." This test part is shown in *Figure 10-1*. It is a set of nine (9) two-inch square (50 mm x 50 mm) boxes with wall thicknesses varying from 0.1" to 0.5" (2.5 mm to 12.5 mm). All nine boxes are generated during a single build, and at the same 0.010" (10 mil or 0.25 mm) layer thickness.

The test parts are built on an SLA. Then they are measured "green" with a Coordinate Measuring Machine (CMM). Next, they are postcured for an hour with appropriate actinic fluorescent lamps (10 lamps at 40 watts each), and then measured a second time with the CMM. Four measurements, two in x and two

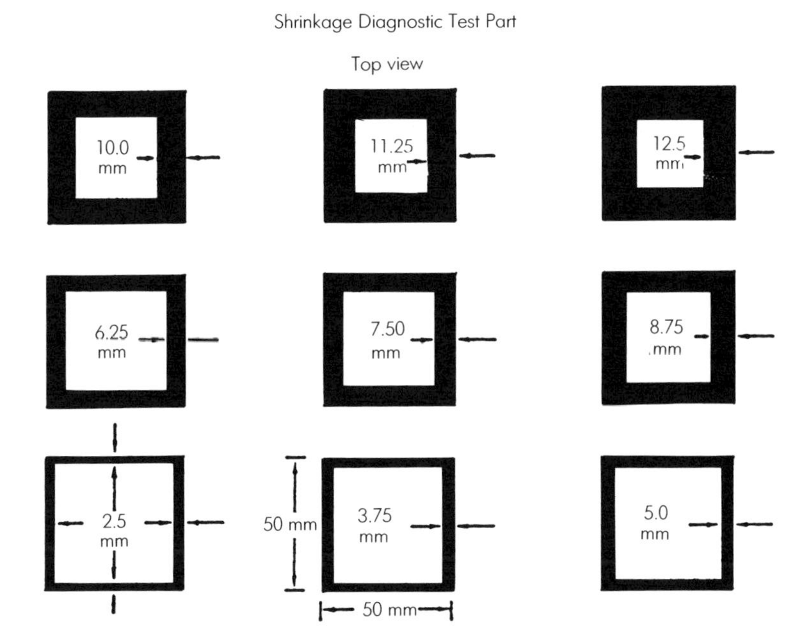

Figure 10-1. "Nine-Box."

in y, are taken for each two-inch dimension (36 points), and 12 readings are taken for *each* wall thickness (three measurements on each of the four walls of a given box). The mean value and standard deviation of S_{pc} are calculated from the "green" state to the postcured state for each of the dimensions mentioned previously. The results are then plotted with shrinkage on the ordinate as a function of linear dimension on the abscissa. The postcure shrinkage value for the two-inch linear dimension is noted.

The test measures both x and y shrinkage. In all of the test runs, the x and y shrinkage values were identical within the standard deviation of the measurements.

It should be noted that the "green" measurement values do *not* include linewidth or shrinkage compensation. A summary of the shrinkage results for the two-inch dimension is shown in Table 10-1.

Table 10-1
Postcure Shrinkage "Nine Box" (% at 2″)

Resin	50 Hatch	25 Hatch	11 WEAVE™	11 STAR™
XB 5081-1	1.20	0.81	0.72	0.56
XB 5134-1	1.34	0.89	0.69	N/A
XB 5139	1.28	0.83	0.73	N/A
XB 5143	1.15	0.63	0.24	0.19
XB 5131	1.19	0.59	0.73	0.28

Shrinkage curves as a function of linear dimension are shown in *Figures 10-2* and *10-3*. Note that, for brevity, in this and subsequent sections of this chapter, only two samples involving the worst case and best case results of the 18 curves available are shown.

The results show that postcure shrinkage is generally *not uniform*, but depends very much on the part building style. All the curves shows a characteristic "airfoil" shape, with the percentage shrinkage being low for thin walls, increasing to a maximum for medium wall thicknesses, and decreasing again for the larger dimensions. With perfectly uniform postcure shrinkage, these curves would be flat.

The dependence upon the part building style shows that the greater the fraction of liquid resin within the laser-cured part, the greater the absolute postcure shrinkage and the more nonuniform this shrinkage becomes. Furthermore, for all resins, WEAVE™ and STAR-WEAVE™ result in much more uniform shrinkage as a function of linear dimension. This is especially important to overall part accuracy.

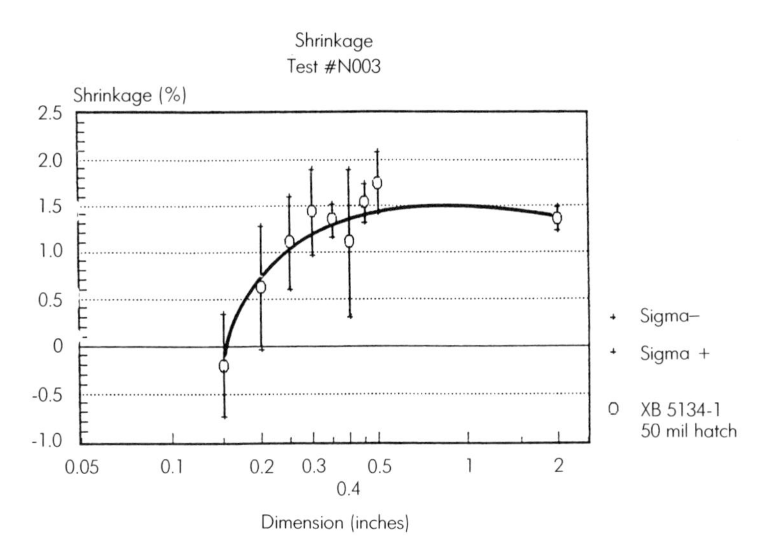

Figure 10-2. Shrinkage versus linear dimension.

Swelling

Swelling, *SW*, is defined as the normalized difference of the measurement of a dimension at 0 resin immersion time, M_0, and the measurement of the same dimension after 24 hours of resin immersion, M_{24}, in mils per inch. Thus:

$$SW = \frac{(M_{24} - M_0)}{M_0}$$

Swelling is the result of liquid resin being absorbed into the outer solidified lines, which makes the part expand. If liquid resin is present on both sides of the line, as in Tri-Hatch, the swelling is more severe.

3D Systems measures swelling by building a part called "Swell-Tower." The part is two inches square by five inches tall (50 mm x 50 mm x 125 mm), with a wall thickness of 0.125″ (approximately 3 mm). The layer thickness used in building this part is 0.010″. The Swell-Tower diagnostic test part is shown in *Figure 10-4*. Swell-Tower is *intentionally* designed to build in a 24-hour period. This is accomplished by adjusting the waiting time between layers (*z*-wait). Building the part in this manner ensures that the bottom of the part is exposed to resin for 24 hours, while the top of the part is exposed for only a few minutes.

The Swell-Tower is measured "green" using a CMM. Measurements are made four times (a to b, c to d, e to f, and g to h) at each of ten equally spaced vertical positions. The average value of the four differential swelling measurements is calculated for each vertical position, and plotted versus depth from the top of the part, which is proportional to the immersion time for any layer.

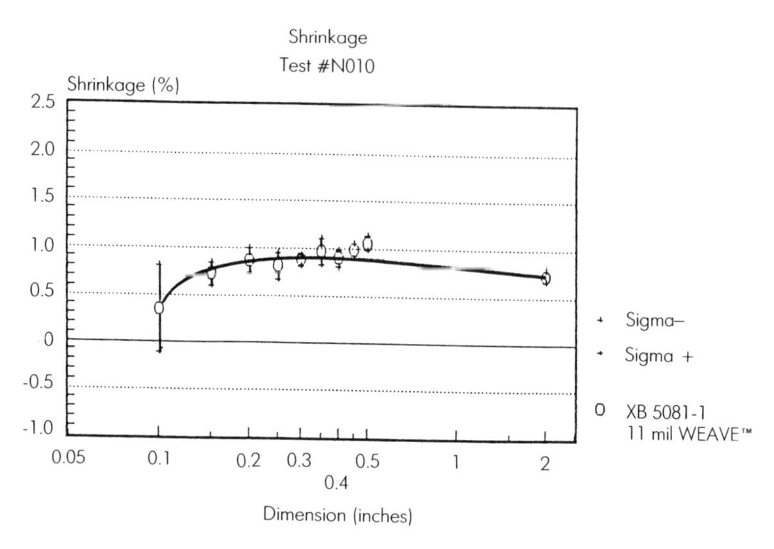

Figure 10-3. Shrinkage versus linear dimension.

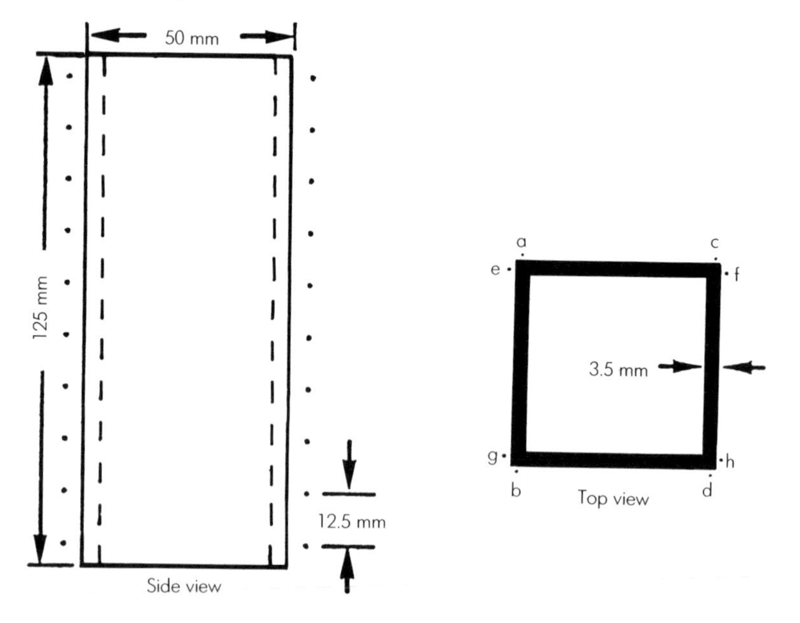

Figure 10-4. Swell-Tower diagnostic test part.

A summary of the swelling results, expressed in mils per horizontal inch, is shown in Table 10-2. Swelling curves as a function of depth from the top of the part are shown in *Figures 10-5* and *10-6*.

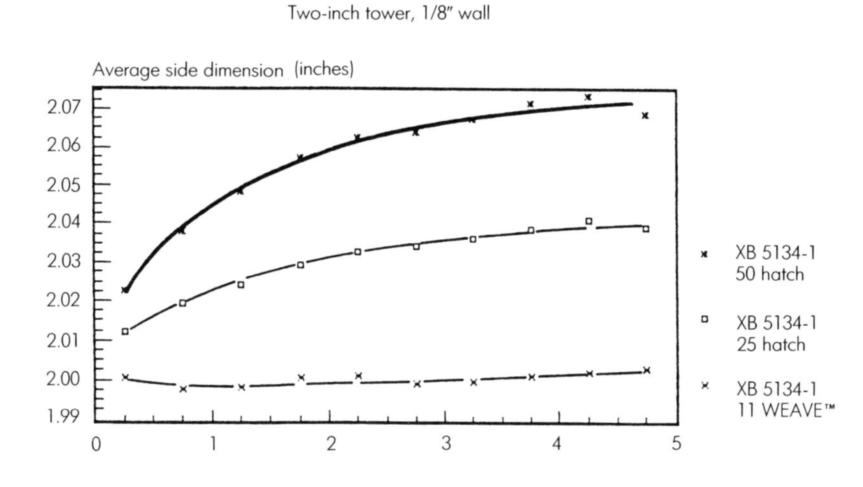

Figure 10-5. Swelling curves as a function of part depth.

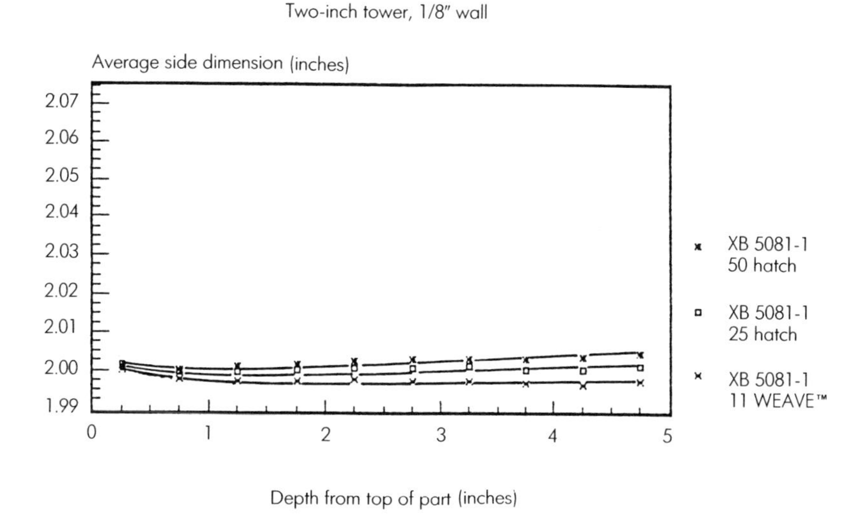

Figure 10-6. Swelling curves as a function of part depth.

The results show that swelling is strongly dependent on both the resin characteristics and the part building style. The dependence upon part building style indicates that the greater the fraction of cured resin within the laser-exposed part, the smaller the swelling and, correspondingly, the more uniform the part becomes.

Table 10-2
Swelling "Swell-Tower" (mils/inch)

Resin	50 Hatch	25 Hatch	11 WEAVE™	11 STAR™
XB 5081-1	2.2	0.7	0.3	0.2
XB 5134-1	15.2	9.9	2.6	N/A
XB 5139	3.5	1.2	0.3	N/A
XB 5143	10.8	4.6	0.7	0.4
XB 5131	1.9	1.8	1.6	0.4

For all resins tested, STAR-WEAVE™ results in the smallest magnitude of swelling. Finally, flexible resins (XB 5134-1 and XB 5143) have shown the largest swelling, and XB 5081-1 has shown the smallest swelling of the resins tested.

Cantilever Curl Distortion

Cantilever curl distortion, C_{f6}, is defined as the curl elevation per unit length along the cantilever. It is measured at a 6 mm unsupported cantilever length, and is expressed as a percentage.

$$C_{f6} = \frac{(M_6 - M_0)}{6\text{ mm}} * 100$$

Where M_6 is the elevation, in millimeters, of the bottom of the cantilever at an unsupported length of 6 mm, and M_0 is the elevation, in millimeters, at the base of the cantilever. The choice of 6 mm, while somewhat arbitrary, was based on the observation that this value provides good measurement sensitivity and good repeatability without being subject to delamination effects.

3D Systems measures cantilever curl distortion with a series of diagnostic test parts called "Twin Cantilever" (TC). This test part is shown in *Figure 10-7*. A single measurement set consists of eight TCs. All eight TCs are generated during a single build, at the same layer thickness, 0.010″ (0.25 mm).

The parts are measured "green," using the CMM. Three readings are taken every millimeter (except at zero and 14 mm) along both unsupported lengths of the cantilever (72 data points per TC). The average value of the three measurements at each position is calculated, and plotted as a function of the distance along the cantilever. A curl factor, C_{f6}, is calculated from the total of 576 measurements.

A summary of the of the curl factor results is shown in Table 10-3, and plots of cantilever curl, as a function of unsupported cantilever length, are shown in *Figures 10-8* and *10-9*.

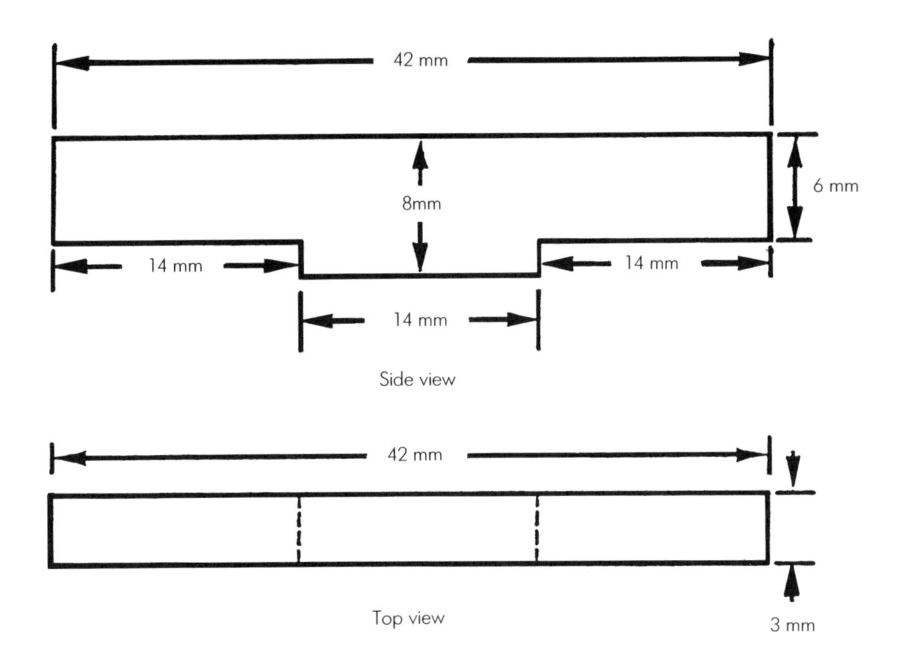

Figure 10-7. Curl diagnostic test part.

The results show that cantilever curl depends on both resin characteristics and part building style. They also show that WEAVE™ does create more cantilever curl, but not enough to cause separation of supports from the platform or the part from the supports. Low viscosity resins generally show more curl, while XB-5131 shows the least curl of the resins tested to date when using WEAVE™.

In all build styles, *curl is directly proportional to the square of the unsupported cantilever length*. Therefore, it is imperative that supports be positioned sufficiently close together to minimize curl distortion.

Table 10-3
Cantilever Curl "TC Part" (C_{f6} %)

Resin	50 Hatch	25 Hatch	11 WEAVE™	11 STAR™
XB 5081-1	9.2	10.0	11.8	9.9
XB 5134-1	7.5	8.5	11.8	N/A
XB 5139	11.1	12.1	17.7	N/A
XB 5143	8.2	7.8	11.7	13.4
XB 5131	8.4	9.4	10.9	12.8

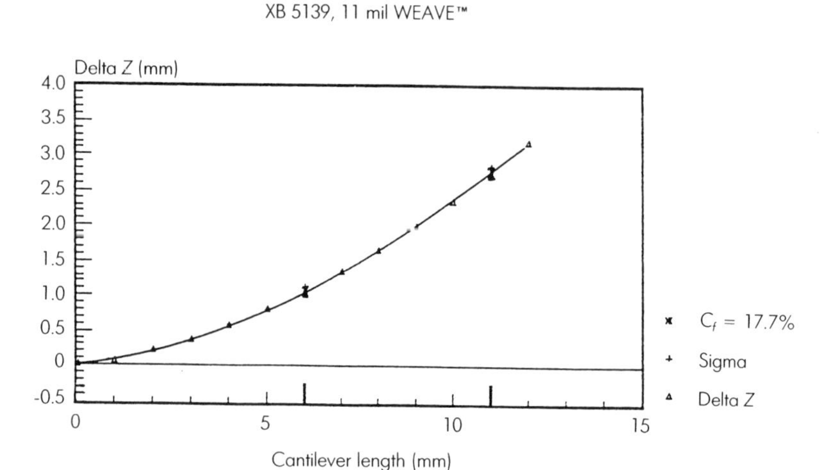

Figure 10-8. Twin cantilever curl diagnostic.

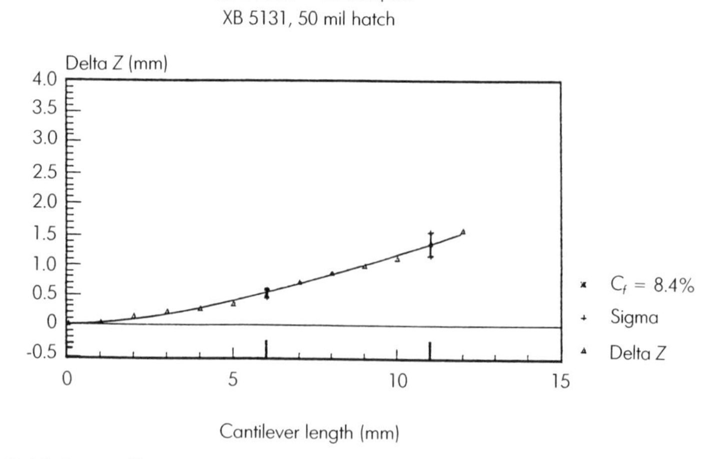

Figure 10-9. Twin cantilever curl diagnostic.

Vertical Wall Postcure Distortion

3D Systems measures vertical wall postcure distortion with a series of diagnostic test parts called "Postcure Walls." This test part is shown in *Figure 10-10*. It consists of two parallel walls, 0.1″ thick, 1″ high, and 4″ long (2.5 mm x 25 mm x 100 mm). The walls are separated by 0.050″ (1.25 mm), and five pairs are built on a single platform. The small separation between the walls is designed to create worst-case postcure distortion since the part is effectively exposed from only one side. The layer thickness of these parts is 0.010″ (0.25 mm).

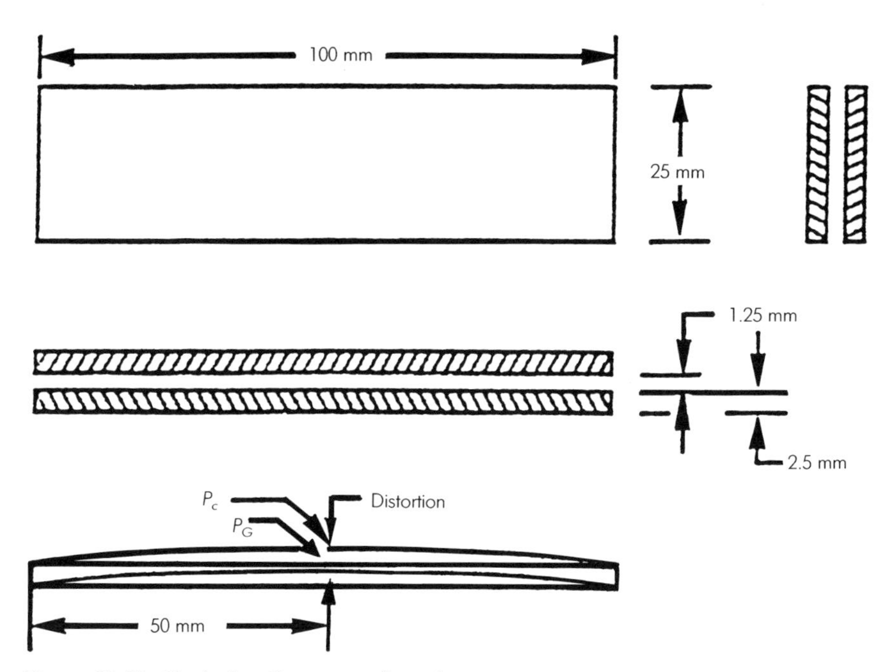

Figure 10-10. Vertical wall postcure distortion.

The walls are first measured "green" with the CMM. They are then postcured for an hour with appropriate actinic fluorescent lamps and finally measured again with the CMM. Three measurements are taken on each wall, one at each end and one in the middle. The distortion of the wall (the maximum departure from flatness) is then calculated. The distortion in the "green" state is then subtracted from the postcure value. The result is the net postcure distortion, summarized in Table 10-4.

Table 10-4
Vertical Wall Postcure Distortion "Postcure Walls" (mils)

Resin	50 Hatch	25 Hatch	11 WEAVE™	STAR™
XB 5081-1	12.5	4.7	1.8	1.1
XB 5134-1	19.8	8.6	1.4	N/A
XB 5139	17.5	6.0	5.8	N/A
XB 5143	4.3	0.6	0.5	1.1
XB 5131	5.7	1.2	1.8	1.1

The results show that postcure distortion depends primarily on part building style, and to a lesser extent on resin characteristics. The greater the fraction of uncured resin remaining within the laser-cured part, the greater the postcure distortion. Since WEAVE™ and STAR-WEAVE™ maximize resin cure in the vat, postcure distortion is correspondingly reduced, especially relative to building with the 50 mil Tri-Hatch method.

Finally, since postcure distortion was previously the greatest source of error in StereoLithography, the improvements resulting from WEAVE™ and STAR-WEAVE ™ are especially significant.

Horizontal Slab Distortion

This diagnostic test involves four measurements: build distortion, postcure distortion, creep distortion, and final distortion.

3D Systems measures horizontal slab distortion with a diagnostic test part called "Slab 6x6." This test part is shown in *Figure 10-11*. Slab 6x6 is a flat part, 6″ square by 0.250″ thick (150 mm x 150 mm x 6.25 mm). The layer thickness is, for the majority of the cases, 0.005″, and also 0.010″ in WEAVE™ and STAR-WEAVE™. The addition of the 0.005″ layer thickness adds a 9 mil hatch spacing WEAVE™ and a 9 mil hatch spacing STAR-WEAVE™ to the matrix.

Slab 6x6 is first measured "green" on the CMM, while *still attached to the platform* (case "A"). After this initial measurement, the part is *removed from*

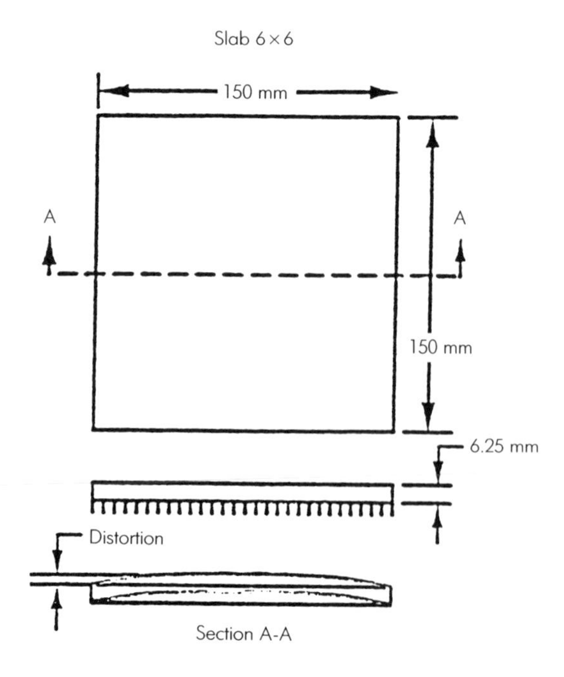

Figure 10-11. "Slab 6X6" diagnostic test part.

the platform and again measured "green" (case "B"). The part is then postcured for an hour with appropriate actinic fluorescent lamps and is measured *postcured* (case "C"). Postcure of the slab is accomplished by intentionally laying the part flat in the PCA, and postcuring it from the top side only. Postcuring the part this way simulates the worst possible situation which can occur with real parts. Measurements are taken a fourth time, *one day later* (case "D"), and a fifth time, *one week later* (case "E"). A total of 576 measurements are taken for each case, or 2880 data points per test. The maximum distortion from a plane is then calculated. A summary of the various slab distortion data is shown in Tables 10-5, 10-6, 10-7, and 10-8.

Horizontal Slab Build Distortion. Horizontal slab build distortion is defined as the difference between the peak distortion of the green part on the platform (case "A") and the green part off the platform (case "B"). This distortion is due to the internal stress developed during the part building process.

Horizontal slab build distortion is significantly reduced for thicker layers, as in 10 mil layer thickness. A summary of the results of this build distortion is given in Table 10-5.

Table 10-5
Horizontal Slab Build Distortion "Slab 6X6" "B-A" (mils)

Resin	50 Hatch	25 Hatch	9 WEAVE™	9 STAR™	11 WEAVE™	11 STAR™
XB 5081-1	28.7	32.4	133.8	79.5	26.5	70.6
XB 5134-1	137.1	28.3	N/C	N/A	N/C	N/A
XB 5139	71.7	80.3	168.2	N/A	26.6	N/A
XB 5143	82.8	50.9	118.9	142.0	162.4	112.1
XB 5131	34.6	N/C	64.3	82.6	28.9	34.4

Horizontal Slab Postcure Distortion. Horizontal slab postcure distortion is defined as the difference between the peak distortion of the green part off the platform (case "B") and the postcured part (case "C"). This distortion is due to postcure-induced stress. Again, the greater the fraction of uncured resin remaining within the laser-cured part, the greater the horizontal slab postcure distortion. Also, thinner layer thicknesses display greater postcure distortion.

Since WEAVE™ increases resin cure in the vat, slab postcure distortion is significantly reduced, especially for 10 mil layer thickness. A summary of the results of the horizontal slab postcure distortion is presented in Table 10-6.

Horizontal Slab Creep Distortion. Horizontal slab creep distortion is defined as the difference between the peak distortion of the cured part (case "C") and the postcured part, one week later (case "E"). This distortion is due to relaxation of both the postcure-induced stresses and internal build stresses.

Table 10-6
Horizontal Slab Postcure Distortion "Slab 6X6""C-B" (mils)

Resin	50 Hatch	25 Hatch	9 WEAVE™	9 STAR™	11 WEAVE™	11 STAR™
XB 5081-1	259.5	88.1	60.2	137.7	1.1	4.8
XB 5134-1	309.5	175.3	N/C	N/A	N/C	N/A
XB 5139	322.1	117.8	112.0	N/A	51.7	N/A
XB 5143	315.9	162.4	26.3	3.8	0.9	11.4
XB 5131	299.1	N/C	133.5	210.8	17.4	16.9

The result shows that creep distortion does not depend as much on part building style as layer thickness. Furthermore, prior results show that creep increases dramatically when parts are cured on the platform. Since postcuring parts off the platform reduces horizontal slab creep distortion dramatically and makes cleaning the platform easier, this procedure is strongly recommended with very few exceptions.

A summary of the results of the horizontal slab creep distortion is presented in Table 10-7.

Horizontal Slab Final Distortion. Horizontal slab final distortion is defined as the difference between the peak distortion of the postcured part, one week later (case "E"), and a flat plane.

The total final distortion is represented in case "E". Note that positive (+) values occur when the center of the slab is elevated (a "dome"). Negative values (-) occur when the center of the slab is depressed (a "bowl").

A summary of the results of horizontal slab final distortion is given in Table 10-8. Distortion as a function of time (build, postcure, and one week later) is shown in *Figures 10-12* and *10-13*.

Table 10-7
Horizontal Slab Creep Distortion "Slab 6X6" "E-C" (mils)

Resin	50 Hatch	25 Hatch	9 WEAVE™	9 STAR™	11 WEAVE™	11 STAR™
XB 5081-1	10.3	4.1	15.4	16.6	11.1	28.3
XB 5134-1	25.3	4.5	N/C	N/A	N/C	N/A
XB 5139	8.9	8.8	91.8	N/A	1.5	N/A
XB 5143	57.5	22.3	1.6	6.1	12.4	19.6
XB 5131	18.3	N/C	21.7	20.9	12.5	25.3

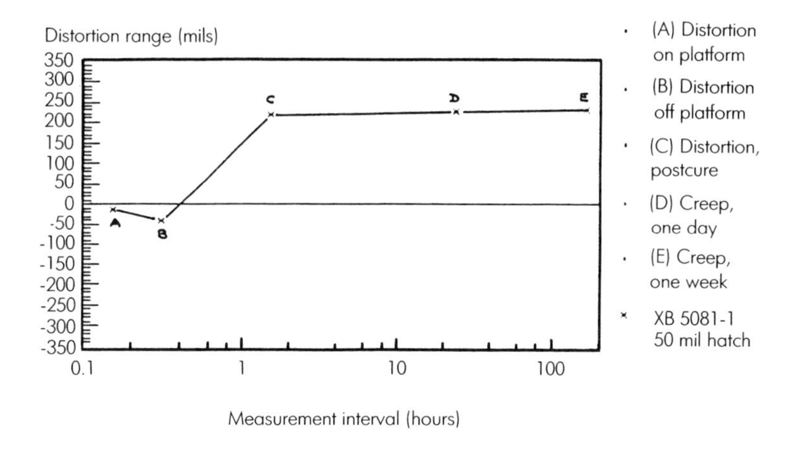

Figure 10-12. Horizontal slab distortion over time.

Slab 6×6, 5 mil layer

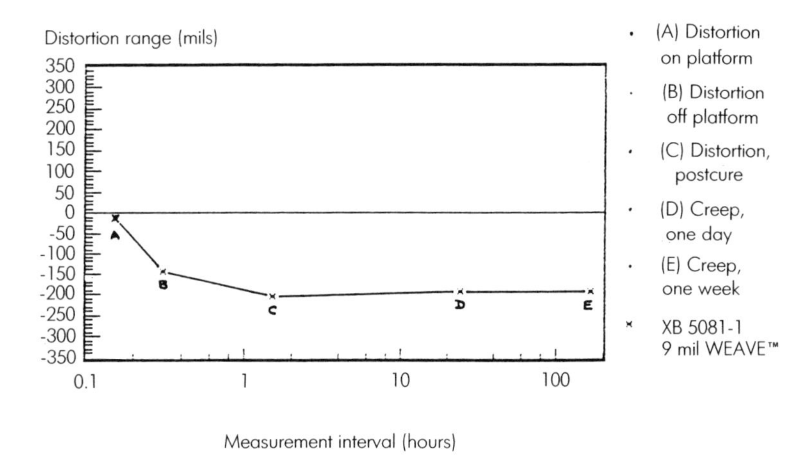

Figure 10-13. Horizontal slab distortion over time.

10.3 The WINDOWPANE™ Technique

Background

We know, from equation 4-33, that the relationship between the cure depth, C_d, and the maximum exposure, E_{max}, also involves the fundamental resin parameters D_p and E_c. However, in actual practice, it is not a simple matter to accurately determine the values of D_p and E_c for a StereoLithography resin.

Table 10-8
Horizontal Slab Final Distortion "Slab 6X6""E" (mils)

Resin	50 Hatch	25 Hatch	9 WEAVE™	9 STAR™	11 WEAVE™	11 STAR™
XB 5081-1	+226.5	+22.5	-194.0	52.9	-65.1	-52.8
XB 5134-1	+166.9	+120.4	N/C	N/A	N/C	N/A
XB 5139	+226.4	+19.9	-214.2	N/A	-44.6	N/A
XB 5143	+166.9	+60.9	-160.4	-145.9	-160.6	-133.7
XB 5131	+225.5	N/C	+71.0	+127.1	-21.5	-42.6

To establish these fundamental resin properties, one must determine the peak irradiance, H_o, of the laser beam, and then calculate the peak exposure on the resin surface, E_{max}. The mathematical relationship between the values of the cure depth of the solidified strings and the appropriate values of the peak exposure, E_{max}, can then be used to determine the resin properties. If the laser beam is Gaussian, H_o has been calculated in Chapter 4, using equation 4-8.

Unfortunately, the irradiance of actual laser beams does not always follow a true Gaussian distribution. As a result, H_o had to be determined experimentally by exploring the power distribution of the laser beam.

The intent of the earlier "banjo-top" technique was to create a "Working Curve," *independent of laser power and beam characteristics*. This was a significant step in improving SL technology. However, handling a solidified "string," and measuring its peak cure depth is very difficult. These tests required skill and specialized equipment. They also demonstrated a significant lack of repeatability from operator to operator.

To make diagnostic part handling and measurement easier, more accurate, and more repeatable, a technique to measure D_p and E_c of an SL resin has been recently developed by 3D Systems. The so-called WINDOWPANE™ procedure involves the *intentional superposition* of the exposure contributions from adjacent scans. The remainder of section 10-3 describes the theory, procedure, and experimental results of the WINDOWPANE™ technique.

Superposition of Adjacent Scans

When a Gaussian laser beam is scanned along the x direction on a resin surface, the exposure in the y direction, orthogonal to the scanned line, can be determined using equation 4-27. If the laser beam is scanned in a "raster" fashion, with line spacing, h_s, the exposure at any point $P(x,y)$ will be equal to the sum of the exposure contributions from those lines within the "zone of influence," as discussed in Chapter 4. For the case of a raster scan, the exposure at various positions in the y direction, can be put into three categories:

Separate Laser-Cured Lines. The laser cured lines on the resin surface are laterally separated and distinct from each other when one of following two conditions occurs:

1. $h_s >> R$ where R is the radius of the zone of influence.
2. $h_s > R$, but the total exposure midway between two adjacent lines is still less than the critical exposure, E_c, so no resin is solidified at these locations.

These conditions are described in *Figures 10-14* and *10-15*.

Uneven Curing. When the line spacing, h_s, becomes small enough, the total exposure at all the intermediate points finally exceeds the critical exposure, E_c, and the whole area is solidified. The top surface is flat, because all the fluid on the planar resin surface has been partially polymerized. In the y direction, the exposure is maximum on the centerlines of each scan vector, while it is minimum midway between any two adjacent lines. From Chapter 4, since the cure depth depends on the exposure incident on the resin surface, it follows that the bottom surface of the solidified area will be uneven. The bottom has "valleys" at the minimum exposure regions and "peaks" at the maximum exposure regions. The maxima and minima in exposure are seen in *Figure 10-16*.

Planar Curing. If the scan lines are positioned sufficiently close together, the difference between the maximum and minimum exposures becomes very small. The result is a correspondingly small difference between the cure depths at the peaks and those at the valleys. In this case, the exposure on the resin

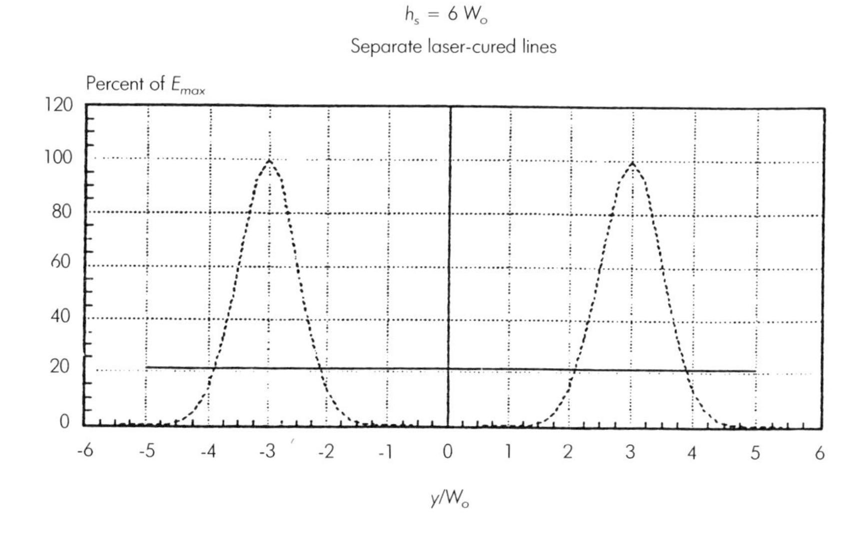

Figure 10-14. Superposition of Gaussian beams, $h_s/W_0 = 6$.

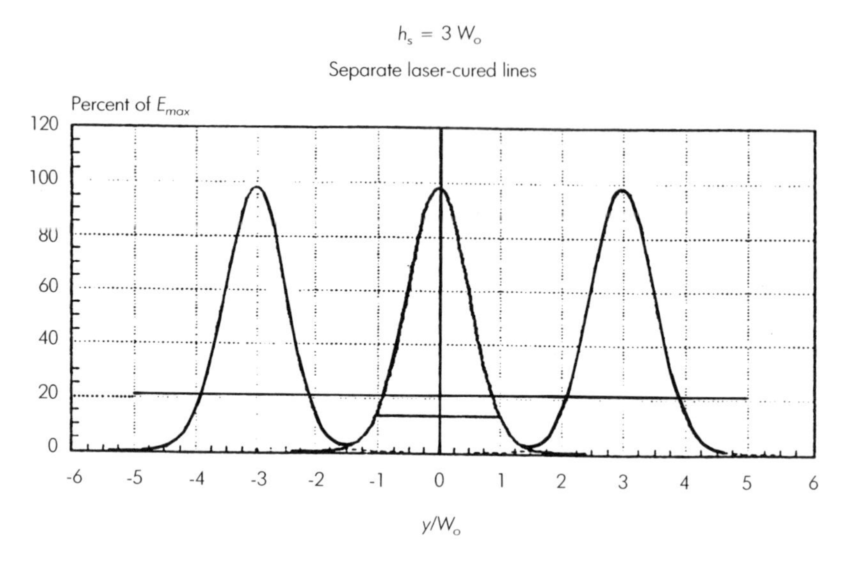

Figure 10-15. Superposition of Gaussian beams, $h_s/W_0 = 3$.

surface is essentially uniform and equal to the average exposure, E_{av}, (Chapter 4, equation 4-59). The bottom surface of the solidified area is therefore very flat. *Figure 10-17* shows the exposure in this case.

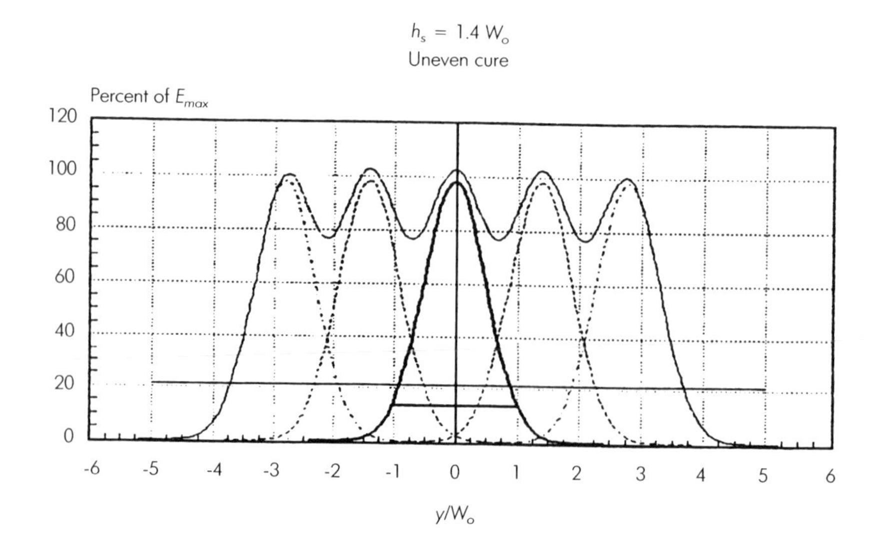

Figure 10-16. Superposition of Gaussian beams, $h_s/W_0 = 1.4$.

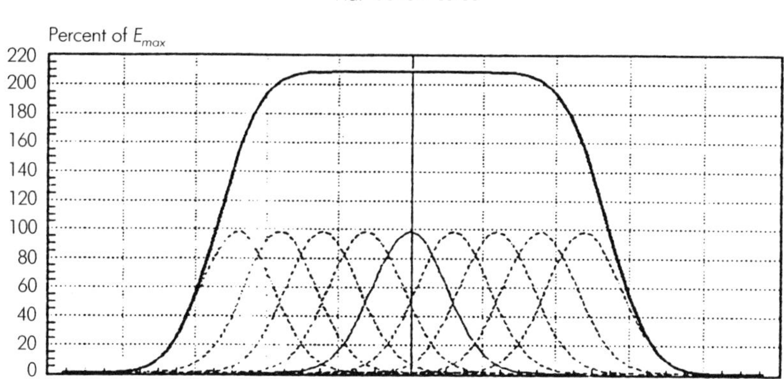

Figure 10-17. Superposition of Gaussian beams, $h_s/W_0 = 0.6$.

Conditions for Planar Cure Depth. From a practical standpoint, an area may be considered to be planar cured when the difference between the cure depth at the "peak" and the cure depth at the "valley" is less than or equal to 0.20 mil or about 5 micrometers. From this definition, we can calculate the maximum line spacing allowed.

From Chapter 4, equation 4-33:

$$C_d = D_p \ln(E_{max}/E_c) \tag{10-1}$$

Let's call the exposure at the "peak" E_p, and the exposure at the "valley" E_v. Substituting these values for E_{max} in equation 10-1, we calculate the corresponding values of the cure depths Cd_p and Cd_v at the "peaks" and the "valleys."

$$Cd_p = D_p \ln(E_p/E_c) \tag{10-2}$$

$$Cd_v = D_p \ln(E_v/E_c) \tag{10-3}$$

The difference is :

$$Cd_p - Cd_v = D_p [\ln(E_p/E_c) - \ln(E_v/E_c)] \tag{10-4}$$

or

$$Cd_p - Cd_v = D_p \ln(E_p/E_v) \tag{10-5}$$

To have $Cd_p - Cd_v \leq 0.20$ mil, this requires that:

$$D_p \ln(E_p/E_v) \leq 0.20 \text{ mil}$$

or

$$E_p/E_v \leq \exp(0.20/D_p) \tag{10-6}$$

Typical D_p values of SL resins range from about 5 mils to about 7 mils. In the most demanding case, $D_p = 7.1$ mils for resin XB 5081-1. Therefore, $E_p/E_v \leq 1.029$. Thus, to assure planar curing according to our 0.2 mil definition, we cannot allow variations in exposure greater than 3% over the entire area.

Now, consider the case of the exposure resulting from the superposition of multiple parallel lines scanned in the x direction. Let the line spacing, h_s, from the centerline of an arbitrary line to the centerlines of adjacent raster scan lines remain constant. The exposure distribution on the resin surface ($z = 0$), as a function of the surface coordinate orthogonal to the scan direction (y), due to that line only is given, for a Gaussian laser, by

$$E_0(y,o) = E_{max} \exp(-2y^2/W_0^2) \tag{10-7}$$

where the subscript, 0, denotes the fact that the centerline of this arbitrary scan is at zero distance from itself.

The exposure distributions resulting from the two neighboring lines, separated by $+nh_s$ and $-nh_s$ respectively, are given by

$$E_n(y,o) = E_{max} \exp[-2(y-nh_s)^2/W_0^2]$$

and

$$E_{-n}(y,o) = E_{max} \exp[-2(y+nh_s)^2/W_0^2]$$

The total exposure, E_{total}, at any distance y, is then given by the sum of the individual exposure distributions at that location, due to all the lines, or

$$E_{total}(y,o) = [E_n(y,o) + E_{n-1}(y,o) + \ldots + E_{-n+1}(y,o) + E_{-n}(y,o)] \tag{10-8}$$

with n approaching infinity. However, from Chapter 4, using the concept of the "zone of influence," only those scan lines located such that absolute value of $y + (-)\, nh_s < 2.146\, W_o$ need be considered. Using this concept, for the case $h_s = W_o$, only the exposure distribution of the lines that are *within* $2W_o$ from the arbitrary line are used to calculate E_{total}. Thus:

$$E_2(y,o) = E_{max} \exp[-2(y-2W_o)^2/W_o^2] \tag{10-9A}$$

$$E_1(y,o) = E_{max} \exp[-2(y-1W_o)^2/W_0^2] \tag{10-9B}$$

$$E_0(y,o) = E_{max} \exp[-2(y)^2/W_o^2] \tag{10-9C}$$

$$E_{-1}(y,o) = E_{max} \exp[-2(y+1W_o)^2/W_o^2] \tag{10-9D}$$

$$E_{-2}(y,o) = E_{max} \exp[-2(y + 2W_o)^2/W_o^2] \qquad (10\text{-}9E)$$

At the position of "peak" exposure, $y = 0$. Substituting this value in equations 10-9A through 10-9E, we obtain :

$$E_p = E_{max} [\exp(-8) + \exp(-2) + \exp(0) + \exp(-2) + \exp(-8)]$$
$$= 1.271\ E_{max}$$

The valley occurs at the point halfway between two adjacent lines. Substituting $y = 0.5\ W_o$ to calculate the total exposure, which is equal to the sum of the exposure of the lines $n = 2, 1, 0, -1$, we obtain:

$$E_v = E_{max} [\exp(-4.5) + \exp(-0.5) + \exp(-0.5) + \exp(-4.5)]$$
$$= 1.235\ E_{max}$$

From the above results, when the line spacing $h_s = W_o$, we find $E_p/E_v = 1.271/1.235 = 1.029$. This ratio decreases when h_s/W_o decreases. Interestingly, this is precisely the "peak" to "valley" exposure ratio required for our criterion that the variation in cure depth should not exceed 0.2 mil. Thus, it is clear that *h_s must be less than W_o* for planar curing. *Figure 10-18* shows the result for Gaussian laser beams with $h_s = W_o$.

Thus, in raster scanning, three conditions must be met to have planar curing:

1. The line spacing, h_s, must be less than W_o.
2. The scanned dimensions of the test sample must be large compared to the size of the laser beam to avoid cure depth variations near the edges of the sample.
3. Conversely, if the sample dimensions are too large, curl effects can lead to distortions which also increase errors and reduce repeatability.

These conditions still apply even when the beam is not Gaussian.

Average Exposure and Drawing Speed

Provided that the laser raster scanning conditions meet the above requirements, the average exposure E_{av} is calculated in Chapter 4, equation 4-59:

$$E_{av} = P_L t_d / A_s \qquad (10\text{-}10)$$

Assume that the laser beam is scanned by moving it in the x direction at an constant velocity, V_s, and that these vectors then propagate in the y direction with line spacing h_s. We also assume that the scanned area is a rectangle of dimensions L_x and L_y in the x and y directions. Thus: $A_s = L_x L_y$.

The time required to complete one scan *line* in the x direction is simply

$$t_L = L_x/V_s$$

The *number* of lines to be drawn in y direction is, within integer round-off,

$$N = L_y/h_s$$

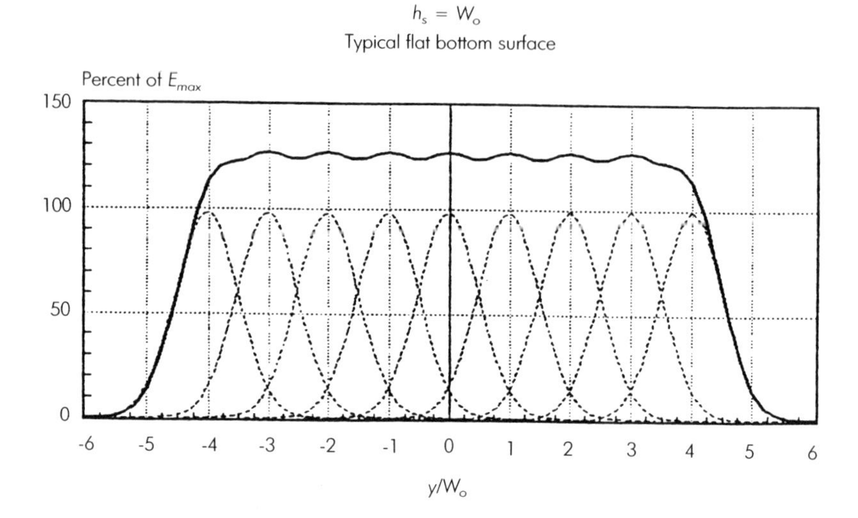

Figure 10-18. Superposition of Gaussian beams, $h_s/W_o = 1$.

The *total drawing time* for a complete raster scan of the sample is then

$$t_d = Nt_L = L_xL_y/V_sh_s$$

Substituting the values for A_s and t_d into equation 10-10, we obtain the important result:

$$\boxed{E_{av} = P_L/V_sh_s} \tag{10-11}$$

Where P_L = laser power (mW)
V_s = laser drawing speed (cm/sec)
h_s = line spacing (cm)
E_{av} = average exposure (mJ/cm^2)

Equation 10-11 is the basic equation used to calculate the exposure for the WINDOWPANE™ method.

WINDOWPANE™ Method

The WINDOWPANE™ technique was developed for measuring the fundamental properties D_p and E_c of an SL resin. In designing the diagnostic part, WINDOWPANE™, the laser scanning characteristics are carefully selected to meet the requirements for a planar cure, previously described, and are based upon the average exposure expression of equation 10-11. *Figure 10-19* shows the top view as well as a projection view of the diagnostic part.

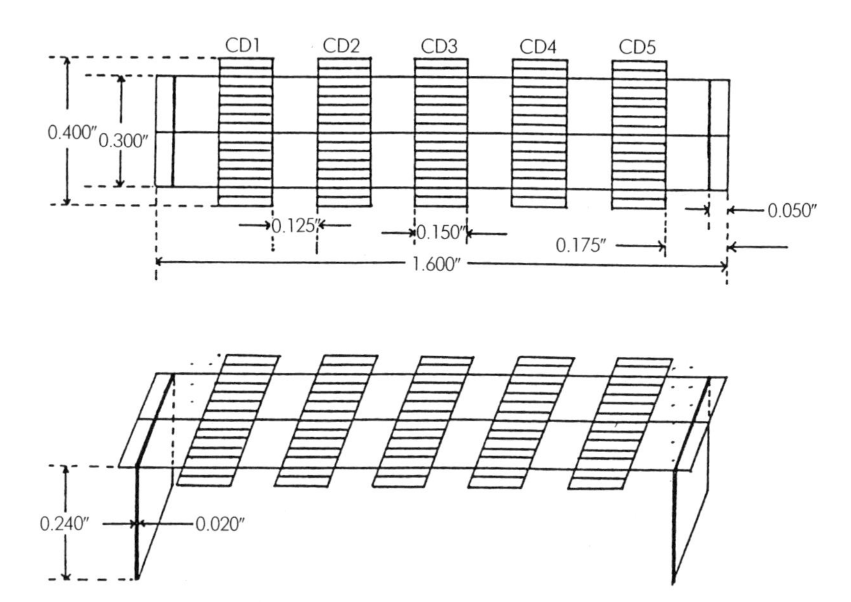

Figure 10-19. WINDOWPANE™ diagnostic part.

The WINDOWPANE™ consists of two supports, one frame, and five separate double-panes labeled from CD1 to CD5. Each double-pane is formed with a different cure depth. Each double-pane is a *single-layer part*, 0.150″ × 0.400″ in size. This size meets the requirements for planar cure, providing that the laser total beam width is less than 12 mils (0.012″) or W_o is less than 6 mils. Also, this size is selected so that the thinnest pane can still be physically handled. The thinnest pane must have a value of cure depth greater than D_p.

One must also consider the acceleration and deceleration of the laser beam. To achieve highly uniform exposure, the laser spot must be moving at constant speed, V_s, for a length of about 50 mils at the middle section of a drawn vector. This span occurs after the acceleration period and before the deceleration period. The 0.150″ × 0.400″ dimensions are suitable for an SLA-250 system. With an SLA-500 system, test data has shown that optimum results occur when the two thinnest panes have dimensions of 0.250″ × 0.400″.

Each pane is separated by a gap of 0.125″ from its nearest neighbors. This distance is far enough that while drawing one pane, there will be no effect on the others.

The supports and frame are designed and built to keep the panes from moving on the resin surface. *Figures 10-20* and *10-21* are photographs of WINDOWPANE™ diagnostic test parts built on an SLA-250 and an SLA-500. Note that the term "WINDOWPANE" derives from the observation that these diagnostic parts resemble a tiny set of framed windowpanes.

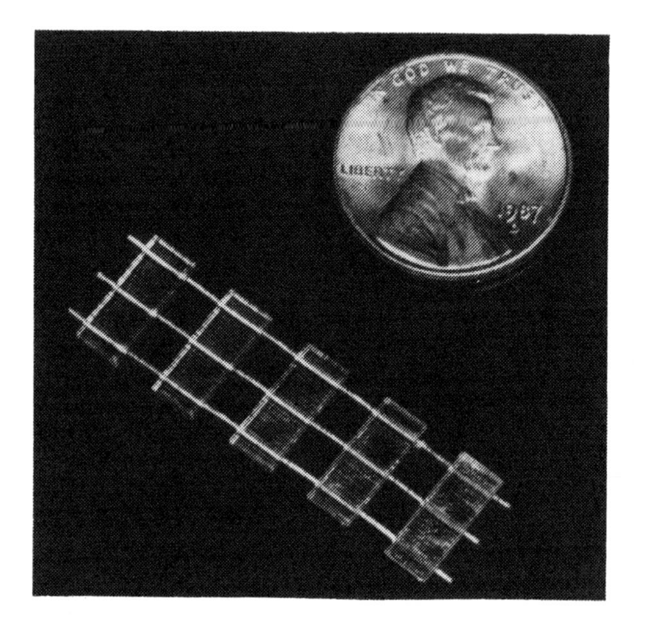

Figure 10-20. WINDOWPANE™ diagnostic part (SLA-250).

Required Equipment

The equipment, instruments, and materials required for the WINDOWPANE™ experiment are as follows:

- SLA to generate a precise raster scan.
- Calibrated UV laser power meter, accurate to 5%.
- Absorption material. (Scott C-fold towels are fine for all approved SL resins.)
- Thermal chamber, capable of reaching 80°C.
- Custom ground point micrometer with 0.05 mil (approximately 1 micrometer) resolution and a rachet type clutch. The micrometer points are then precision ground to produce flat, parallel, circular measuring surfaces. The diameter of each flat area should be between 25 and 30 mils (0.6 mm to 0.7 mm). The measuring force is the activation force of the clutch. This force must be weak enough that the tips will not penetrate into the test parts during the measurement and yet strong enough to provide reproducible results. With all currently approved resins, the optimum force is about 3 Newtons.

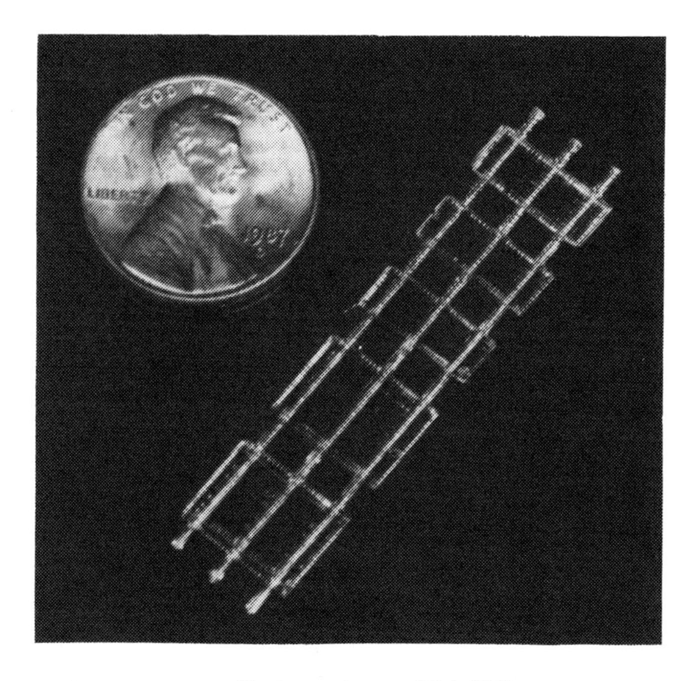

Figure 10-21. WINDOWPANE™ diagnostic part (SLA-500).

- Flat glass plate, approximately 6″ X 6″ X 0.25″.
- Razor blade and tweezers.
- Actinic lamp postcuring apparatus (PCA).
- A vat of the resin to be tested.

Building WINDOWPANES™

In building test parts, the exposure is controlled by the laser drawing speed, V_s. Supports are built using a layer thickness of 20 mils (0.5 mm). Finally, at the last layer on the free resin surface, the frame and the panes are built. To ensure that the test parts can be physically handled, the exposure for each component of the WINDOWPANE™ must be carefully chosen. The supports and the frame must also be sufficiently strong. Since each double-pane has a different cure depth, the thinnest must be able to be handled, and the thickest must still be in the linear region of the ''working curve.'' From experience, the values of cure depth should be between D_p and $4D_p$ for most types of SL resin. *To achieve these requirements, a few iterations must be performed.*

The perimeter is drawn first and is strongly solidified. The panes are drawn with the whole area of each pane being scanned before the next is started. The laser beam diameter, B, should be maintained at approximately 10 mils (0.25 mm). The line spacing, h_s = 4 mils (0.10 mm) is chosen, since this value has produced excellent, highly repeatable results. For this case, $h_s/W_o = 0.8$, which more than satisfies the requirements for planar curing.

For statistical purpose, five test parts should be built. One may use standard multiple part building techniques to generate all five of them at one time. The time required to build the complete ensemble of five WINDOWPANES™ is typically only about 20 minutes.

Postprocessing

Draining all excess resin from the test parts is critical. If this is not done properly, one will obtain incorrect cure depth measurements and, consequently, misleading results. Different ways to drain the test parts have been studied. The best method realized to date is to absorb the excess resin using an appropriate absorption material at an elevated temperature. In handling the test parts, great care must be taken to keep the top side of the WINDOWPANE™ dry.

After being built, the test parts are then elevated and maintained in this position for about five minutes to allow excess resin to drain. Next, they are removed and placed "wet side down" on the absorbing material. Both ends of the WINDOWPANE™ frame are then cut off so that the "panes" can lay flat. After removing the ends, all five WINDOWPANES™ are transferred to a new piece of the absorbing material and placed into a temperature-controlled chamber. For existing approved resins, 30 minutes at 80°C is excellent. When the drainage is completed, the test parts are put "dry side down" on a clean glass plate for postcuring. Postcure in an appropriate PCA for one hour to ensure that the sample will be tack free.

Measurement

Each double-pane of a WINDOWPANE™ is divided into two halves by the center line of the frame. One measurement is taken at the center of each individual pane. There are a total of five WINDOWPANES™ per experiment each consisting of five double-panes. A total of 10 measurements are made for each of the five different values of cure depth (CD1, CD2,...CD5). Therefore, 50 measurements are taken for the complete experiment. When making a cure depth measurement, center the pane to be measured and adjust the point micrometer to allow the tips to move toward the pane slowly until the clutch is activated. Because of the clutch, the measuring force will be constant from one measurement to the next. During the measurements, the "zero reference" of the point micrometer should be frequently checked.

The values of the measured cure depth are best recorded in a tabular format. The average value, and the range of each cure depth are then calculated. With the specified values of P_L, V_s and h_s, the values of the exposure are determined from equation 10-11. A typical record of actual measurements is shown in Table 10-9.

Extensive testing at 3D Systems has shown that the *WINDOWPANE™ technique is capable of achieving a standard deviation of 0.15 mil or about 4μm.* Thus, if any range exceeds 1.00 mil, or almost seven standard deviations, inconsistency among the measurements is probable. The reasons may be:

- Excessive resin has not been completely absorbed.
- The "zero reference" of the point micrometer may have changed during the measurement.

In this case, the experiment, or at least that measurement, must be repeated.

Linear Regression Analysis

The "working curve" equation, 4-33 can be written as:

$$C_d = -D_p \ln(E_c) + D_p \ln(E_{max})$$

In the case of the WINDOWPANE™ procedure, E_{max} and C_d are replaced by the calculated E_{av} and $C_{d,av}$ respectively.

$$C_{d,av} = -D_p \ln(E_c) + D_p \ln(E_{av}) \quad (10\text{-}12)$$

Table 10-9
WINDOWPANE™ Cure Depth Measurements
P_L = 23 mW; B = 9.8 mils;
h_s = 4 mils Resin: LMB 5086, Lot #KDD-043

C_d identification	CD1	CD2	CD3	CD4	CD5
E_{av} (mJ/cm^2)	22.70	39.80	70.10	121.20	214.00
Measurement 1 (mils)	9.50	12.50	15.65	18.75	22.05
Measurement 2 (mils)	9.20	12.45	15.75	18.80	22.20
Measurement 3 (mils)	9.55	12.45	15.90	19.25	22.10
Measurement 4 (mils)	9.25	12.20	15.70	18.65	22.10
Measurement 5 (mils)	9.20	12.35	15.85	18.80	22.25
Measurement 6 (mils)	9.40	12.55	15.80	18.75	21.90
Measurement 7 (mils)	9.55	12.65	15.85	18.70	21.95
Measurement 8 (mils)	9.25	12.55	15.75	18.75	22.20
Measurement 9 (mils)	9.55	12.65	15.75	18.80	22.30
Measurement 10 (mils)	9.50	12.60	15.80	18.80	22.10
$C_{d,av}$ = Average (mils)	9.40	12.50	15.78	18.82	22.12
Range (mils)	0.35	0.45	0.25	0.65	0.40

We can re-write equation 10-12 as:

$$y = A + B\,x \qquad (10\text{-}13)$$

Where $y = C_{d,av}$; $x = \ln(E_{av})$; $A = -D_p \ln(E_c)$; and $B = D_p$.

From the experimental data, A and B can be calculated using the standard linear regression analysis formulas (see References 1 and 2).

3D Systems has developed a software package known as "Diagnostic Disk," which will perform all the necessary linear regression analysis of a set of WINDOWPANE™ measurements. The output of this software is the value of the resin penetration depth, D_p, the resin critical exposure, E_c, and the linear regression correlation coefficient, R.

The correlation coefficient indicates how well the experimental data points actually fit a straight line. R always lies in the interval $-1 \leq R \leq 1$. Values of R close to $+(-)1$ indicate a good linear correlation. For current SL technology, R must be greater than 0.99 to ensure no worse than a 1 mil error in predicting values of cure depth. *Figure 10-22* shows a semilogarithmic plot of the working curve for the experimental data previously tabulated.

Experimental Results

Experiments were performed using the WINDOWPANE™ method to determine the *repeatability* of data collection. In this experiment, three field test personnel from 3D Systems were chosen to determine the properties of XB 5081-1 resin. The tests were performed using an SLA-250. The HeCd laser power was $P_L = 20.3$ mW. Each operator built four WINDOWPANES™ and

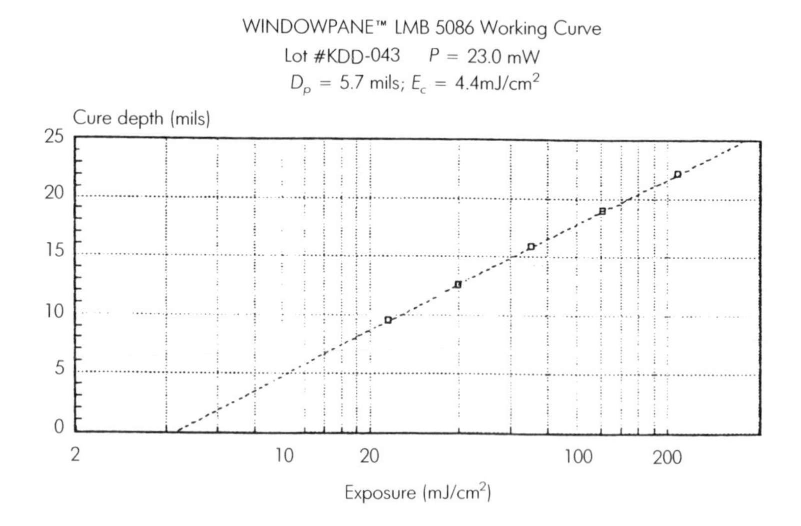

Figure 10-22. Working curve for experimental data.

made eight measurements for each cure depth. The data and measured test results are detailed in Table 10-10.

In this experiment, *the average standard deviation of all the cure depth measurements is about 0.13 mil or about 3.3 μm!* Thus, with about 99.9% confidence limits, the measurements can be repeated within ±0.4 mil. The values of D_p and E_c were calculated by computer using linear regression. The results were as follows:

	D_p (mils)	E_c (mJ/cm^2)	R
Operator 1	**7.2**	**5.7**	**0.9999**
Operator 2	**7.3**	**5.8**	**0.9997**
Operator 3	**7.1**	**5.6**	**1.0000**

R values indicate that data fits the form of equation 4-33 extremely well. Figure 10-23 shows that all the experimental data points from three separate working curves determined by three different individuals are well within 0.4 mils of the mean line.

It is especially noteworthy that *none* of the three service personnel who performed these tests had ever done so before. Further, they were intentionally given *no training whatsoever*. They were simply handed the procedure and told to "follow the instructions very carefully, in a step-by-step manner."

This is certainly the "acid test" for any written experimental test procedure. Nonetheless, all three individuals obtained results comparable with those achieved by one of this chapter's authors, who actually developed the WINDOWPANE™ technique and has built and measured hundreds of these samples. This strongly suggests that the WINDOWPANE™ diagnostic technique is now essentially independent of the finesse of the experimenter. Of course, accurate results definitely require that the test be performed very carefully and in accord with the procedure described herein.

Including the time in the thermal oven and the PCA, the entire procedure takes about three hours. This is time very well spent whenever the user needs to check D_p and E_c. Accurate parts demand accurate cure depths, which in turn require accurate values of D_p and E_c.

10.4 Reverse WINDOWPANE™

The cure depth, C_d, strongly affects the quality of parts. If we do not cure the resin deeply enough, layers do not attach to each other, resulting in layer delamination. If the cure depth is too great, curl will occur, resulting in distortion and reduced part accuracy.

Thus, controlling cure depth, C_d, is very critical in part building. The

Table 10-10
WINDOWPANE Cure Depth Measurements
P_L = 20.3 mW; B = 9.8 mils; h_s = 4 mils
Resin: XB 5081-1, Lot #EFF-081

Operator 1					
C_d identification	CD1	CD2	CD3	CD4	CD5
E_{av} (mJ/cm^2)	26.7	46.8	81.9	143.7	250.7
Measurement 1 (mils)	11.15	15.25	19.15	23.25	27.65
Measurement 2 (mils)	11.15	15.15	19.20	23.20	27.25
Measurement 3 (mils)	11.25	15.10	19.15	23.15	27.35
Measurement 4 (mils)	11.20	15.15	19.20	23.20	27.40
Measurement 5 (mils)	11.25	15.30	19.15	23.20	27.45
Measurement 6 (mils)	11.20	15.45	19.10	23.20	27.65
Measurement 7 (mils)	11.30	15.05	19.10	23.20	27.50
Measurement 8 (mils)	11.25	15.25	19.05	23.25	27.45
Average (mils)	11.22	15.21	19.14	23.21	27.46
Range (mils)	0.15	0.40	0.15	0.10	0.40
Operator 2					
C_d identification	CD1	CD2	CD3	CD4	CD5
E_{av} (mJ/cm^2)	26.7	46.8	81.9	143.7	250.7
Measurement 1 (mils)	11.15	15.05	19.05	23.25	27.85
Measurement 2 (mils)	11.50	15.35	18.85	23.05	27.85
Measurement 3 (mils)	11.25	15.35	19.15	23.35	27.55
Measurement 4 (mils)	11.20	15.20	18.95	23.20	27.55
Measurement 5 (mils)	11.20	15.05	19.15	23.05	27.40
Measurement 6 (mils)	11.15	15.10	18.95	23.15	27.35
Measurement 7 (mils)	11.20	15.15	19.10	23.20	27.30
Measurement 8 (mils)	11.20	15.15	19.15	23.15	27.55
Average (mils)	11.23	15.18	19.04	23.18	27.55
Range (mils)	0.35	0.30	0.30	0.30	0.55
Operator 3					
C_d identification	CD1	CD2	CD3	CD4	CD5
E_{av} (mJ/cm^2)	26.7	46.8	81.9	143.7	250.7
Measurement 1 (mils)	11.30	15.00	18.95	23.00	27.50
Measurement 2 (mils)	11.10	15.30	19.15	23.35	27.40
Measurement 3 (mils)	11.30	15.35	19.40	23.35	27.05
Measurement 4 (mils)	11.25	14.85	19.15	23.10	27.10
Measurement 5 (mils)	11.30	15.20	19.10	23.05	26.90
Measurement 6 (mils)	11.00	15.20	19.00	23.05	27.20
Measurement 7 (mils)	11.00	15.50	19.20	23.10	27.35
Measurement 8 (mils)	11.20	15.45	19.20	23.10	26.95
Average (mils)	11.18	15.23	19.14	23.14	27.18
Range (mils)	0.30	0.65	0.45	0.35	0.60

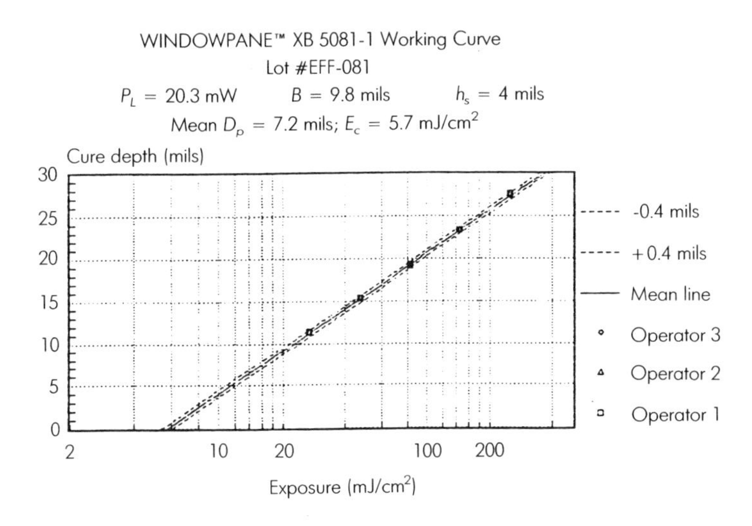

Figure 10-23. Experimental data points from three working curves.

previous section described how to accurately determine the fundamental resin parameters D_p and E_c. With these two key parameters, the cure depth can be calculated using equation 4-33. However, in practice, there are still some uncertainties. For example:

- How close is the laser beam to Gaussian mode?
- How well does the system determine the laser power?
- How consistent is the resin from batch to batch?
- How well is the system calibrated?

Reverse WINDOWPANE™ Method

With the WINDOWPANE™ technique, we control the laser power and scan speed to build the test parts. Cure depths are measured and the values of D_p and E_c are calculated. In the "reverse WINDOWPANE" method, we specify the cure depths for the test parts. The RP&M system reads the laser power and uses the values of P_L, D_p, E_c, and the desired cure depth, C_d, to determine the value of the scan speed for part building. The cure depth of the test part is then measured and compared to the specified cure depth.

Test Part. The diagnostic test part is again the WINDOWPANE™. In the early days of SL technology, most of the parts were built using either 10 or 20 mils layer thickness. The resin was cured 6 mils deeper than the layers. Therefore the values of cure depth were typically 16 or 26 mils.

Currently, the advanced part building techniques, WEAVE™ or STAR-WEAVE™ are often used to build parts with 10 mils layer thickness. Here 9 mils cure depth is selected for the hatch vectors and 16 mils for the borders. Thus, the appropriate values of the cure depths to be checked are CD1 = 9, CD2 = 16, CD3 = 21, and CD4 = 26 mils.

Test Procedure. The Reverse WINDOWPANE™ test procedure involves part building, postprocessing, and measurement. The required equipment and processes were discussed in section 10.3.

Tolerance

The deviations of the measured cure depths compared to the predicted cure depths depend on part building technique. In the Tri-Hatch building technique, every vector is cured 6 mils deeper than the layer thickness, and a cure depth error of ±2 mils is allowed. However, with WEAVE™ or STAR-WEAVE™, the actual hatch cure depth is 9 mils for a 10 mil layer thickness. The maximum cure depth error allowed in this case is ±1 mil.

Test Results

Typical diagnostic test results are tabulated in Table 10-11.

Table 10-11
Reverse WINDOWPANE™ Cure Depth Measurements
P_L = 24.2 mW; B = 10.1 mils; h_s = 4 mils
SLA-250 Serial number: 90-0076
Resin: XB 5081-1, Lot#: Mixed

C_d identification	CD1	CD2	CD3	CD4
Measurement 1 (mils)	9.30	15.85	21.20	26.35
Measurement 2 (mils)	8.95	16.20	21.35	26.05
Measurement 3 (mils)	9.05	16.15	20.90	26.45
Measurement 4 (mils)	9.15	15.95	21.45	26.05
Measurement 5 (mils)	9.05	16.25	21.20	26.20
Measurement 6 (mils)	8.95	16.15	21.15	26.05
Measurement 7 (mils)	9.15	15.95	21.25	26.00
Measurement 8 (mils)	9.00	16.25	21.15	25.90
Measurement 9 (mils)	9.20	15.95	21.00	26.25
Measurement 10 (mils)	9.00	16.05	21.15	26.00
Range (mils)	0.35	0.40	0.55	0.55
C_d Average (mils)	9.08	16.08	21.18	26.13
C_d Specified (mils)	9.00	16.00	21.00	26.00
Error (mils)	0.08	0.08	0.18	0.13

Discrepancy

In this experiment, discrepancies become readily apparent. The measured C_d averages are outside the specifications when the difference between the actual cure depth and the specified cure depth is greater than 1 mil. The data is regarded as inconsistent when *any* of the range values are greater than 1 mil. Thus, there are two cases of discrepancy:

1. The range is outside the specification. The reasons can be either that excessive resin has not been completely absorbed, or that the "zero reference" of the point micrometer may have changed during the measurement. In this case, that measurement must be repeated.
2. The C_d average is outside of the specification. The causes of this discrepancy lie with the calibration of the RP&M system. Typically:

- The power reading is out of calibration.
- The laser beam profiling is out of calibration.
- The resin parameters D_p and E_c are out of specification.

If the third item is the cause, the remedy is to perform the WINDOWPANE™ diagnostic test as previously described. If either the first or second item is the cause, then system maintenance is required.

10.5 CHRISTMAS TREE™ Method

3D Systems investigated potential diagnostic parts that might provide linewidth compensation and shrink factor in a simple, rapid, accurate, and repeatable manner. Initially, it was assumed that multiple diagnostic parts would be required. After some iteration, it became clear that linewidth compensation (LWC) and shrinkage compensation factor (SCF) could be determined from a single diagnostic part.

Figure 10-24 is a photograph of this diagnostic part, known for obvious reasons as CHRISTMAS-TREE™. *Figure 10-25* is a schematic drawing of this part.

The basic concepts behind the CHRISTMAS-TREE™ diagnostic part are as follows:

1. Build five simple, geometrically scaled rectangular slabs, all having thickness to width to length ratios of 1:2:10.
2. Let the range of dimensions extend from 0.200″ to 5.00″, as shown in *Figure 10-25*. In this way, we achieve a wide span in linear dimension, while keeping the size small enough that it can be easily measured using digital calipers (for example, Mitutoyo Digimatic Model 500-351).
3. Connect the five rectangular slabs together to ensure that their orientation with respect to one another is preserved. However, be certain that the connection itself has minimal impact on each slab.

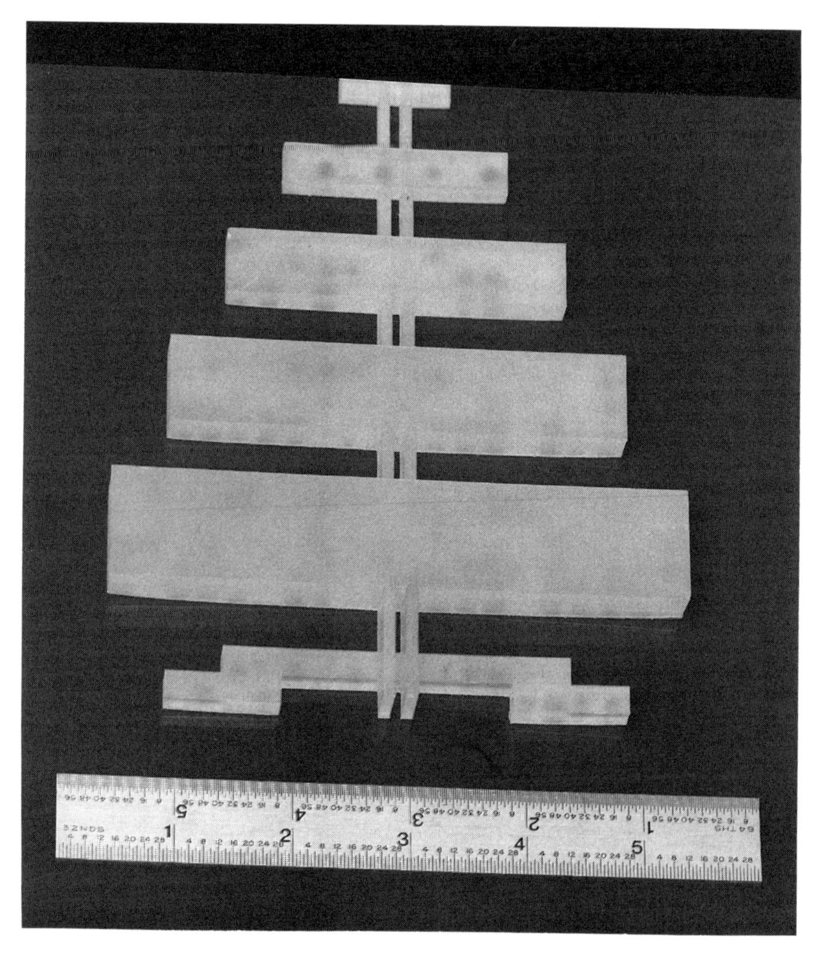

Figure 10-24. CHRISTMAS-TREE™ diagnostic test part.

4. Intentionally build all the slabs with zero LWC and zero SFC since the purpose of this diagnostic part is to experimentally *measure* these quantities, rather than assigning estimated values.
5. Measure the *length* of each slab five times. Since there are five slabs, this will involve 25 measurements. Fortunately, each measurement takes only about five seconds, so this step requires only a few minutes.
6. Measure the *width* of each slab five times. This will also involve another 25 measurements and a comparable amount of time.

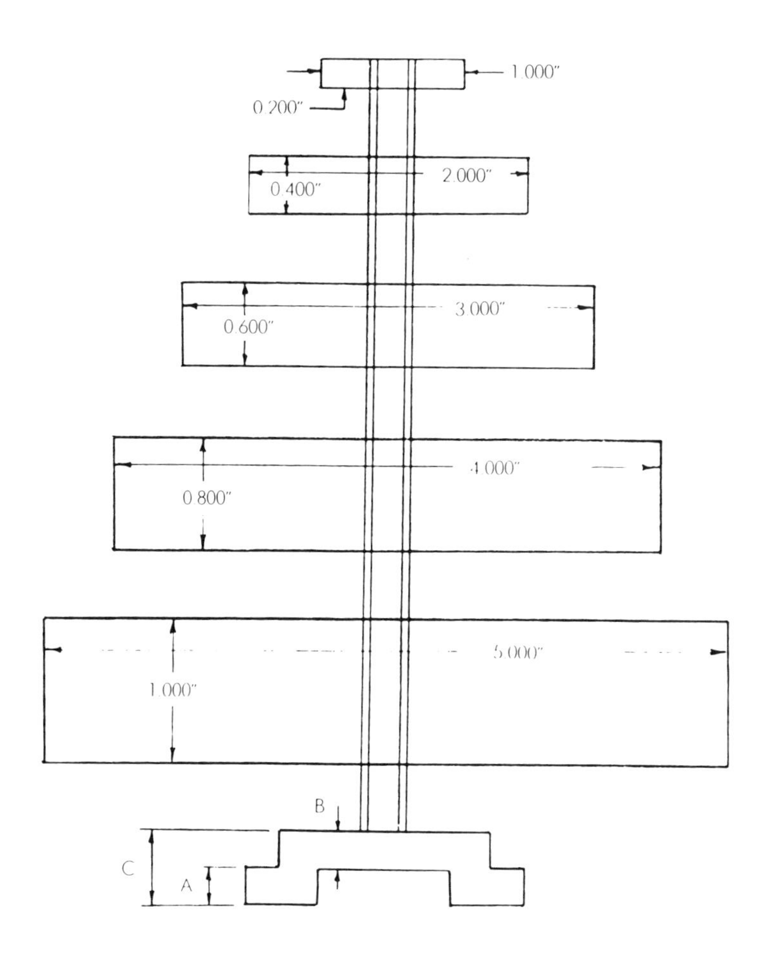

Figure 10-25. CHRISTMAS-TREE™ diagnostic test part.

7. The "base" of the CHRISTMAS-TREE™ diagnostic test part is intentionally shaped in the form of an ascending stair-step, a flat section, and a descending stair-step, as shown in *Figure 10-24*. We define the width of the outermost step on the left and right as *A*, the width of the flat section as *B*, and the width of the double step on each side as *C*, as shown.
8. Measure the values of *A*, *B*, and *C* five times on the right and five times on the left. This will require 30 measurements. Hence, the total number of caliper measurements to be performed on this diagnostic part is 80.

This should take under 10 minutes, especially after one has done it a few times.

9. Input each of the measurements for the 10 dimensions into the special "Diagnostic Disk" linear regression software analysis program.
10. For each of the measurements, the program will subtract the nominal CAD dimension. The result will be 50 error values.
11. The computer program will then automatically calculate the mean value of each set of five errors for a single dimension, will perform a linear regression analysis on the 10 mean values, and will compute the "best fit" values for LWC and SCF.
12. If only one CHRISTMAS-TREE™ is built, the computer program will output the best mean value SCF for both x and y shrink factors. However, if two CHRISTMAS-TREES™ are built *orthogonally* on the same platform, the computer can then separate the independent x values from the y values, and thereby output the optimum values of SCF(x) and SCF(y) separately.

As a final check on the value of LWC, input the 10 sets of data obtained for each of the three quantities A, B, and C discussed previously in steps 7 and 8 into the "Diagnostic Disk" software. The program will calculate

$$L_{w,i} = [A_i + B_i] - [C_i] \qquad \text{for } i \text{ from 1 to 10.}$$

Then, the software will calculate the mean value L_w, compare the value of L_w with that of LWC, as determined by the computer from the ordinate intercept method. If these two quantities are within 2 mils, then the program will output the value of LWC for future linewidth compensation on this particular SL machine. If, on the other hand L_w and LWC differ by more than 2 mils, the entire experiment should be repeated as there is a reasonable probability that either a procedural or experimental error has been made.

Results

If properly performed, the combination of the three diagnostic parts, WINDOWPANE™, Reverse WINDOWPANE™, and CHRISTMAS-TREE™ should provide the user with accurate values of the following six resin and build parameters, as well as confirmation of three additional system parameters for their particular SLA, resin, laser, and optics.

Resin Parameters	Build Parameters	System Parameters
D_p	C_d	P_L
E_c	LWC	W_o
	SCF(x), SCF(y)	V_s

We recommend that users learn to perform these tests rapidly and accurately. Each diagnostic sequence should only take a few hours. Once the system has been properly characterized, these tests need be repeated only whenever one of the following events occurs:

1. A new laser or laser head is installed.
2. A new resin is installed.
3. Something suddenly "changes" regarding the quality or accuracy of the parts being generated.

The time spent performing these diagnostic tests will be well rewarded in the form of improved part quality.

REFERENCES

1. Taylor, J.R., *An Introduction to Error Analysis*, University Science Books, Mill Valley, CA, 1982, pp. 153–157.
2. Downie, N.M. and Heath, R.W., *Basic Statistical Methods*, Harper & Row, New York, 1970, pp. 129–137.

chapter 11

Accuracy

The combination of the causes of phenomena is beyond the grasp of the human intellect, but the desire to find those causes is innate in the soul of man. The human mind, with no inkling of the immense variety and complexity of circumstances conditioning a phenomenon, any one of which taken separately may seem to be the cause, snatches at the first and most easily understood approximation, and says: This is the cause!

—Count Leo Tolstoy
War and Peace, Part XIII
Russia, 1869

11.1 Introduction

Accuracy is the cornerstone of RP&M. Within this book, we discuss in some detail each of the following: resin properties, enhanced laser performance, the development of photomodulus, software algorithms, solid modeling CAD systems, basic part building, advanced hatching techniques, improved post processing methods, diagnostic measurements, productivity gains, and financial benefits. Every one of these is not only possible, but has been achieved by a wide array of users. However, without accuracy, they are often meaningless.

Developing a new impact resistant resin is only significant if the resulting prototype is accurate. Improved laser performance would be a hollow victory if the resulting parts were distorted. Enhanced photomodulus of inaccurate parts is worthless. An improved software algorithm generating invalid results is almost an oxymoron. The whole point of part building with the new advanced methods

By ***Jan Richter***, *Senior Research Engineer, and* ***Paul F. Jacobs, Ph.D.***, *Director of Research and Development, 3D Systems, Inc., Valencia, CA.*

is to generate accurate parts. The benefits of the process are virtually moot if the prototypes are not accurate.

Having said this, one must first decide exactly *how accurate* a prototype must be to be useful. Second, a reliable method of *measuring* part accuracy is needed. And third, sufficient data must be gathered to be able to establish the *repeatability* of the process. As the reader will soon discover, from a simple example, the real foundation for establishing the accuracy of a process is conceptually straightforward. Unfortunately, it is also demanding, requiring thousands of measurements to just begin to sort out "the immense variety and complexity of circumstances conditioning a phenomenon....", as Tolstoy noted over a century ago.

11.2 Fundamentals of Physical Measurements

About 35 years ago, one of us was involved in a physics course which had a laboratory requirement. The laboratory portion used its own textbook.[1] This text, now long out of print, has since gathered its share of dust. Nonetheless, this book still has one of the simplest and best explanations of the fundamental principles of physical measurements extant. Much of this section is based upon the concepts described in Reference 1. The specific examples which follow have been updated and made relevant to RP&M.

Let us begin by considering a simple cube. Further, let us assume that the cube was designed on a CAD system, with the intent that the length of each side, S, shall be exactly 2.000″. (While this book would normally list dimensions in both English and metric units, for this example, we shall use English units exclusively in the interests of clarity. A similar example could also be done using metric units.) We now build this cube on an SLA, and when completed we take care to mark the x, y, and z directions subsequent to resin stripping, while the part is still attached to the platform. Next, we carefully remove it from the platform, postcure it, and then stand ready to measure the part. Now, what do we measure it with?

We could use a yardstick. However, the smallest markings on most yardsticks are usually one-eighth of an inch. Since we could probably visually estimate quite easily to within half or perhaps a quarter of the smallest divisions, we might estimate the length of one side of our cube to within about 1/32″ using a yardstick. Anything much finer is beyond our ability to resolve. Hence, the *resolution* of a yardstick is about 1/32″, or roughly 0.030″ or 30 mils. If 30 mils is good enough, then a yardstick is appropriate.

Certainly, if one were measuring a room, this level of resolution would be more than adequate. In fact, if you saw a room dimension listed as 20′ 3.031″, you would probably think: "My, what a strange value for the size of a room." Good! You should have this reaction. Why? For two reasons: first, because the last indicated digit (the 1 in 3.031) is *beyond the resolution limits of the measuring instrument*, and second, because *the conditions of the measurement need to be described at a level commensurate with the accuracy implied.*

As an illustration, let us assume that the room was built with an oak floor, and that the linear coefficient of thermal expansion of white oak, with a water content of 10%, is hypothetically 24 microinches/inch/° F. As a result, a temperature rise in the oak floor of only 1°F (0.56°C) would cause an increase in the length of the room of about 0.006″.

Since the temperature of the floor can easily vary by 20°F from the coldest day in winter to the hottest day in summer, the change in the length of the room could amount to as much as 0.120″. Now, how meaningful is 20′ 3.031″? This is the reason that you should be skeptical of such a statement.

At this point, it should be clear that 20′ 3″ is an appropriate measurement for this room, but that 20′ 3.031″ is not, unless more information is given. For example, one could have firmly secured two mirrors on opposite walls, and then used an optical interferometer to make the measurement. Optical interferometers measure distance through the observation of fringe shifts in the interference patterns of light. In this case, the resolution is determined by the ability to distinguish fractions of a fringe shift.

The major difference between an optical interferometer and a yardstick, ignoring for the moment the great differences in cost and complexity, is resolution. While the yardstick might have a resolution of about 0.030″, the interferometer will probably have a resolution in the vicinity of 0.000002″.

At first glance, this looks terrific. Unfortunately, if one were to attempt to measure the dimensions of a room using an optical interferometer, two big problems would occur. First, the measurement would be extremely difficult, and second, the results would be wildly fluctuating from one minute to the next. The reasons are that the system would be hypersensitive to tiny temperature variations, as well as vibrations from cars driving nearby, and probably your own breathing as you made the measurements. The simple action of a thermostat turning on the central heating system would probably look like an earthquake on a strip chart recording of the interferometer output. A person simply walking anywhere in the house would likely drive the instrument to its limits, and the data would take some time to return to equilibrium as the system recorded the tiniest levels of vibration in the walls, floors, and ceilings.

The main point of this discussion is, of course, not intended to suggest that one should measure 2″ cubes with yardsticks, or living rooms with optical interferometers. The real point is to make the reader aware that such questions as "What was the temperature?," or "What was the humidity?," or even "How recently did somebody walk down the hall?," may indeed be relevant whenever a measurement pretends or implies that it is especially accurate.

Measurement Precision

The concepts of resolution, precision, and accuracy are all close relatives. And, like cousins, they may be related, but they are *not* one and the same. Returning to our little cube, let us assume that you finally decide that a yardstick does not have fine enough resolution. Further, an optical interferometer, while capable of very fine resolution, is also quite expensive, very time consuming to set up and calibrate, and highly impractical.

After thinking about this measurement for a while, you finally decide that the results need to be within 0.001″, or one mil. However, to do this, you realize that the measurements will intrinsically be limited by the resolution of the instrument. Thus, the *accuracy* of the measurements will generally be constrained by the resolution of the equipment.

As a result, you finally decide to use a high quality digital caliper with a resolution of 0.5 mil (the smallest value that is displayed). At this point, you make your first measurement and get a reading of 2.060″. You decide to measure it again along the same side. This time your reading is 2.061″. You then decide to measure it eight more times, for a total of 10 measurements. The measured values are as follows: 2.060″, 2.061″, 2.061″, 2.0615″, 2.0625″, 2.0615″, 2.0605″, 2.061″, 2.0605″, and 2.060″. You now pull out your calculator and determine the *mean value*, which is simply the sum of all 10 readings divided by 10, or in general:

$$\overline{S} \equiv \frac{1}{N} \sum_{i=1}^{N} S_i \tag{11-1}$$

where S_i is the ith measurement value,

N is the total number of measurements,

and $\overline{S}$ is the mean value of the measurements.

For our example, with $N = 10$, we obtain $\overline{S} = 2.06095''$. However, no matter how many digits the calculator may blindly produce, *the significant figures cannot exceed that of the input data itself.* This is so because beyond that point, the raw data is not trustworthy and hence is not "significant." Therefore, we take $\overline{S} = 2.061''$. Next, the standard deviation, σ, is defined as

$$\sigma = [\sum_{i=1}^{N} (S_i - \overline{S})^2 / (N\text{-}1)]^{1/2} \tag{11-2}$$

For this data, the standard deviation turns out to be 0.0007454″. Again, considering the resolution limit of the instrument as well as the raw data, the

only significant figures would be 0.0007″, or 0.7 mil. The standard de
which is a measure of the *precision* of the measurements, is small. In fact, it is comparable to the instrument resolution (0.5 mil). This data would generally be referred to as originating from a precise set of measurements. Therefore, this would be an accurate result, right? **Wrong!**

Three days ago, somebody accidentally knocked the caliper off a lab bench onto a concrete floor and then put it back without reporting what had happened. As a result, the caliper is now out of calibration. Your data is precise, but it is **not** accurate due to the presence of a *systematic error*. In making physical measurements there will always be unavoidable *random errors*, but we should strive to eliminate systematic errors.

We may now define *precision as a measure of how tightly clustered a set of measurements are about their mean value*. The more tightly clustered the data, the more precise the result. *The indicator of precision is the standard deviation of the data*. The smaller the value of σ, the greater the precision. Does this tell us anything about accuracy? Not directly.

The reason is that *accuracy is an indication of how close a measurement is to the true value of something*. The Heisenberg Uncertainty Principle (see Reference 2), which is fundamental to quantum mechanics, tells us that no measurement can ever be perfectly accurate. However, when using the very finest digital caliper in the world, one is still many orders of magnitude from reaching Uncertainty Principle limitations.

A much closer physical measurement limit involves the molecular structure of matter. With recent advances in instrumentation, some laboratories are actually approaching the molecular limit. For SL resins involving so-called long chain polymer molecules, this could be as little as a few tenths of a nanometer across the width of such a molecule, to as many as about 7 nm along its length. At this level, the measurement would actually depend upon the local molecular orientation! To our eyes, the surface may seem smooth, but on a molecular level it is very bumpy.

Even in the extreme case, 7 nm is about equal to 0.3 microinch, or about 0.0003 mil. If 0.3 microinch was your definition of acceptable accuracy, you would literally have to worry about the surface orientation of individual polymer molecules. The key point here is that *the specification of an acceptable accuracy level establishes the difficulty of the measurement*. The required accuracy level also defines the problems and costs associated with assuring that any particular measurement does, or does not, meet that specification. This is one of the central predicaments of quality assurance. Loose specifications reject few parts, but ultimately may result in sloppy products. Tight tolerances improve product quality, but, initially at least, may result in a flood of rejects. In summary, be careful what you ask for, it may cost you a king's ransom to get it.

Measurement Accuracy

At this point, the reader should begin to appreciate some of the problems associated with making "accurate" measurements, as well as how to judge if they are indeed "precise." For example, we have already seen a case where precise measurements were not accurate (the damaged caliper). Conversely, one can also think of a case in which an accurate result could stem from data that is not precise. Returning to our cube, imagine a series of measurements as follows: 2.040″, 2.021″, 2.000″, 1.980″, and 1.960″. The mean value of this measurement series is 2.0002″, which is an amazingly *accurate* result since the "true" CAD value was intended to be 2.000″.

However, the standard deviation, as a measure of the *precision* of the measurements, would be about 0.032″, or about 32 mils. Clearly, if you were trying to make measurements to within 1 mil, this would be regarded as very imprecise data. In short, the "accuracy" in this case was more a matter of what is sometimes referred to as "compensating errors" than it was the inherent validity of the data.

Now, we are finally getting closer to the heart of the matter. Let us again return to our little cube. Having just been disappointed after finding out about the damage to the caliper, we decide that this time, before we proceed any further, we will *calibrate* the instrument. This is done by using it to measure some set of "standard" test blocks that have been made from a material with a very low linear coefficient of thermal expansion, such as Invar. This is done to minimize expansion or contraction effects. Finally, these standard test objects should come with a specified set of *tolerances*, defining the maximum and minimum values within which the measurements should occur.

Had we done this the first time, we would have quickly discovered that "something was wrong." After testing the original caliper on two or three standard test blocks, it would have been quickly evident that a systematic error existed. At this point, the first caliper should be sent for repair and recalibration, and another instrument should be selected and calibrated. When either the new or the old caliper was able to consistently yield readings within the tolerances of the standard test objects, only then is it ready for use.

We are now ready to make further measurements. However, let us measure the cube **10 times** in each of the three principal directions (x, y, and z). The results of **one set** of 30 measurements on a **single** cube might be as shown in Table 11-1 (dimensional values in inches, as noted previously).

Are the results precise? Well, the standard deviations are all about 1.6 mils. If we intend to measure within one mil, we would at first tend to say that perhaps these measurements were not sufficiently precise, despite the fact that the values of the standard deviations for the x, y, and z dimensions were seemingly quite *repeatable*. But, are the three sets of data sufficient to establish repeatability? In this case, probably not. Why? Because the three mean values are not repeatable, and, as we shall soon see, the number of test points is too small.

Table 11-1
Cube Measurement

No.	*x* dimension	*y* dimension	*z* dimension
1	**2.012**	**2.008**	**2.011**
2	**2.011**	**2.007**	**2.010**
3	**2.012**	**2.009**	**2.010**
4	**2.013**	**2.010**	**2.007**
5	**2.015**	**2.006**	**2.008**
6	**2.013**	**2.011**	**2.006**
7	**2.010**	**2.008**	**2.009**
8	**2.015**	**2.006**	**2.009**
9	**2.013**	**2.007**	**2.006**
10	**2.013**	**2.008**	**2.008**
Mean Value	**2.013**	**2.008**	**2.008**
Standard Deviation	**0.0016**	**0.0016**	**0.0017**

Specifically, the mean value, $\overline{X}$, of the x dimension measurements is definitely larger than either of the mean values $\overline{Y}$ or $\overline{Z}$ of the y or z data. The difference between $\overline{Y}$ and $\overline{Z}$ would not be considered statistically significant since to the nearest mil they are equal. However, the difference between $\overline{X}$ and $\overline{Y}$ is significant, according to Chauvenet's criterion[3]. Specifically, $(\overline{X} - \overline{Y})/\sigma = 0.005''/0.0016'' = 3.13$. From the standard tables of Gaussian probabilities (see Reference 3), it can be shown that a difference as large as this could be expected to occur, due to random error, only about once in 500 measurements!

Therefore, we are left with a number of questions.

1. Is there a "real" difference between the performance of our system in the x direction relative to that in the y or z directions?
2. Is the data repeatable?
3. Is the data accurate? If so, how accurate?

To determine the answers to these questions, we need to do more work. The information necessary to accomplish these tasks follows.

The Measurement Distribution Function

An important test of physical measurement data is to plot the number of measurements whose values fall within a preset "bin" on the ordinate, against the values of the dimensions themselves on the abscissa. By a "bin," we mean a small increment of the measurement. For example, one might create bins of 1 mil width. In this case, one would count all the measurement values that lie between 2.000″ and 2.001″ as part of the total for the first bin. Then one would count all those values between 2.001″ and 2.002″ in the second bin, etc. for all bins necessary to span the extremes of the data. Those values that land "on" the dividing value between bins (2.001″) may be assigned by an arbitrary convention, such as "count measurements landing on the divider as part of the lower absolute value bin."

This plot is known as a "Measurement Distribution Function" (MDF), since it graphically shows the distribution of all the various measurements of a given dimension. For the case of the data in the previous table, since it all occurs to the nearest 0.001″, or 1 mil, we could conveniently choose 1 mil wide bins that "straddle" the integer mil values (from 2.0005″ to 2.0015″). In this way, every measurement unambiguously falls in a bin, with none landing "on" the dividers. *Figure 11-1* shows the MDF for the data of our example. Also, to see if there is any statistically significant difference between the results in the x, y, and z directions, the data for each of these three subgroups should be plotted separately. Here N is the number of measurements within a given bin and S is the actual length of a given measurement.

Note that in *Figure 11-1*, a total of 10 bins are required to encompass the data. Also, when points from the x, y, or z distributions would otherwise land "on top of one another," they are simply placed laterally adjacent, with the small horizontal gap between the symbols not intended to have any physical significance.

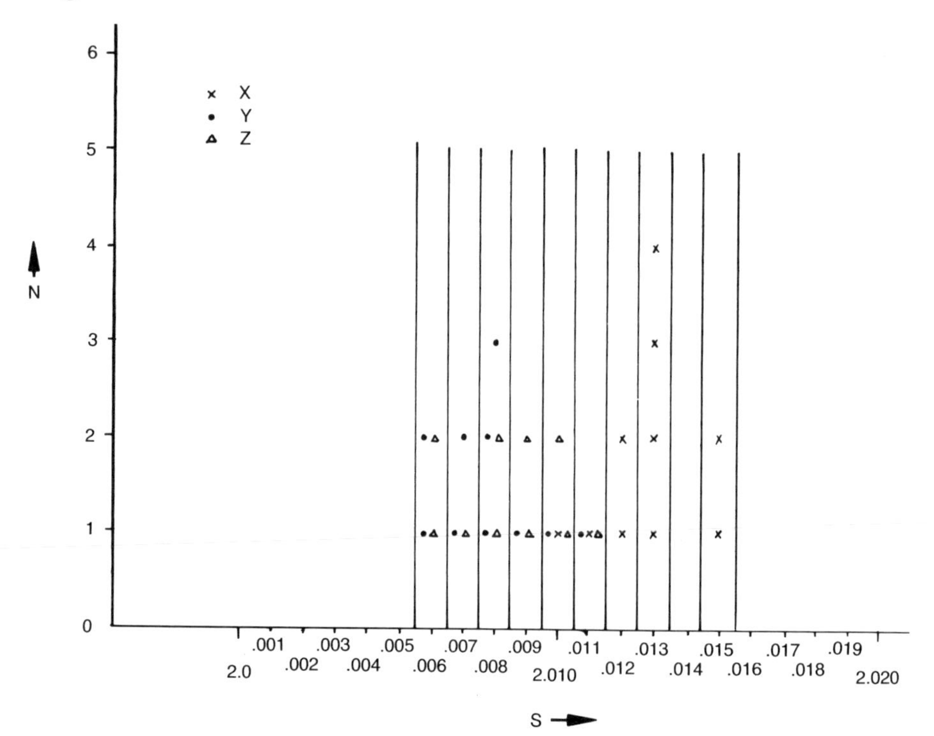

Figure 11-1. Measurement distribution function.

Unfortunately, about the only conclusion one could draw from *Figure 11-1* is that "the cube is probably a little bigger than 2″ on each side," and "the three sides are about the same." The reason these conclusions are so vague is because there is insufficient data to draw stronger conclusions. While it would "seem" as though there "might" be a "tendency" (note the vague wording) for the x dimensions to be greater than the y or z dimensions, there is just enough overlap in the data that you cannot draw specific conclusions with a high *confidence level*.

It is at this point, unfortunately, that the world tends to divide into two camps. We may refer to one group as the "data takers," and the other as the "conclusion jumpers." The data takers believe that what is needed is much more data. This will take time, money, and lots of hard work. Conversely, the conclusion jumpers take a glance at the original data and quickly decide that the "average" measurement is about 2.010″, so there must be roughly a 10 mil error, and since this occurred on a 2″ dimension, then the accuracy of the system must correspond to an error of about "five mils per inch." Also, the x, y, and z dimensions are "close enough."

Meanwhile, the data takers go back to the lab and painstakingly build *one hundred* 2″ cubes. This takes many weeks, uses almost half a vat of SLA-250 resin, logs hundreds of hours on the laser, begins to load up a tank full of TPM, and involves multiple calibrations of the caliper as well as one for the SLA. Finally, these stoic fellows gather their data and begin to analyze it mathematically.

The Error Distribution Function

Once our data takers carefully examine their results, they will quickly notice that most of the data is reasonably close to the desired dimension. Therefore, rather than plotting the measurements of a single dimension separately, they might as well plot the error for all the measurements on the same abscissa. Thus, we may define ϵ_i as the *error* of the ith measurement, such that

$$\epsilon_i \equiv S_i - S_d \qquad (11\text{-}3)$$

where S_i is the value of the ith measurement

and S_d is the *design* value of that measurement

Note that since S_i may be either larger or smaller than S_d, then ϵ_i may be either positive or negative.

Since the design value, S_d, for our cubes was 2.000″, we may now subtract 2.000″ from all the data. Using a modern digital computer, this is done very rapidly. Next, we plot N, the number of measurements per bin, on the ordinate, versus the error, ϵ, on the abscissa. This type of plot, involving N versus ϵ, is *absolutely fundamental* to any study of accuracy, and is known as the **Error**

Distribution Function, or EDF. A carefully generated EDF is an essential tool for determining system accuracy and repeatability. The major advantage of the EDF relative to the MDF is that a whole range of measurements can be plotted simultaneously since you do not need to select variable abscissa values for each different dimension.

Figure 11-2 shows the results of 1000 measurements on 100 cubes, just in the *x direction* (10 measurements per cube). We will get to the data for the *y* and *z* directions shortly.

Upon inspection of *Figure 11-2*, we notice a number of things. First, the data peaks near zero error. Second, the data, in the form of a histogram, involves a series of discontinuous horizontal line segments rather than a smooth curve. It would only become a fairly smooth curve if we had many more bins, with each bin being much narrower. Thus, we take as much data as possible until sufficiently confident to draw a smooth curve between the data points, as seen in *Figure 11-2*.

The central concept here is that the greater the number of measurements, the higher the confidence that interpolating a smooth curve will not lead to erroneous conclusions. For example, if the errors are truly *random*, the Error Distribution Function should approach a classical Gaussian distribution. Under these conditions, the function would be symmetric, and centered about zero error. Physically, this implies that positive and negative errors are equally

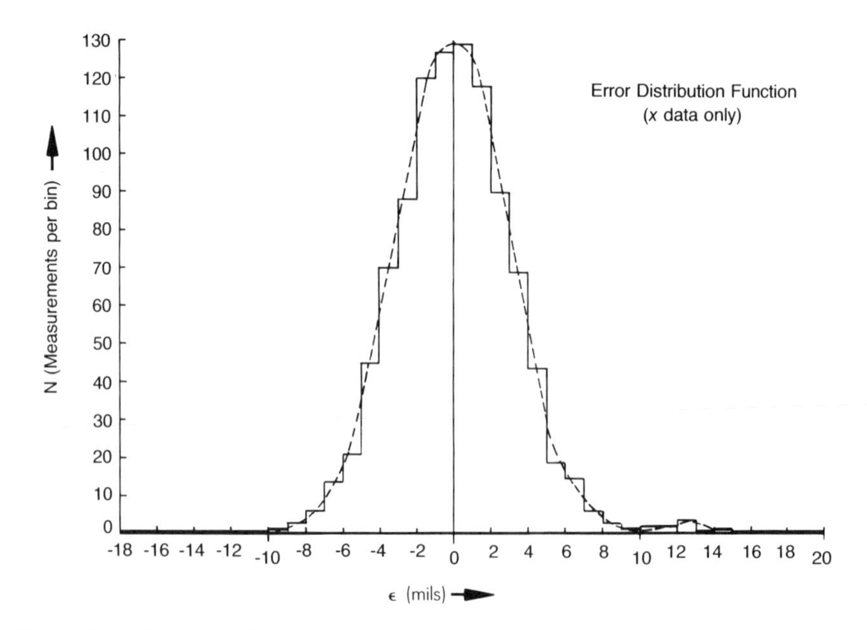

Figure 11-2. Error distribution function in *x*.

likely, and their magnitudes are equally distributed.

In fact, any departure of the EDF from symmetry is usually a good indicator that some sort of *systematic error* is present. This is often a clue that system performance can still be improved by identifying and ultimately reducing or eliminating the source of the systematic error.

Inspection of *Figure 11-2* also reveals a small, secondary "bump" centered near +13 mils error. Looking at the raw data, it becomes clear that almost all of the values associated with the "bump" seem to stem from the x dimension measurements of the first cube. After you walk over to the lab and speak to the SLA operator who built all the cubes, he checks his records and discovers that he *forgot to include linewidth compensation on the first cube*, but then corrected this omission on the subsequent 99 cubes. Sorting this kind of thing out is not trivial.

In this case, our conclusion jumpers would definitely have fallen into Tolstoy's trap. While their first conclusion, that the part was about 10 mils large, certainly seemed reasonable at the time, the main point is that it was not due to a machine error, but rather an "operator error." Also, their conclusion that the system accuracy corresponds to an error of about "five mils per inch" is no longer supported by the data. Finally, their conclusion that x, y, and z are "close enough" remains to be proved or disproved.

To do this, we return to the data for all 3000 measurements on 100 cubes (now actually 2970 measurements on 99 cubes, since we need to discard the data from the first cube, as this was clearly built under different conditions than the other 99). Thus, 990 measurements are available in each direction (x, y, and z). The results for all the measurements are plotted in *Figure 11-3*.

This is much more substantial data and affords a number of important conclusions:

A. The standard deviation of these measurements may be determined from the EDF. We shall define the Gaussian function in a manner consistent with Reference 3, such that:

$$f(x) = A \exp\left[-(x-\overline{X})^2/2\,\sigma^2\right] \tag{11-4}$$

For the case $\overline{X} = 0$ (an EDF centered at zero), when $x = 0$, $f(0) - A$. Furthermore, it is clear that when $x = \sigma$, $f(\sigma) = A \exp[-1/2] = 0.6065\ A$. Therefore, taking a simple ratio, $f(\sigma)/f(0) = 0.6065$. In other words, the value of the standard deviation of the entire data ensemble can be determined directly from the EDF by drawing a horizontal line at roughly 61 % of the maximum EDF value, and then graphically establishing the two values where this line intersects the EDF curve. The standard deviation of the data is then the half-width of the EDF function at 61% of its maximum value.

For our specific example, the results are as follows: $\sigma(x) = 3.03$ mils, $\sigma(y) = 3.02$ mils, and $\sigma(z) = 3.03$ mils. Remembering our earlier point about significant figures, we take $\sigma = 3.0$ mils for all three curves.

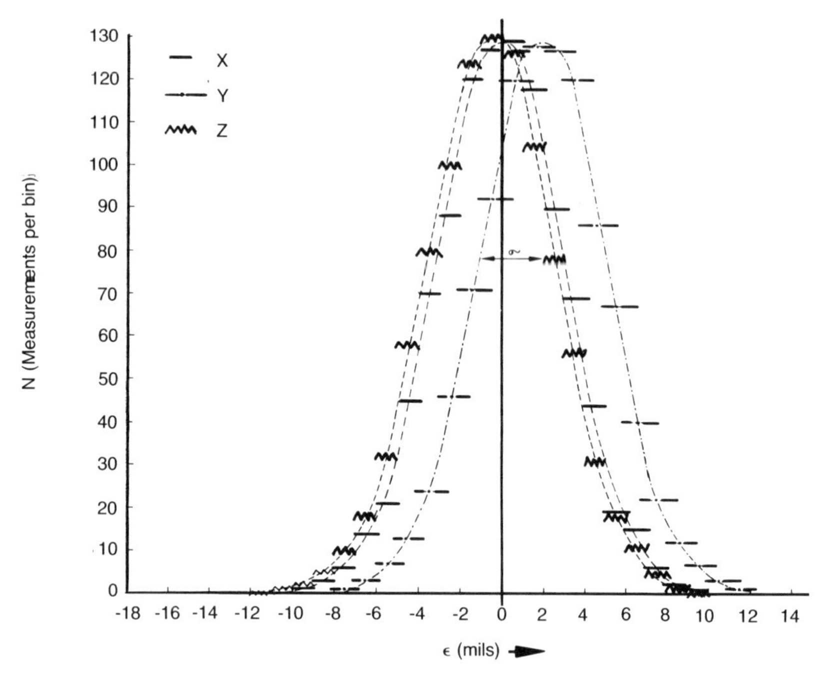

Figure 11-3. Error distribution functions in *x*, *y*, and *z*.

B. Since the standard deviation of each group of measurements is 3 mils, it will **not** be possible to set realistic tolerances at ±1 mil. From Reference 3, we find that for $t = 1$ mil/3.0 mils $= 0.33$, only about 26% of all dimensions built in either x or y or z could be expected to lie within ±1 mil. Even for $t = 2/3 = 0.67$, only about half of the dimensions will lie within ±2 mils. Listing the percentage of measurements that can be expected to lie within a given *tolerance* band, assuming that the EDF is Gaussian with a standard deviation of 3 mils, we obtain the results displayed in Table 11-2.

Table 11-2
Tolerance Results

Tolerance Band	% Measurements in Tolerance
± 1 mil	26% (nearest percent)
± 2 mils	50% (nearest percent)
± 3 mils	68% (nearest percent)
± 4 mils	82% (nearest percent)
± 5 mils	91% (nearest percent)
± 6 mils	95% (nearest percent)
± 7 mils	98% (nearest percent)
± 8 mils	99.2% (nearest 0.1%)
± 9 mils	99.7% (nearest 0.1%)
±10 mils	99.9% (nearest 0.1%)

The user must now decide: "Can I live with ±5 mils for about 90% of the part dimensions?" If the answer is yes, fine; the data will support such a set of tolerances. If the user still insisted upon ±1 mil tolerances, the inescapable conclusion, based on the data of *Figure 11-3*, is that only about 26% of the part dimensions could be expected to lie within this specification.

C. Furthermore, inspection of the data shows that the agreement between the measurements in the x and z directions is quite close. Specifically, the *most probable error* in x (the value of ϵ for the x data at that location where the EDF peaks) happens to be very near zero. What this means is that positive and negative errors are equally likely. This is an excellent result, and indicates that the sources of error are predominantly random, with little evidence of any systematic errors. Also, although the z data does show a most probable error of about -0.5 mil, this is sufficiently small relative to the standard deviation (3.0 mils) that according to Chauvenet's criterion, as noted earlier, one could quite reasonably conclude that any difference in the performance of the machine in the x and z directions is **not statistically significant**.

D. However, there is a greater difference between the y and x data, or the y and z data. For the y data, the most probable error is about +2 mils. This means that the y dimensions are more likely to be oversized rather than undersized. This indicates that the y dimensions exhibit a systematic displacement, which biases them towards positive errors. This could be due to asymmetry in the laser beam, an incorrect shrinkage compensation factor, SCF(y), problems with the scanning system, or something else. Whatever the cause, this bias in the y direction is almost certainly statistically significant based on Chauvenet's criterion, since t = 2/3 = 0.67. Hence, from the Gaussian tables, there is about a 50% probability that this effect is real since about 1470 of the 2970 measurements would be expected to show differences this great.

Cumulative Error Distribution

As we have seen, the EDF can provide much important information regarding accuracy and repeatability. However, an additional tool that is very helpful is the Cumulative Error Distribution (CED). What this curve provides, in a very handy format, is a straightforward method of establishing practical tolerance levels as well as realistic accuracy expectations.

The CED is based upon the Error Distribution Function (EDF), so one must develop the EDF first. From a mathematical standpoint, the CED is a graphical representation of the normalized integral of the EDF from zero error to infinity. Physically, what this means is really quite simple.

Let us return to *Figure 11-2* for thc purposes of illustration. First, count the number of measurements falling between -1 mil and +1 mil error. Add these together and divide by the total number of measurements. This normalizes the data. The result is then the fraction of all measurements *within* one mil of their design value. This result is then multiplied by 100 and expressed as a percentage. For our example, from *Figure 11-2*, this would be 129 + 127 = 256 measurements out of 990, or about 26% within one mil of their design value.

Second, count the number of measurements falling in the next higher bin for positive errors as well as the next lower bin for negative errors. In our case, this would be the value of N for the bin from +1 mil to +2 mils (118) plus the value of N for the bin from -1 mil to -2 mils (120). This subtotal would then be 118 + 120 = 238. Next, add this value (238) to the original value from -1 mil to +1 mil (256), forming a subtotal of 494. Then, renormalize and convert to percent. The result is 49.9%. Thus, about half of the measurements could be expected to lie within two mils of their design value.

Obviously, this procedure is continued bin by bin until 100% of all the measurements have been counted. *A plot of the percentage of measurements falling within a given error versus the magnitude of the error is known as the CED*. The origin of the term "cumulative" should now be evident since the CED is formed by accumulating data from the EDF on a bin by bin basis.

Figure 11-4 shows the CED that results from the EDF of *Figure 11-2*. Note the two handy features of the CED. First, it is very easy to see what kind of results you could expect if you set different tolerance levels. Clearly, the tighter the tolerance, or maximum allowable error, the smaller the percentage of measurements that one could expect to stay within that specification. This helps guard against unrealistic expectations.

Second, the CED will actually establish the *confidence limits* for a given specification. Because of the Gaussian nature of random errors, there is always a very small but finite probability that one may encounter an error many times larger than the standard deviation of the entire ensemble of data. In other words, strictly speaking, to insure that **every** measurement will fall within the tolerance band, the tolerance would have to be set at infinity! This is obviously of no value.

However, if the user is willing to accept, say 99% of all measurements within a given tolerance, the allowable tolerance value very rapidly becomes quite finite. Unfortunately, the problem with setting the confidence limits at 99% is that the CED is usually quite flat in this region, and a small error in plotting the results can cause a big error in determining the appropriate tolerance value.

On the other hand, the CED is quite steep near 50% probability. Unfortunately, few people are willing to settle for a tolerance level where there is only a "fifty-fifty" chance that any measurement will lie within the specification. In short, this gives a sharp and well defined intersection that is relatively insensitive to small plotting errors, but which establishes far too low a confidence limit.

Further, by drawing a horizontal line at 68% on the ordinate, one can determine the standard deviation of the data ensemble, assuming the original

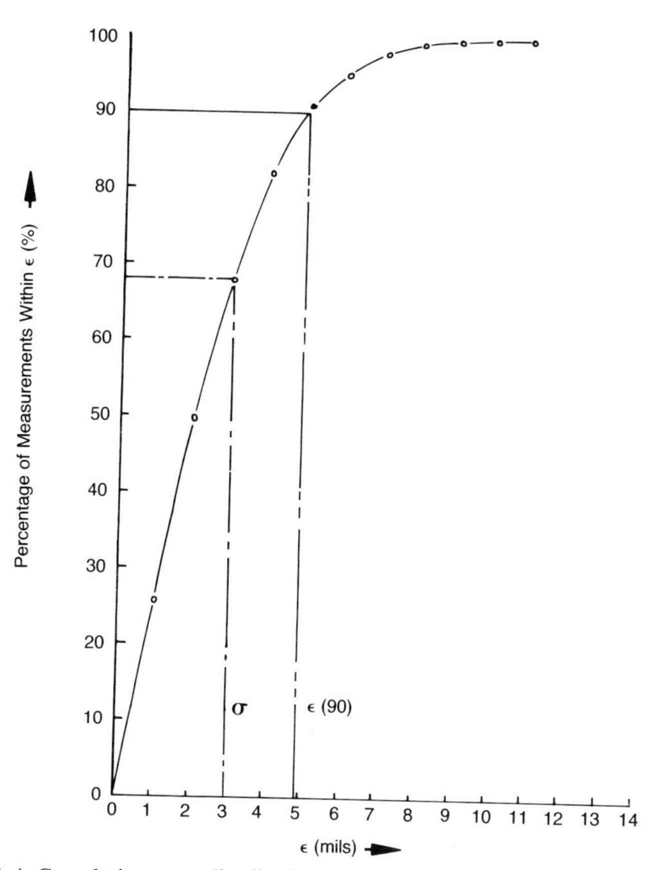

Figure 11-4. Cumulative error distribution.

EDF was Gaussian. In this case, the CED provides a simple way to establish the value of σ. However, if the EDF was not Gaussian, this method does **not** yield the standard deviation.

Our experience indicates that an excellent compromise between these extremes involves a quantity that we may define as $\epsilon(90)$, to be read as "epsilon ninety." *The value of $\epsilon(90)$ is specifically defined as that tolerance value which will result in a 90% confidence limit for all measurements.* Simply stated, one draws a horizontal line on the CED plot at a value of 90% on the ordinate, and then graphically determines the value of the error at the intersection of that line with a "smoothed" version of the CED. This value is then exactly $\epsilon(90)$, and establishes the tolerance level within which 90% of the measurements can be expected to exist. For our example, this would be about 4.9 mils.

Finally, provided that the EDF is very close to a Gaussian function, the value of $\epsilon(90)$ can then be related to the standard deviation of all the measurements by the simple result:

$$\epsilon(90) = 1.645\ \sigma \tag{11-5}$$

Nonetheless, $\epsilon(90)$ has important significance even if the EDF is **not** Gaussian. *Specifically, $\epsilon(90)$ will still provide the 90% confidence limit tolerance. This result is always useful no matter what the EDF may be like.*

All this from some little cubes...and thousands of physical measurements and a lot of data reduction. The important thing for the reader to realize is how much more specific the answers can be when the data is statistically significant. The uncertainty of the conclusions is dramatically reduced, and real decisions can be made regarding practical tolerances that can be achieved. Before spending hundreds of thousands of dollars on an RP&M system, you should do at least the following:

1. **Decide what tolerances you really need!** But be careful, if you pick unreasonably tight tolerances (± 1 mil for 99% of all measurements), it is possible that no system, including a numerically controlled milling machine, will be able to satisfy these specifications.
2. **Ask tough questions!** Demand to see actual data, EDFs, CEDs, and especially $\epsilon(90)$. Also, check *repeatability* (the extent that the mean value and standard deviation of any subset of the data agrees with the ensemble average from the EDF). This is vital if you want to be able to build parts on Tuesday that are very nearly identical to the ones you built on Monday. Such data should exist and be available for your examination.

In conclusion, we referred earlier to a group of individuals as data takers. Other names for such people are scientists, engineers, and technicians. We also referred to a different group of people as conclusion jumpers. An alternate name for these people is simply human beings. Every once in a while, human beings will "jump to a conclusion" which happens to be correct. They will congratulate themselves and tell others how much time was saved. Unfortunately, human beings also jump to the wrong conclusion, and may waste considerable time and money in the process. One of the reasons that science and engineering have advanced significantly since the time of Newton, is that, although their work may seem slow and plodding, the data takers improve the odds.

Finally, a word of caution. Remember that scientists, engineers, and technicians are also human beings. They definitely have human failings. Ultimately, RP&M as a technology will be best served by scrupulously reporting the most extensive accuracy results possible. The work leading to the results presented in the following sections of this chapter was a definite attempt in this direction.

11.3 Machine Accuracy

In this section, we will discuss the inherent accuracy of the SLA itself. This is separate and distinct from part accuracy, which we shall discuss in further detail in sections 11.4 and 11.5. To determine machine accuracy, all other distortions should be uncoupled as much as possible. What we want to know are the answers to the following questions:

1. How repeatably does the system position the laser beam? If one were to draw the same object many times, how much variation would occur?
2. Is the machine calibration algorithm working?
3. Is the machine leveling system functional?
4. Is the scanning system working correctly?
5. Is the software issuing the correct machine instructions?

To answer these questions, a standard test part had to be designed. The ideal machine accuracy test part would have the following properties:

A. It would be large enough in the x and y dimensions to insure that the system is capable of building accurately near the extremes of the platform as well as near the center.
B. It would have a substantial number of small, medium, and large dimensions.
C. It would have individual and separated areas, so shrinkage would have a minimal influence on the measured dimensions.
D. It should not take too long to build.
E. It should not consume a large amount of resin.
F. It should be easily measured on a CMM.

3D Systems designed such a part in April of 1990. The part is known as the "Acceptance Test Part" (AT Part) and is shown in *Figure 11-5*. This part consists of five equal boxes, one in the middle of the platform and one in each of the four corners. *Figure 11-5* also shows the measurements that are taken on this test part. Note that there are no dimensions on the drawing. This part was designed to suit all SLA models, so an individual part was created for each SLA.

The aspect of the AT Part that will be covered in this section involves the development of a database to establish **accuracy** and **repeatability** on an SLA machine. The plan was to select one SLA-250, have 3D Systems Field Service calibrate and tune the machine as per their regular specifications, and finally build **30 AT Parts**. During the building of these 30 parts, every parameter is held constant with the only "variable" being the day the part was built.

The following step-by-step procedure was used:

A. Pick a build style. Tri-Hatch was chosen since this build style results in minimal build stresses in the green state.
B. Pick a resin. XB 5081-1 was selected.
C. Make sure that *no* shrink factor or linewidth compensation is employed.
D. Build the AT Part. Depending on laser power, this normally takes 2.5 to 3 hours.
E. Clean the part. This is done in TPM.

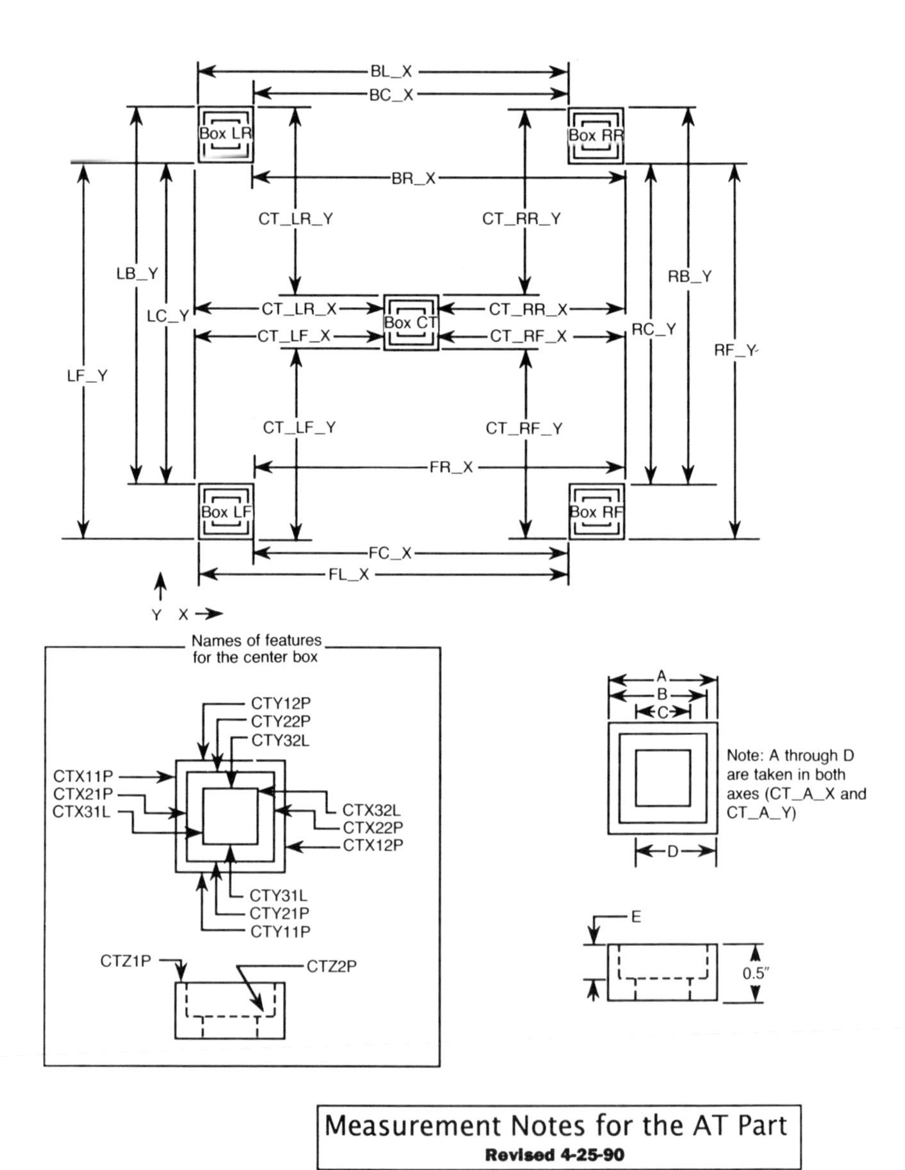

Figure 11-5. Drawing of the AT Part machine accuracy standard.

F. Measure the part. A Brown & Sharpe "Validator" CMM was used to take the measurements. The complete measurement series requires about 10 minutes. Note that the part is measured "green" and while still attached to the platform.
G. From these measurements, determine the actual values of each of the dimensions.
H. Group the dimensions into their nominal values (the 0.100″ dimension, the 0.250″ dimension, etc.)
I. Calculate the mean value and the standard deviation for each group.
J. Build 30 AT Parts. This was done on SLA #90076 at 3D Systems, Inc., and was completed in May 1990.
K. Calculate the mean value and the standard deviation for each subset of dimensions for all 30 parts.
L. Plot the standard deviation as a function of scale for the data from all 30 parts.

The results of this study gave us considerable insight into the accuracy capabilities of the StereoLithography machine itself. *Figure 11-6* shows the standard deviation for the data from all 30 parts. A number of important features of this data are worth noting:

1. The standard deviation of the measurements is roughly proportional to the length of the dimension. Thus, **we can characterize machine accuracy in terms of mils per inch or percentage of length**. Note that this is true only for machine accuracy and *not* for overall RP&M accuracy.

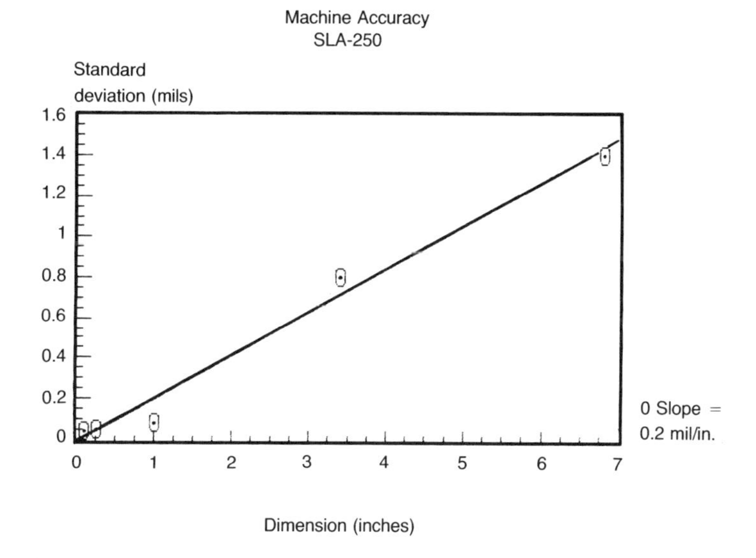

Figure 11-6. Machine accuracy: standard deviation versus dimension.

2. The machine accuracy is about an order of magnitude better than the overall RP&M accuracy.
3. Because the machine itself is intrinsically *capable* of high accuracy, it is especially important that all other variables (build styles, resin temperatures, calibrations, and cleaning techniques) are maintained at their optimum values.

Although we have only discussed the SLA-250 in this section, an equivalent part exists both for the SLA-500 and the SLA-190. Both the SLA-500 AT Part test data and the SLA-190 AT Part test data are reported in the same manner as the test data for the SLA-250 AT Part.

11.4 The User Part

Previously we discussed resin and build style characteristics (see section 10.2) and machine accuracy (see section 11.3). Now it is time to put all this information together to determine how accurately parts can be built on a SLA system. With recent improvements to the machine accuracy it was apparent that **quantitative results** were needed to establish SL part accuracy.

To provide such results, selection of a particular accuracy test standard is necessary. The ideal accuracy test part would have the following properties:

A. It would be large enough in the x and y dimensions to insure that the system is capable of building accurately near the extremes of the platform as well as near the center.
B. It would have a substantial number of small, medium, and large dimensions.
C. It would have both ''inside'' and ''outside'' dimensions. This is important in verifying that linewidth compensation is working properly.
D. It should not take too long to build.
E. It should not consume a large quantity of resin.
F. It should be easily measured with a CMM.
G. It would have many of the features of ''real'' parts (thin walls, flat surfaces, holes, etc.).
H. Ideally, it would be a part not designed by 3D Systems, to ensure complete impartiality.

Fortunately, such a part does exist. During 1990, the StereoLithography User Group, consisting of about 150 industrial companies, service bureaus, U.S. government laboratories, and universities, developed exactly such an accuracy test standard. Known simply as the ''User-Part,'' a photograph of this accuracy test standard is shown in *Figure 11-7*.

A drawing of the User-Part is also shown in *Figure 11-8*, which indicates some of the many dimensions to be measured on a CMM. In all, a total of 170

Figure 11-7. Photograph of User-Part.

measurements are made on each User-Part. Of these, 78 are in the x direction, 78 are in the y direction, and 14 are in the z direction.

3D Systems Inc. subsequently agreed to adopt the User-Part as the RP&M standard for establishing the accuracy levels achieved by an SLA when building parts. The User-Part has since become a metric for part building optimization as well as repeatability studies of accuracy.

11.5 User Part Accuracy Study

The aspect of the User-Part that will be covered in this section involves the development of a database to establish the **accuracy** and **repeatability** of the SL process. The plan was to select one SLA-250, have 3D Systems Field Service calibrate and tune the machine as per their regular specifications, optimize the machine and resin parameters, and finally build 15 User-Parts. During the building of these 15 parts, every parameter is held constant.

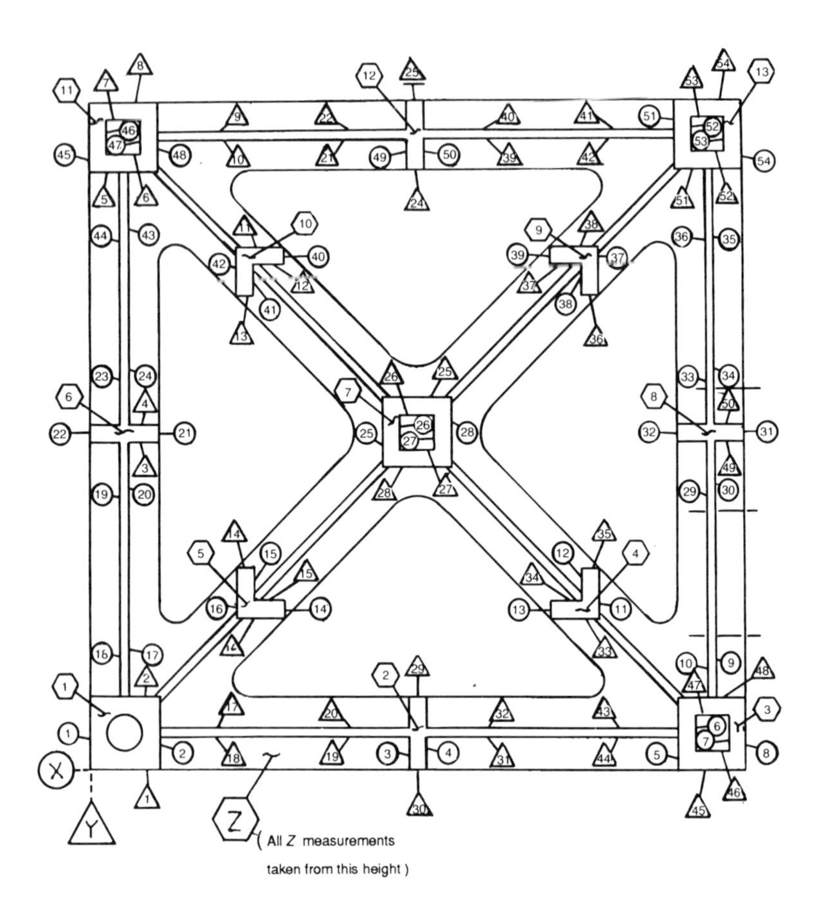

Figure 11-8. Drawing of User-Part.

The step-by-step procedure 3D Systems used was as follows:

A. Pick a build style. STAR-WEAVE™ was chosen since it has proven to be the most accurate build style to date.
B. Pick a resin. XB 5081-1 was selected.
C. Pick a resin shrinkage factor and a linewidth compensation value. The values from a CHRISTMAS-TREE™ test should be employed for the first User-Part to be built.
D. Build the User-Part. Depending upon laser power, this normally takes 8 to 10 hours and can be conveniently done overnight.
E. Clean the part. This was done using TPM, which is effective for all resins and has the least deleterious effect upon part accuracy of all solvents tested to date.

F. Postcure the part. This was done using the new actinic-fluorescent based PCA. Postcure takes about one hour. Also the part should be positioned vertically (standing on edge) to minimize postcure distortion.
G. Measure the part. A Brown & Sharpe "Validator" CMM was used to take the measurements. The complete measurement series requires about 20 minutes.
H. From these measurements, determine the actual values of each of the 170 dimensions.
I. Compare each of these actual dimensions with the appropriate CAD dimension. Calculate the difference, which is the **error** for that specific measurement.
J. Tabulate the errors for all 170 measurements.
K. Plot the errors as a function of length. Follow the same procedure used for the CHRISTMAS-TREE™, as described in Chapter 10.
L. Determine the **shrinkage factor** and **linewidth compensation**.
M. Separately determine the root-mean-square (RMS) error for the 78 x dimension measurements, for the 78 y dimension measurements, and for the 14 z dimension measurements.
N. Determine the RMS error for all 170 measurements.
O. Modify the values of the shrinkage factor and linewidth compensation to reduce the errors.
P. Repeat the **entire procedure** using the new shrinkage factor and linewidth compensation value.
Q. Continue iterating until the optimum results are achieved (minimum RMS errors).
R. Plot the Error Distribution Function over the whole range of errors.
S. Plot the Cumulative Error Distribution. This plot describes the total number of measurements within any desired tolerance level.
T. Determine $\epsilon(90)$, the 90% confidence limit tolerance level, from the CED data.
U. Plot the 99% confidence limit (three standard deviations) error level as a function of scale. This plot describes how the errors are distributed for small, medium, or large dimensions.
V. Build 15 User Parts. This was done on SLA #90076 at 3D Systems, Inc., and was completed in June 1991.
W. Plot the EDF for the data from all 15 parts.
X. Plot the CED for the data from all 15 parts.
Y. Establish the value of $\epsilon(90)$ for the data from all 15 parts.
Z. Plot the 99% confidence limit error as a function of scale for the data from all 15 parts.

The results of this study (literally A to Z) gave us considerable insight into the accuracy capabilities of StereoLithography.

Figure 11-9 shows the Error Distribution Function for the data from all 15 x 170 = 2550 measurements. Note that:

1. Within the limitations of finite sampling, the EDF is quasi-Gaussian.
2. Over 60% of the errors are within ±4 mils.
3. The overall RMS error is about 5 mils.
4. Nonetheless, a few errors are as great as 18 mils.

Figure 11-10 shows the Cumulative Error Distribution for all 2550 points. Note that:

1. About 30% of the dimensions of this part are within ±1 mil of their CAD value.
2. About 58% of the dimensions are within ±3 mils of their CAD value.
3. About 73% of the dimensions are within ±5 mils of their CAD value.
4. However, about 2% of the dimensions have errors between 14 and 18 mils.

Figure 11-11 is a plot of the 99% confidence limit error as a function of the length of the dimension. This plot is generated by taking all the dimensions of the User-Part that lie within a certain range of lengths (from 0 to 1″, 1″ to 2″, etc.), and then calculating the 99% confidence limit error (±3 standard deviations) for each sub-set. Inspection of *Figure 11-11* shows the following:

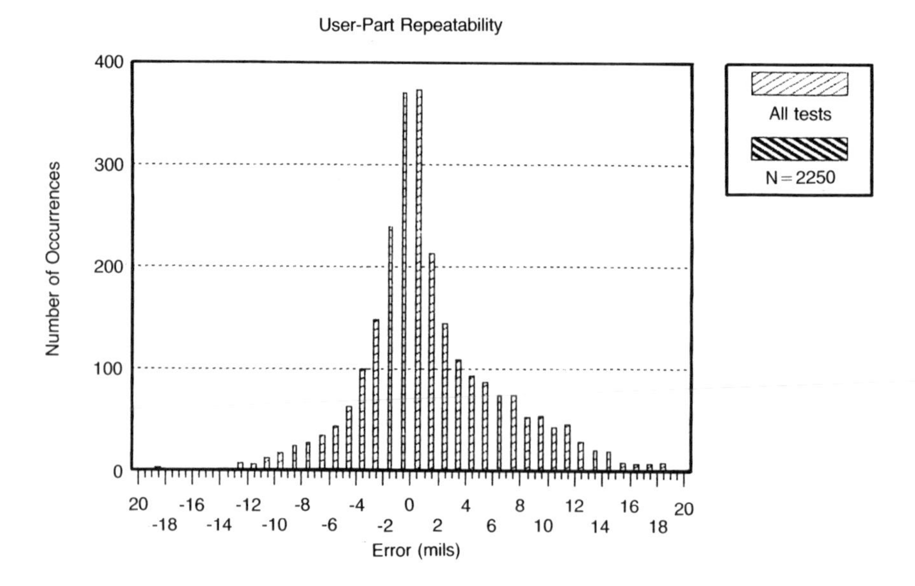

Figure 11-9. Error distribution function.

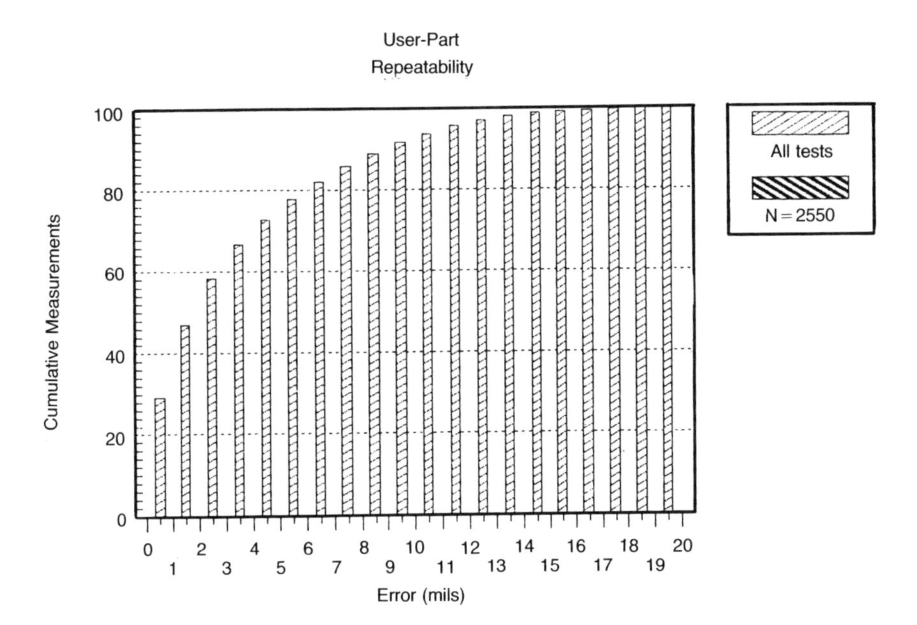

Figure 11-10. Cumulative error distribution.

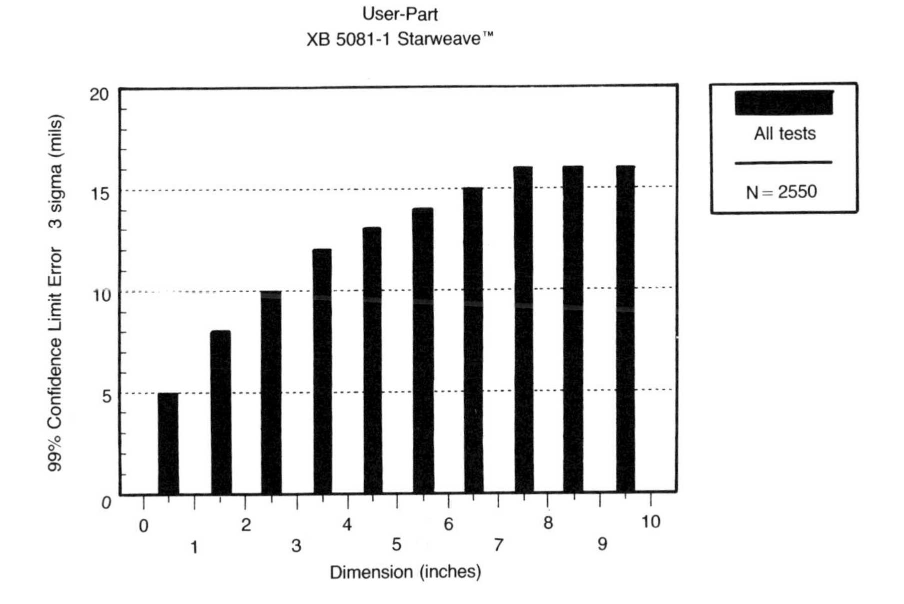

Figure 11-11. 99% confidence limit error versus dimension length.

1. The 99% confidence limit error increases with increased length.
2. However, the increase of error with length is **nonlinear**. Thus, characterizations of RP&M accuracy based on a ''percentage of length,'' or ''mils per inch,'' both of which imply a linear function, *are simply not in agreement with experimental results.*
3. The actual results suggest that the 99% confidence limit error scales very nearly as the **square-root** of the length. Thus, for example, 9″ dimensions will tend to have errors about three times as great as 1″ dimensions, not nine times as great! *Figure 11-11* can also be used as a tolerance standard for StereoLithography when using the SLA-250 with XB 5081-1 and STAR-WEAVE™.
4. As a footnote to these results, it should be evident that characterizing the ''accuracy'' of StereoLithography is not as simple as referring to a particular number. At this time, we strongly believe that the best measure of accuracy for the entire SL process is the Error Distribution Function itself. This function, as well as the Cumulative Error Distribution, which is derived from it, allows one to generate realistic tolerance standards for StereoLithography.

11.6 Summary

To appreciate the remarkable progress that has been made during the past two years with respect to the accuracy of parts built using StereoLithography, one should examine *Figure 11-12*. This is a plot of $\epsilon(90)$ for all 170 dimensions of the User-Part, built using SL resin XB 5081-1. The data is shown for three different part building techniques.

1. 50 mil, 60°/120°/X Tri-Hatch. This was by far the most common SL part building method through the fall of 1990.
2. WEAVE™. This build method, developed and tested within 3D Systems throughout 1990, was introduced to the general SL user community in late 1990.
3. STAR-WEAVE™. This latest build method was developed and tested within 3D Systems during the first half of 1991, and subsequently announced to the SL user community in September 1991.

It is important to note that the data presented in *Figure 11-12* are for User-Parts generated after completing the full A to Z iteration cycle described in detail in section 11-5. Thus, linewidth compensation as well as shrinkage compensation have not only been applied, but they have been independently optimized for each part building method.

This procedure is best accomplished by initially building the CHRISTMAS-TREE™ diagnostic part, and then determining the linewidth compensation value

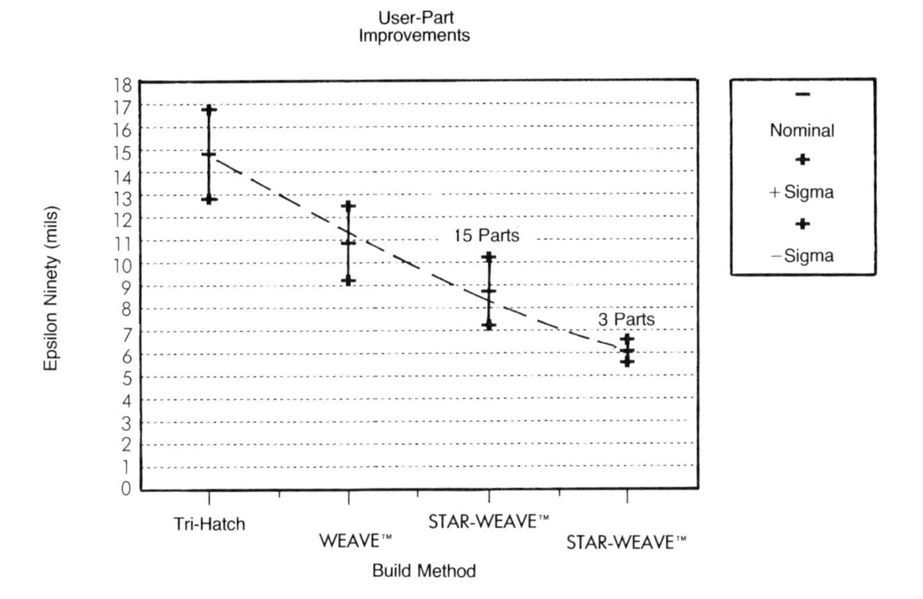

Figure 11-12. Improvements in User-Part accuracy.

and the shrinkage compensation factor for your specific machine and for the actual resin in your vat. Next, one should build a second CHRISTMAS-TREE™, but this time using the values of LWC and SCF(x) and SCF(y) determined from the first build as input for the second build. This is an excellent "cross check" of these values. If they are correct, the second CHRISTMAS-TREE™ should be very close to its CAD dimensions. Continue iterating the diagnostic part until the best results are achieved. This will save both time and resin since the CHRISTMAS-TREE diagnostic part consumes much less resin and builds in about one-fifth the time of a User-Part.

Once the optimum values of LWC and SCF(x) and SCF(y) have been established, they should then be used to build a User-Part. By following the procedure described above, even the first User-Part should produce very good results. Nonetheless, by carefully studying the raw CMM data, or by plotting an Error Distribution Function, one may still detect small systematic errors, which will generally manifest themselves in the form of an asymmetric EDF. Adjustments to the linewidth or shrinkage compensation values can then be used to complete the iteration cycle. If all this is done in a careful and systematic manner, the end result should establish the true accuracy capability of your system.

The reader may be thinking at this point, "Why should I go to all this trouble?" The answer is twofold. First, you have purchased an expensive RP&M system. Going through the aforementioned procedure will establish in your own mind what your particular system is capable of in the way of part accuracy. Furthermore, simply *doing* all of the indicated steps will cause you to gain confidence that you too can produce excellent results. The confidence gained is well worth the time spent.

Some additional observations regarding *Figure 11-12* follow. The reader should note the considerable reduction in ϵ(90) in going from Tri-Hatch to WEAVE™ to STAR-WEAVE™, to the very best optimized results for STAR-WEAVE™. This improvement in system accuracy has occurred within the past two years. *Furthermore, it should be clear from this data that StereoLithography is currently approaching the point where 90% of all the measurements on the agreed accuracy standard test part will be within five mils (0.005") of their CAD value.*

Finally, the reader may still be thinking: "Okay, the data does look pretty good, and I know that it is fast, but how does the accuracy compare with what I could get using NC milling?" Some very recent work at the University of Nottingham, Nottingham, England[4] may shed some light on this issue. To answer precisely this question, drawings of the User-Part were submitted to various precision machine shops in the United Kingdom, with the intent of having this part generated using state-of-the-art numerically controlled milling machines. Also, the part was to be produced in both aluminum and perspex (an acrylic engineering plastic in common use in the U.K. and generally very similar to Plexiglas). The generation of these parts was accomplished during late 1991, and the parts were subsequently measured by E. Gargiulo of DuPont.[5] The results of these measurements are shown in *Figure 11-13*.

These results are quite remarkable. They show that the best *SL-produced User-Parts in STAR-WEAVE™ are more accurate than the NC milled versions in a well-known engineering plastic!* Finally, the SL results show a value of ϵ(90), which is within 0.0007" of the value for the NC milled User-Part made from aluminum. This is truly phenomenal for such a young technology. Even allowing for the inevitable slowing in the rate of advance of accuracy improvement, it is also evident from *Figures 11-12* and *11-13* that within the near future RP&M may equal or surpass the accuracy level of NC milling in aluminum.

Hopefully, the detailed experimental results presented in this chapter will help the reader to appreciate that this new technology has made great strides in accuracy during the period from 1989 through early 1992. Work continues. As we learn more about the fundamental processes, and develop still more advanced part building techniques, the technology can only advance further. Certainly, we have no intention of going backwards. The progress that has been made to date has enabled many remarkable applications leading to significant cost and time saving benefits.

Nonetheless, it is important to remember that the cornerstone of RP&M is still accuracy. It is the responsibility of the RP&M system manufacturer to

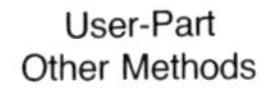

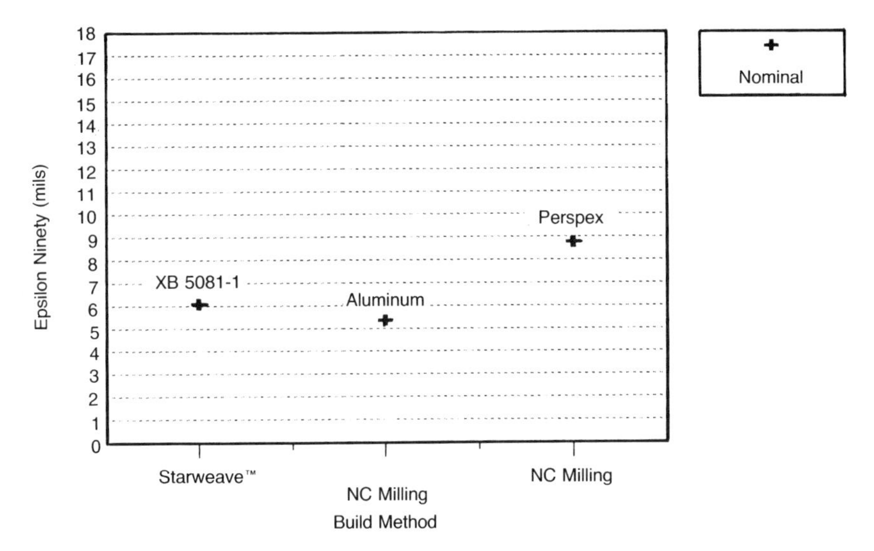

Figure 11-13. Accuracy comparisons: StereoLithography and NC milling.

demonstrate the accuracy of the system in a reliable, repeatable, and conclusive manner. It is, however, the responsibility of the user to learn how to get the very best performance from their system. The information presented in this chapter is intended to provide users with optimum results. *If a wide array of RP&M users reach these proven levels of performance, we are more than confident that this technology will provide some of the impetus needed to generate real and significant gains in manufacturing productivity.*

REFERENCES

1. Ingersoll, L.R., Martin, M.J., and Rouse, T.A., *Experiments in Physics*, McGraw-Hill, New York, 1955, pp. 1-17.
2. Rushbrooke, G.S., *Introduction to Statistical Mechanics*, Oxford University Press, London, 1962, Chapter IV, pp. 54-58.
3. Taylor, J.R., *An Introduction to Error Analysis: The Study of Uncertainties in Physical Measurements*, University Science Books, Mill Valley, CA, 1982.
4. Dickens, P., University of Nottingham, England, private communication, November, 1991.
5. Gargiulo, E., E.I. du Pont De Nemours and Company, Wilmington, Delaware, private communication, November, 1991.

chapter 12

Texas Instruments: An Aerospace Case Study

Education is learning about
how little we know.

—Unknown

12.1 Introduction

Over the past 25 years, the computer-aided engineering/computer-aided design/computer-aided manufacturing (CAE/CAD/CAM) industry has developed technological advances that allow industrial users to increase productivity and decrease the cycle time necessary for product development. These technologies include basic wireframe and surface design, specialized application software packages, finite element analysis, numerical control, solid modeling, and, recently, rapid prototyping.

Each of these technologies plays a significant role in industry today. The Defense Systems & Electronics Group of Texas Instruments currently uses these technologies in the mechanical design engineering process. This chapter discusses the two specific technologies of solid modeling and rapid prototyping (specifically, StereoLithography) including their advantages, benefits, and practical applications. Case studies are provided to demonstrate the increase in productivity that may be gained by utilizing these technologies. Finally, the use of RP&M parts within the solid mold investment casting process will also be discussed.

---

*By **Paul Blake**, Design Center, and **Owen Baumgardner**, Manager, Fab Producibility Engineering, Texas Instruments, Inc.*

12.2 Solid Modeling

Solid modeling is an advanced design capability that offers the mechanical design engineer numerous advantages over traditional wireframe design methods. Visual and analytic capabilities are improved and required inputs for rapid prototyping are provided.

As its name implies, solid modeling represents an object as a solid part. These parts are typically shown as shaded images on the computer screen (*Figures 12-1* and *12-2*). On the other hand, wireframe design systems represent edges as lines and faces as surfaces. As the complexity of a wireframe design increases, the ability to visualize and understand the design decreases. *Figure 12-3* demonstrates the visual ambiguity that wireframe designs create. Solid modeling alleviates this ambiguity. Solid modeling is able to generate cross sections or cut-away views with minimal effort (*Figure 12-4*). These capabilities allow the design engineer to visualize and understand the part's actual appearance and function within an assembly by clearly displaying the critical relationship of part layouts, interferences, and clearances.

The numerical and material information intrinsic to a solid model offers two unique capabilities. First, the solid model permits the engineer to derive mass property data (weight, center of gravity, moments of inertia, surface areas, etc.)

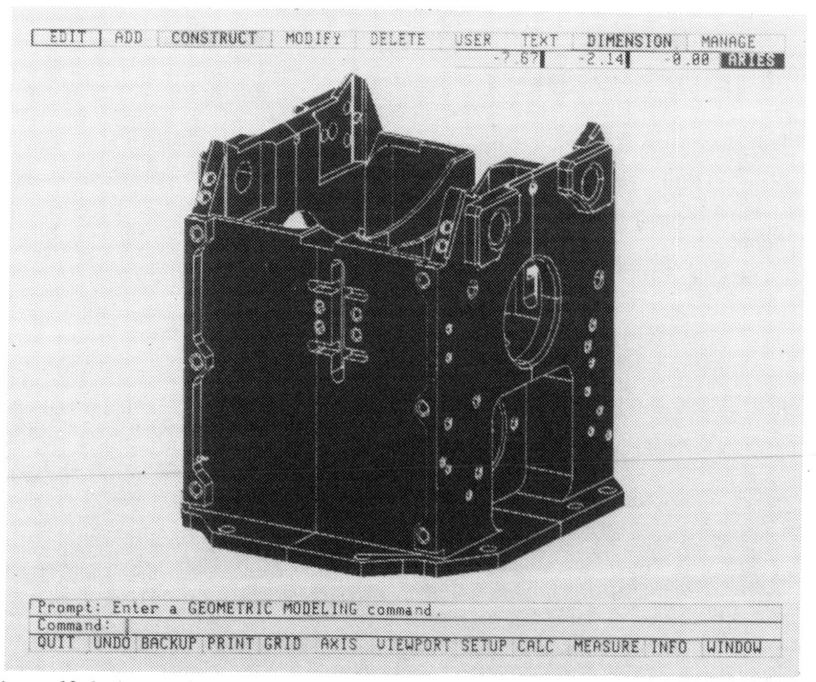

Figure 12-1. Assembly of solid-model parts.

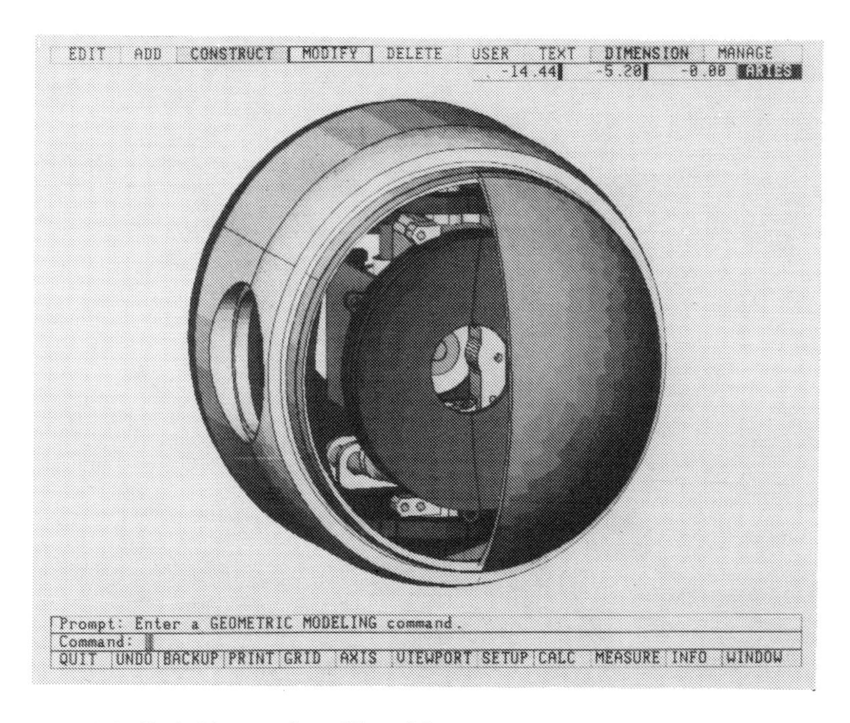

Figure 12-2. Shaded image of a solid model.

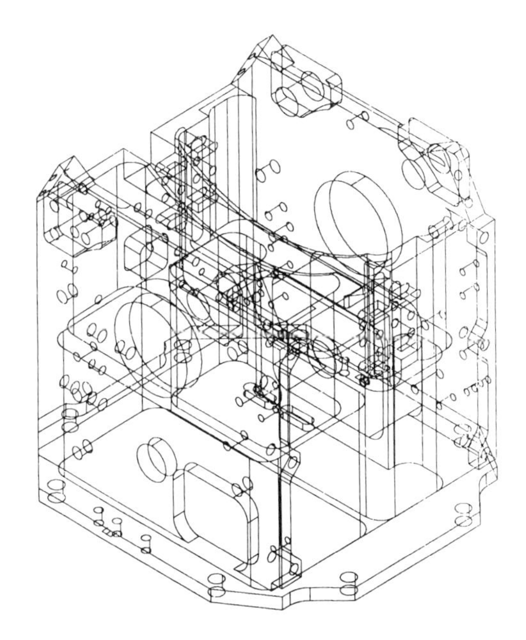

Figure 12-3. Wireframe representation.

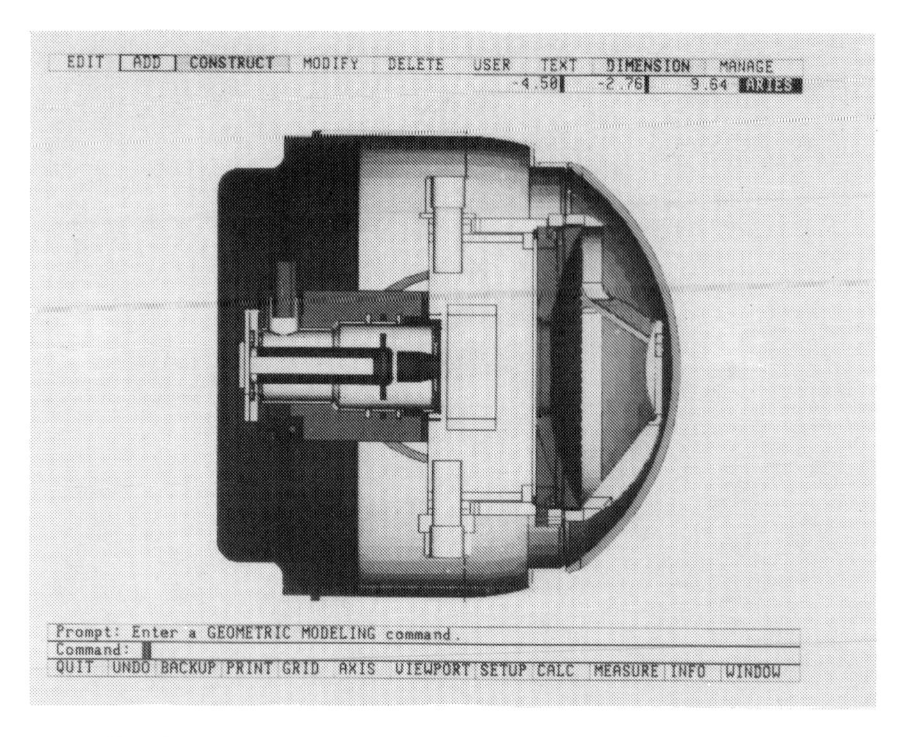

Figure 12-4. Section view of solid-model assembly.

of parts and assemblies (*Figure 12-5*). This eliminates the need for hand calculations to estimate these values from drawings or sketches. Second, the mechanical integrity of the design may be investigated using various finite element methods to determine structural, thermal, kinematic, and aerodynamic properties. These capabilities allow the engineer to design, analyze, and iterate on the CAE computer using the same solid model. Automating the design process helps reduce the need for, but does not eliminate, the costly iterations of fabrication and functional testing of the hardware.

Finally, solid model databases serve as direct input to rapid prototyping systems. However, it should be noted that surfaced part databases may also be used as inputs, but with reduced efficiency. The input is in the form of a tessellation that is generated from the solid model or surface database.

12.3 Benefits of Solid Modeling and Rapid Prototyping

The use of solid modeling and the SL process has resulted in numerous benefits for the design engineering community at Texas Instruments. These include increased visualization capability, detection of design flaws before

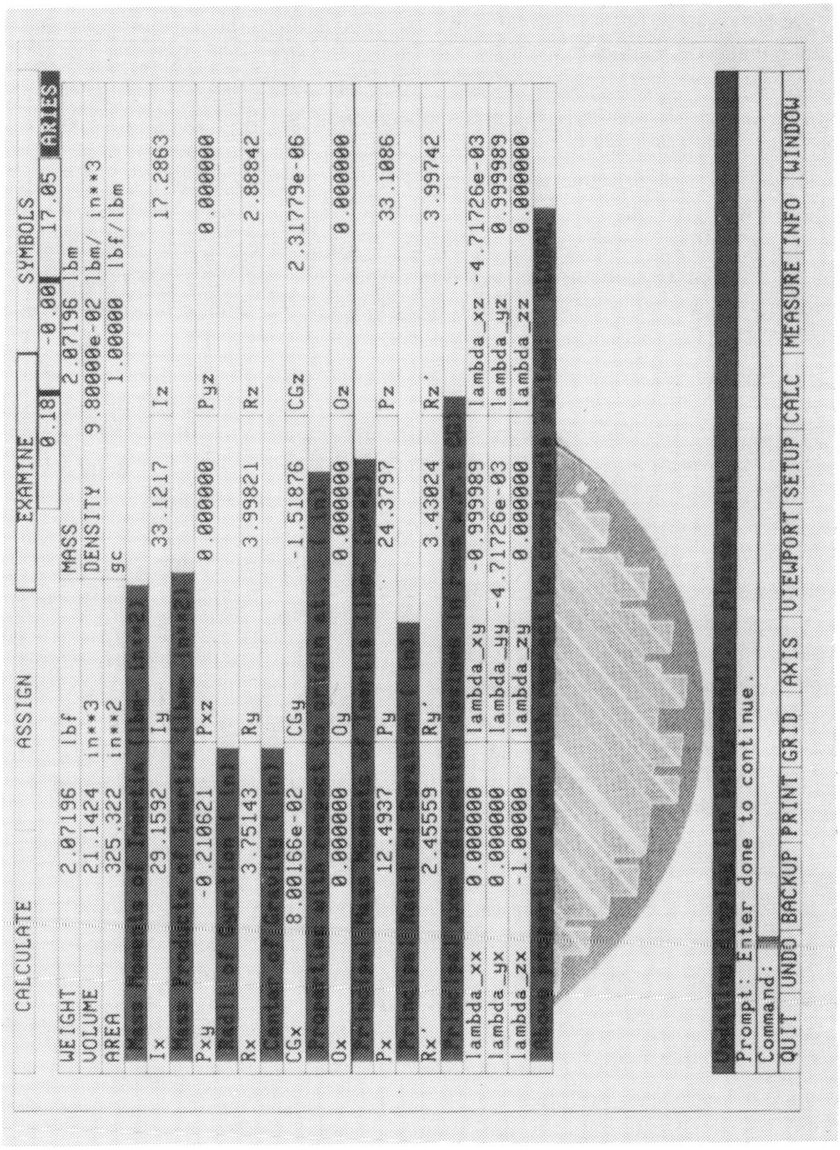

CALCULATE ASSIGN EXAMINE SYMBOLS

0.18 -0.00 17.05 ARIES

WEIGHT	2.07196 lbf	MASS	2.07196 lbm		
VOLUME	21.1424 in**3	DENSITY	9.80000e-02 lbm/ in**3		
AREA	325.322 in**2	gc	1.00000 lbf/lbm		
Ix	29.1592	Iy	33.1217	Iz	17.2863
Pxy	-0.210621	Pxz	0.000000	Pyz	0.000000
Rx	3.75143	Ry	3.99821	Rz	2.88842
CGx	8.00166e-02	CGy	-1.51876	CGz	2.31779e-06
Ox	0.000000	Oy	0.000000	Oz	0.000000
Px	12.4937	Py	24.3797	Pz	33.1086
Rx'	2.45559	Ry'	3.43024	Rz'	3.99742
lambda_xx	0.000000	lambda_xy	-0.999989	lambda_xz	-4.71726e-03
lambda_yx	0.000000	lambda_yy	-4.71726e-03	lambda_yz	0.999989
lambda_zx	-1.00000	lambda_zy	0.000000	lambda_zz	0.000000

Prompt: Enter done to continue.

Command:

QUIT | UNDO | BACKUP | PRINT | GRID | AXIS | VIEWPORT | SETUP | CALC | MEASURE | INFO | WINDOW

Figure 12-5. Mass property data of solid-model assembly.

hardware fabrication, increased ability to calculate mass properties, design optimization, and decreases in cost and cycle time associated with prototype part fabrication.

The first benefit of these two technologies, and perhaps the most obvious one, is the enhanced *visualization* capability. Engineers, designers, technicians, and managers now discuss scale plastic prototype parts, not two-dimensional drawings. The SL prototype parts and assemblies are also used for fit check verification (where tolerances permit), assembly methods and tooling planning, cable routing, weight reduction studies, customer communication models, and vendor quoting tools.

An additional benefit is the *decrease of cycle time* required to produce prototype parts. The current SL prototype turnaround cycle, defined as the time difference between database receipt to part fabrication completion, is between two and 10 days, depending on engineering requirements. This is a dramatic decrease over conventional machining methods, which typically take months. Therefore, the engineering team may visually, dimensionally, and functionally inspect and *verify* critical and high-risk parts and assemblies earlier in the design phase of the overall design cycle. This benefit also results in the increased ability of the design engineer to explore and *iterate* new concepts or improvements into the design. The engineer may accept or reject the new concepts leading to design *optimization*. Another cycle time benefit is the increased time available to explore less expensive casting methods for part *fabrication* over traditionally more expensive machine processes.

Texas Instruments has experienced typical cycle time reductions of two weeks for simple parts and 20 weeks for complex machined or cast parts in the fabrication of prototype parts. Also, customer concept communication models have been fabricated for half the cost and in half the time of conventional methods. Specific examples of cycle time and cost reductions are found in *Figures 12-6* through *12-9* and *Figure 12-18*.

The method chosen to produce prototype parts varies according to engineering requirements. ''Paper dolls'' (*Figure 12-10*), computer data bases, rapid prototyping parts, SL pattern castings, and machine parts provide the engineer with a wide spectrum of choices. The relative differences between these methods are outlined in Table 12-1. The part accuracy, cycle time, and cost vary dramatically according to which process is used.

Another important benefit of solid modeling and SL rapid prototyping is cost reduction associated with eliminating design flaws early in the design cycle. The product development and fabrication cycle is constantly at the mercy of continuous requirement and design changes. These changes coupled with the inability to easily verify overall part layout configurations contribute to part assembly problems such as interferences and misaligned hole patterns. Previously, these problems often went undiscovered until piece parts were fabricated and assembled. Costs rose as parts were scrapped or modified, and the amount

Figure 12-6. Machined motor housing. (Material: 6061-T6; time required: 104 hours; cycle time: 4 months.)

of required touch labor also increased in order to assemble and disassemble the piece parts. Engineering change notices had to be incorporated into the design, significantly adding another element to the overall cost.

Solid modeling permits engineers to visually inspect the designed piece parts and assemblies for mating hole patterns and part clearances. As the design is completed and verified, the engineer may use the SL rapid prototyping process to build a three-dimensional model as a final review before committing the design to fabrication. These capabilities serve as cost avoidance steps that help meet proper fabrication and assembly requirements (*Figure 12-11*). Solid modeling also significantly increases the ability to *calculate mechanical properties* of the parts and assemblies throughout the design process. For example, in many aerospace and military systems, weight is a critical design requirement. Solid modeling mass property calculations assist the engineer by calculating part and assembly weights. This information allows the engineer to optimize the design for weight requirements before part fabrication. These capabilities serve as additional cost avoidance steps that help meet the proper fabrication and assembly requirements.

Figure 12-7. Time required: 8 hours; cycle time: 3 days. (Time required is for SLA rapid prototype touch labor only, not machine run time.)

12.4 StereoLithography and Investment Casting

A significant and immediate impact on the cost and cycle time associated with the investment casting process has been achieved by incorporating the RP&M part into the process. The ability to use the SL modeled part as a wax substitute is the key to supplying castings for low production runs without the high cost of foundry pattern tooling. In addition, providing machine shops with a near net-shaped part reduces the number of required machining steps to produce a completed part.

The two types of investment casting processes are solid mold (sometimes known as "flask" casting) and shell casting. They differ only in the method used to form the ceramic mold. Both processes require a pattern, gating to a central sprue, ceramic mold (either flask or shell), removal of the pattern by melting, pouring metal into the cavity left by the melted pattern, removal of the model material from the cast cluster, and cutting of the castings from the sprue.

Currently, the most successful procedure is to use the RP&M part as a wax pattern substitute for investment casting in the solid mold process. The SL part expands more than the traditional wax pattern as it is heated in order to be melted from the mold. This increased expansion causes the RP&M part pattern to crack

Figure 12-8. SLA model of a missile. Outside quote was $10,000 and two months cycle time. Time required was 75 hours and one month cycle time. (Time required for SLA rapid prototype touch labor only, not machine run time.)

the weaker mold in the shell process, but not the metal reinforced mold in the solid mold process. It should be noted that newer resins are becoming available that are showing evidence of eliminating this problem with the shell mold process.

The ceramic solid mold is created by placing the RP&M part, with the gate and runner system attached, into an open-end metal flask. Investment slurry is poured into the flask, completely surrounding the SL part with its gate and sprue system. Typically, extra SL patterns are sent to the foundry to allow establishment of the gating requirements. Before the binder in the slurry solidifies, the flask is placed under a vacuum to remove all entrapped air. When the investment is set, the RP&M part pattern is melted out of the mold and the process is completed by following the same procedures used for any normal investment casting.

There are four classes and grades for castings: classes 1, 2, 3, and 4; and grades A, B, C, and D. The class is established by the engineer according to design requirements for functionality, safety, reliability, and mechanical properties. The grade, also established by the engineer, is used as the accept/reject

Figure 12-9. SLA sweep volume model (seeker assembly). Time required: 54 hours; cycle time: 8 weeks. The lessons learned were that (1) the primary mirror had interference with gimbal in azimuth sweep, and (2) the secondary mirror was strengthened in its attachment legs. (Time required is for SLA rapid prototype touch labor only, not machine run time.)

criteria for radiographic inspections. Material available for the investment casting process includes all ferrous and nonferrous alloys.

The following nonferrous (aluminum) alloys have been successfully poured to date: A356-T6, 356-T6, A357-T6, 357-T6, A201-T7, and a matrix metal composite of A357-T6 with 20% silicon carbide. Most of the castings have been class 3 and grade D, meeting typical engineering requirements. Grade C may also be obtained, but requires extra gating models for the foundry to determine the correct gating systems. The average cycle time has been four weeks. *This is a four month decrease in fabrication time*! See *Figures 12-12* and *12-13* for examples of parts cast using this process.

12.5 Case Studies

The entire product development and production cycle is complex. Solid modeling, RP&M parts, and castings made directly from RP&M patterns are used throughout the process to produce better designs by facilitating concurrent

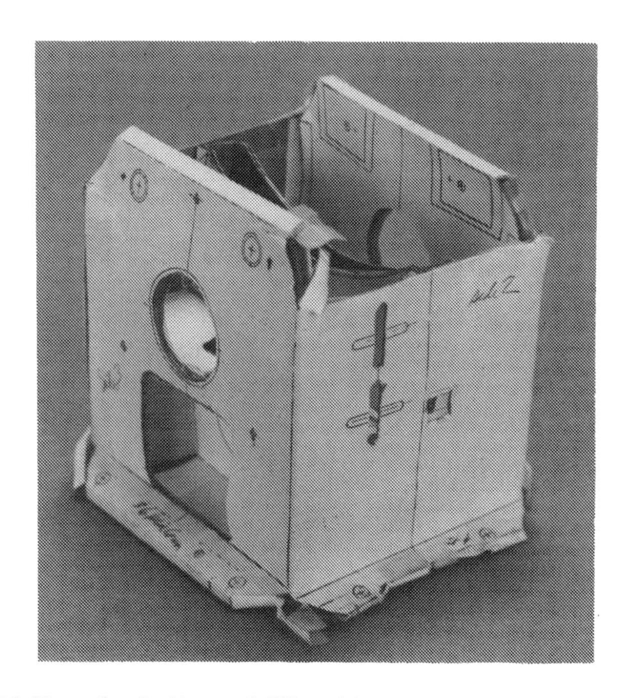

Figure 12-10. Example of a "paper doll" model.

Table 12-1
Engineering Prototype Comparison

	Cost	Cycle Time	Accuracy	Visualization	Part Strength
1. Paper Doll	Low	Days	Very Poor	Poor	Poor
2. Wire Frame[1]	Low-Med[2]	Days-Weeks[2]	Very Good	Poor/Good	N/A
3. Solid Model[1]	Low-Med[2]	Days-Weeks[2]	Excellent	Good	N/A
4. Rapid Prototype Part	Low	Days[3]	Very Good	Excellent	Good[4]
5. Rapid Prototype Casting	Low-Med	Weeks[5]	Very Good	Excellent	Excellent
6. Machine Hogout	High	Weeks-Months[6]	Excellent	Excellent	Excellent

1. Computer-generated models versus physical prototypes.
2. Length depends upon design complexity.
3. Typically two to 10 days (part configuration dependent).
4. Resin dependent and resins are improving.
5. Typically four weeks (part configuration dependent).
6. Typically four to six weeks (part configuration dependent).

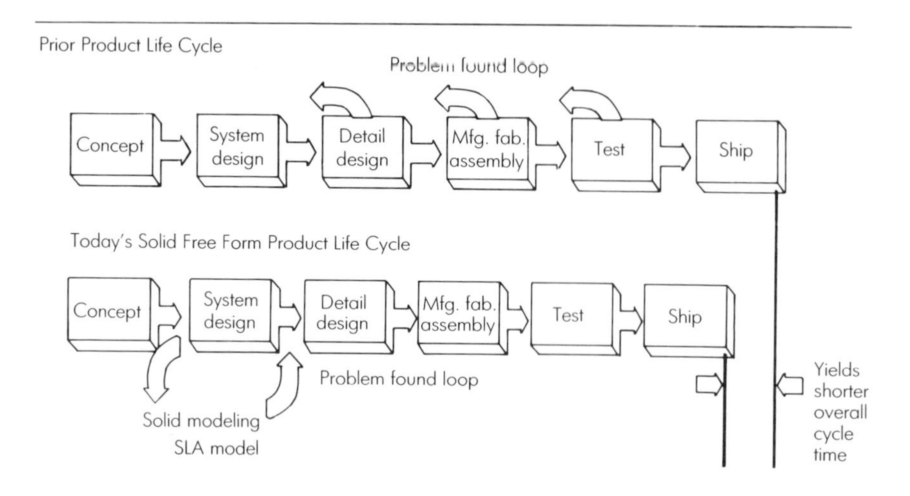

Figure 12-11. Typical product life cycle for mechanical parts.

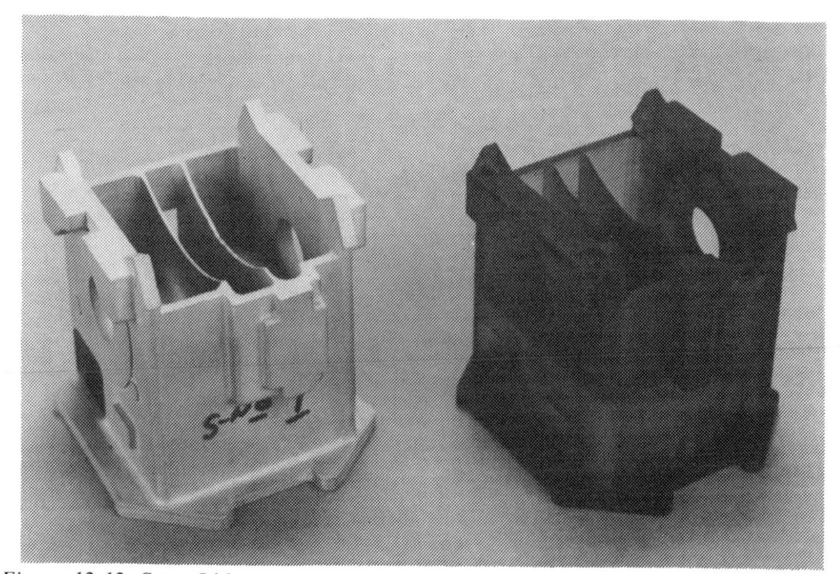

Figure 12-12. StereoLithography part with casting.

Figure 12-13. SL part with casting.

engineering and promoting quality. Table 12-2 outlines where each of these technologies is used and the typical applications in the different steps of the cycle.

Table 12-2
Solid Modeling and Rapid Prototyping
Product Development Cycle Applications

	Concept/ Proposal	System Design	Detail Design	Analysis/ Test	Manufacturing Assembly
Solid Modeling	• Geometry definition	• Geometry design/ definition • Assembly verification • Drawing production • N/C Hand-off		• Weight/CG • Finite element	• Vendor quotes
Rapid Prototyping Parts	——— Design and assembly verification ——— ——— Concept communication ———	• Human factor studies		• Pressure drop • Flow (air/liquid) • Radar cross section	• Vendor quotes • Cable routings • Potting molds • M&T process development
Rapid Prototyping Castings				• Static • Vibration • Thermal	

Figure 12-14 demonstrates the progress a typical product undergoes through its initial design phase. Six different models have been produced over a period of six months for various applications ranging from customer reviews to vendor quote models. The parts were built in series starting from left to right in the figure. The following is a brief explanation of the different applications for which each part was used:

Part #1
Date 3/91

- Initial Customer Review
- Design Approach Validated with
  A. Customer
  B. Management
  C. Human Factors

Part #2
Date 6/91

- Design Verification Model

Part #3
Date 7/91

- Design Verification Model
  A. Battery Vent Design Check
  B. Hardware Mount/Adapter Assembly Check
  C. Potentiometer Interference Detection

Part #4
Date 9/91

- Assembly Fit Check *Figures 12-15* and *12-16*
- Customer Review
  A. Lens Focus Mechanism
  B. Human Factors

Parts #5 & 6
Date 9/91

- Casting Models
- Quotes from Vendors
  A. 4 Vendors
  B. 1 Casting Returned (See *Figure 12-17*)
  C. No clarification required from vendors
- Methods and Tooling Process Development
- Dimension Scheme Definition

A second case study extensively used rapid prototyping technology for various applications ranging from engineering design verification, cable routing analysis, and assembly fit checks to the production of SL patterns for the

Figure 12-14. Progession of a product through the initial design phase.

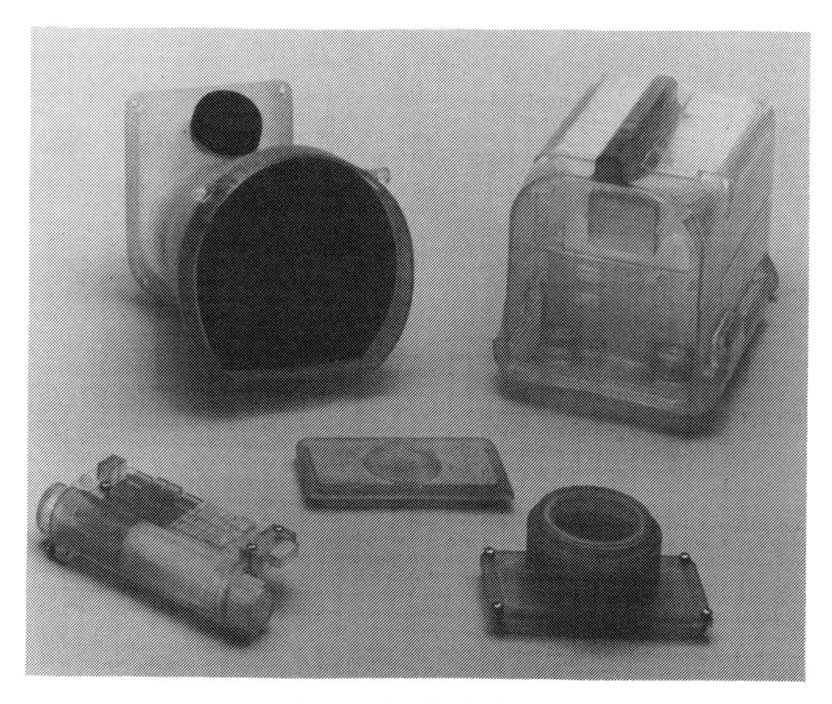

Figure 12-15. Parts to be used for assembly fit check.

Figure 12-16. Completed assembly fit check.

Figure 12-17. Casting returned from vendor.

investment casting process. The various project requirements dictated that eight unique parts be designed, analyzed, and manufactured (quantity of 16 per part) within a seven-month cycle. Part size ranged from 3″ (7.6 cm) to 36″ (91.4 cm) and shape from simple to very complex.

A cost and cycle-time graph is shown in *Figure 12-18* comparing the three different manufacturing methods of SL pattern casting: RP&M, traditional wax tool/pattern casting, and machining from stock material. *The SL pattern casting clearly demonstrated a significantly lower cost and shorter cycle time.*

The appendix to this chapter provides a detailed examination of the costs and cycle-time data for the above example.

As the program progressed, numerous additional advantages became evident as the prototype parts and SL pattern castings were used.

1. The engineer was able to make additional changes/iterations up to eight weeks later in the design cycle without negatively impacting the manufacturing lead time and cost requirements (See *Figure 12-18*).
2. Prototype parts were used for *verification* of the part and assembly design. As the design progressed, approximately one of every four design *iterations* resulted from problems that the prototype models clearly demonstrated. These ranged from part interference to cable connector misalignment.
3. A complete assembly of the parts was used to assist in *optimization* of the cable routing paths and required lengths. Typically, this process is done later in the design cycle with drawings, computer data bases, and/or the actual parts. Utilizing the rapid prototype assembly, an estimated 30-40 engineering change notices (ECNs) were avoided, and three months of cycle time was saved.

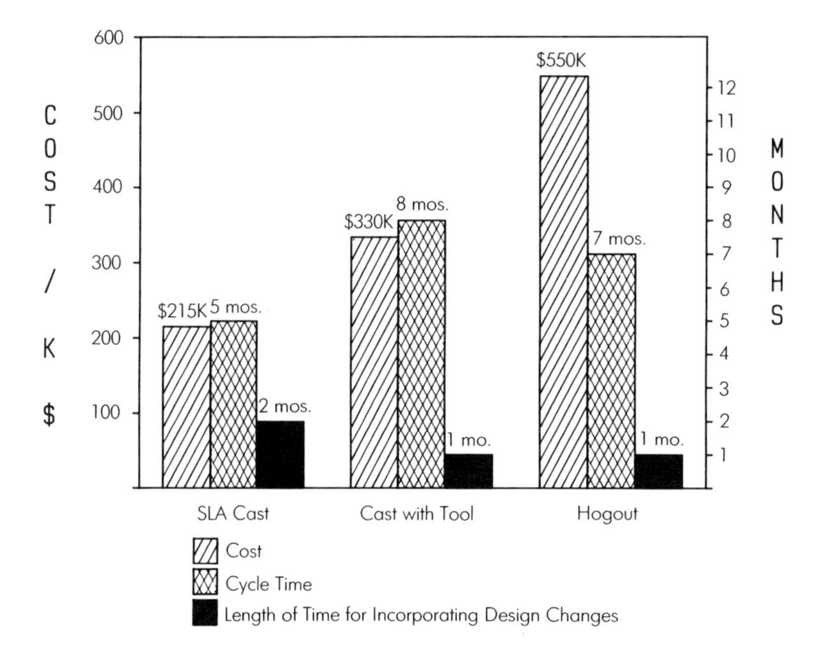

Figure 12-18. Comparison of manufacturing cost and cycle time.

4. The requalification cost of the castings was eliminated. Typically, the cycle time is too long and costly to produce casting in the initial production phase. Previously, machined parts were used to meet the cycle time requirements at a much higher cost. As the project progresses into full-scale production, castings are integrated into the system to reduce the overall cost. However, these parts must be structurally tested to ensure proper strength. This testing was eliminated as the castings were integrated into the systems from the first production phase. This new method resulted in over $20,000 in savings in this step alone.
5. Typically, the foundry requires four test SL patterns to establish the gating requirements. Of the four test RP&M patterns that were used at the foundry, one casting was returned to the program. This part served as a tooling proof part for the subsequent machining operation, resulting in shorter machining cycle times and reduced scrap for the production RP&M pattern castings.

12.6 Guidelines and Practical Limitations

Five practical guidelines can be used to assist in the implementation of solid modeling and rapid prototyping in the industrial environment.

First, there is no substitute for good engineering. CAE/CAD/CAM, computers, and rapid prototyping assist engineers in performing their tasks. These design tools cannot replace engineering judgement, discipline, design creativity, and practical design applications. Solid modeling and SL are not panaceas, but are excellent tools to assist engineers in accomplishing their tasks.

Second, as with all CAE/CAD tools, there is a learning curve associated with solid modeling. To receive maximum benefit, sufficient time should be allowed for the user to become acquainted with the capabilities and intricacies of the software.

Third, designing with a solid modeling system is the most effective way to take advantage of rapid prototyping. If designs are maintained in wireframe mode and an SL rapid prototype part is desired, the part must be remodeled or converted into an acceptable database format. This adds to the cost and cycle time required to produce a "rapid prototype" part.

Fourth, the solid mold or flask investment casting process will eventually be supplanted by the more traditional shell mold investment casting process. As the chemical, casting, and rapid prototyping industries are successful in the development of an acceptable RP&M part for the shell mold investment process, applications for the solid mold process will diminish.

Fifth and finally, the SL process does have limitations which need to be balanced against the part requirements. These limitations include: surface finish, tolerance, wall thickness, and feature configurations.

Surface Finish

The general finish is acceptable for engineering evaluation requirements. However, for display model, surface finish flaws are noticeable at five to six feet, but, at present, may be corrected by a fiberglass filler, sanding, and painting. Ultimately, further research and development is needed to improve the intrinsic surface finish of RP&M parts.

Tolerance

Typical part dimensions can be held to ±0.005″ per inch. Tighter tolerances can be realized with experimentation and the use of some of the newer and more accurate part building techniques. Part accuracy is also strongly influenced by part configuration.

Wall Thickness

As a general rule, 0.050″ (1.27 mm) is the minimum wall thickness that can be consistently reproduced. Thinner sections may suffer distortion during the postcuring cycle.

Feature Configuration

Part position and orientation should be aligned with the important engineering design features for the best possible surface finish and accuracy. The layering (or "staircase") effect dictates that all critical part features be built in the *X* and *Y* plane of the SLA, not in the *Z* plane.

Conclusion

Solid modeling and rapid prototyping offer the mechanical design engineer a more cost effective means of designing and building prototype parts than traditional methods. Texas Instruments has experienced productivity increases as these processes have been used in the design and fabrication areas of the product development life cycle. Both solid modeling and RP&M technology will play important roles in the future as companies strive to decrease design costs and cycle times.

APPENDIX

		CASTING COST		MACHINING COST								
				SET-UPS			MACHINE		TOOLING			
	QTY	CAST COST	SET UP COST	#	HOURS	COST	HOURS	COST	HOURS	COST	MATL$	
Upper housing	15	$1,350	$1,500	4	60	$12,000	40	$2,000	120	$6,000	$500	
Lower housing	15	1,200	$1,500	4	48	9,600	32	1,600	120	6,000	500	
Forward blkhd	15	91	500	2	16	1,600	4	200	60	3,000	300	
Intermed. blkhd	15	188	750	4	16	3,200	4	200	60	3,000	300	
PS bulkhead	15	95	500	2	12	1,200	4	200	60	3,000	200	
Mid bulkhead	15	61	350	2	8	800	4	200	40	2,000	100	
Low structure	15	133	500	2	16	1,600	6	300	60	3,000	300	
Fairing	15	122	500	4	12	2,400	4	200	30	1,500	200	
SUB TOTAL		$3,240						$4,900				190,500
SLA COST												25,000
Total $		$48,600	$6,100			$32,400		$73,500		$27,500	$2,400	$215,500
												Total $

Notes:

- Tooling hours include N/C programming and tool fabrication.
- Tooling is based on simple, temporary "shop-aid" tooling.
- Machining cost is based on a labor rate of $50/hour.

Appendix Figure 12-1. SLA pattern investment casting cost.

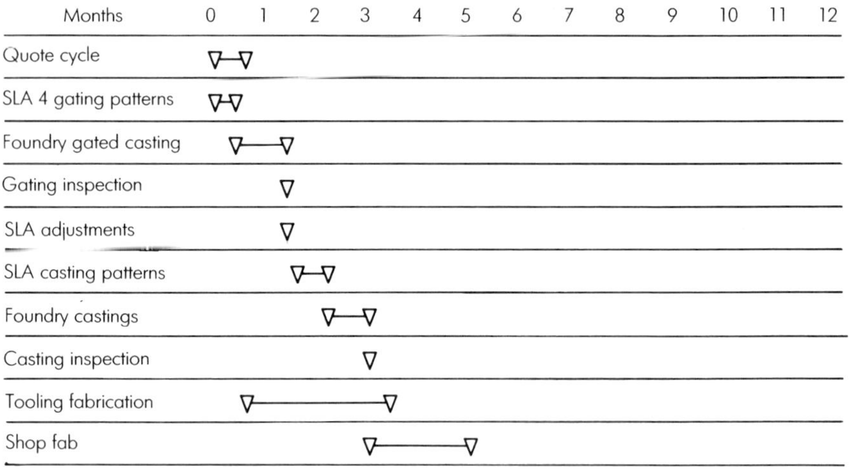

Months 0 1 2 3 4 5 6 7 8 9 10 11 12

Quote cycle

SLA 4 gating patterns

Foundry gated casting

Gating inspection

SLA adjustments

SLA casting patterns

Foundry castings

Casting inspection

Tooling fabrication

Shop fab

Appendix Figure 12-2. Prototype manufacturing cycle via SLA pattern investment casting.

	Qty.	Material Cost	MACHINING COST: SET-UPS #	SET-UPS HOURS	SET-UPS COST	MACHINE HOURS	MACHINE COST	TOOLING HOURS	TOOLING COST	TOOLING MATL$	
Upper housing	15	$500	4	150	$30,000	200	$10,000	160	$8,000	$300	
Lower housing	15	500	4	120	24,000	150	7,500	120	6,000	300	
Forward blkhd	15	50	2	24	2,400	16	800	100	5,000	100	
Intermed. blkhd	15	150	4	60	12,000	80	4,000	100	5,000	100	
PS bulkhead	15	50	2	32	3,200	20	1,000	100	5,000	100	
Mid bulkhead	15	30	2	12	1,200	12	600	40	2,000	100	
Low structure	15	150	2	40	4,000	36	1,800	100	5,000	100	
Fairing	15	250	4	20	4,000	24	1,200	60	3,000	100	
Sub total		$1,680					$26,900				
Total $		$25,200			$80,800		$403,500		$39,000	$1,200	$549,700 Total $

Notes:

- Tooling hours include N/C programming and tool fabrication.
- Tooling is based on simple, temporary "shop-aid" tooling.
- Machining cost is based on a labor rate of $50/hour.
- Cost and hours are estimates based on similar parts.

Appendix Figure 12-3. Machining hogout cast.

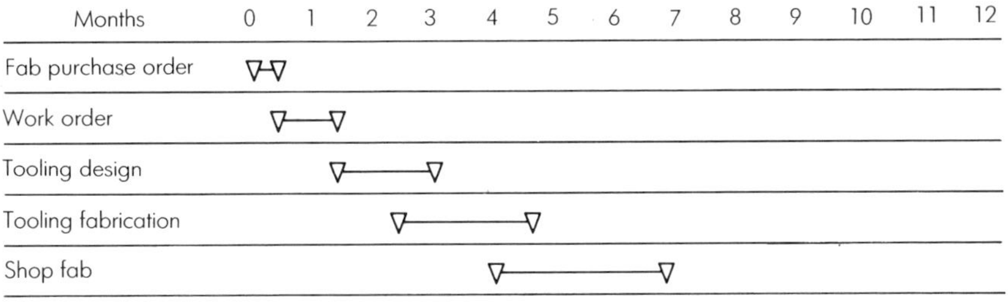

Months	0	1	2	3	4	5	6	7	8	9	10	11	12
Fab purchase order													
Work order													
Tooling design													
Tooling fabrication													
Shop fab													

Appendix Figure 12-4. Prototype manufacturing cycle via machining hogout.

Prototype Manufacturing Cycle Via Wax Investment Casting

		CASTING COST		MACHINING COST							
				SET-UPS			MACHINE		TOOLING		
	QTY	CAST COST	TOOLING COST	#	HOURS	COST	HOURS	COST	HOURS	COST	MATL$
Upper housing	15	$1,350	$30,000	4	60	$12,000	40	$2,000	120	$6,000	$500
Lower housing	15	1,200	24,240	4	48	9,600	32	1,600	120	6,000	500
Forward blkhd	15	91	8,400	2	16	1,600	4	200	60	3,000	300
Intermed. blkhd	15	188	13,680	4	16	3,200	4	200	60	3,000	300
PS bulkhead	15	95	8,400	2	12	1,200	4	200	60	3,000	200
Mid bulkhead	15	61	7,080	2	8	800	4	200	40	2,000	100
Low structure	15	133	9,360	2	16	1,600	6	300	60	3,000	300
Fairing	15	122	9,840	4	12	2,400	4	200	30	1,500	200
SUB TOTAL		$3,240	$111,000					$4,900			
Total $		$48,600	$144,300			$32,400		$73,500		$27,500	$2,400

$328,700 Total $

Notes:

- Tooling hours include N/C programming and tool fabrication.
- Tooling is based on simple, temporary "shop-aid" tooling.
- Machining cost is based on a labor rate of $50/hour.
- Cost and hours are estimates based on similar parts.

Appendix Figure 12-5. Wax investment casting cost.

Months 0 1 2 3 4 5 6 7 8 9 10 11 12

Quote cycle

Vendor tool and part

First lot inspection

Fab purchase order

Work order

Tooling design

Tooling fabrication

Shop fab

Appendix Figure 12-6. Prototype manufacturing cycle via wax investment casting.

chapter 13

Chrysler Corporation: An Automotive Case Study

Businessmen go down with their businesses because they like the old way so well that they cannot bring themselves to change...Seldom does the cobbler take up with a new fangled way of soling shoes and seldom does the artisan willingly take up with new methods in his trade.

—Henry Ford
My Life and Work, 1922

13.1 Perspective

"Launching new technology is fraught with corollaries of Murphy's Law: What can go wrong, will, and at the worst possible time. We at Chrysler have lived through power failures, laser burnouts, spilled resin, imperfect CAD data, and stray laser vectors. Approaching the anniversary of Chrysler's first CATIA™ part (a 5/16th scale Viper oil pan) built on February 9, 1990, it is important to keep things in perspective. First, StereoLithography works as promised, with only slight exaggerations as to ease of use. Second, good CAD models are required. As the SLA team at Jeep and Truck Engineering (JTE) learns from our experiences, we will share our education with you."

---

*By **Lavern D. Schmidt**, P.E., Manager, Design Aids and Packaging Development, and **William L. Phillips**, Supervisor, Wood, Plaster, and StereoLithography Labs, Chrysler Corp., Jeep and Truck Engineering, Detroit, MI.*

Thus began our editorial in "SLA Today," a StereoLithography newsletter, written to the Chrysler community on our progress with Rapid Prototyping & Manufacturing technology.[1,2,3,4] The article continued, "The SLA Team has to take additional steps to fully develop the process at Chrysler:

1. Upgrade our file viewing capabilities to spot erroneous data. This was accomplished in mid-August 1990, allowing our personnel to find errors *before* we build a part.
2. Upgrade the SLA build computers to 386 CPUs to speed the rate of build in the vats. This step was also accomplished in August, 1990, effectively doubling our build capability to the equivalent of four SLA-250s with the older 286 computers.
3. Publish an SLA reference manual to inform designers of proper CAD system procedures, as well as techniques to eliminate errors. A major portion of our education is contained in this manual.

To date, the biggest hurdle to fully utilizing this technology has not been technical but *cultural*. Quite simply, we need solid models to build a part. It takes time and training to gain the experience necessary to do solid design properly, commodities that are in short supply.

The SLA team will do our part, striving to build all our models as fast as possible. You'll get trained to fully utilize all the advantages of 3D solid modeling and StereoLithography. Together we'll push this rapidly developing technology, driving down our costs and time to market." After our first year, we were confident and proud to be on the cutting edge.

The purchase of two 3D Systems SLA-250s at Chrysler was funded by our Accelerated Quality Improvement Program (AQIP) in 1989. Our proposal was "sold" to the Chrysler Investment Committee on the basis of improving vehicle quality and reducing cycle time.

13.2 Background

Packaging problems are often discovered too late in the pilot and program vehicle build process to allow correction without excessive time, effort, and dollars being expended. These late changes invariably affect the quality of the vehicle produced during V1 (Production Vehicle Launch).

AQIP Proposal

Utilize two SLA-250 units to create three-dimensional plastic components directly from CATIA™ CAD information. These systems will allow the JTE Design Aid and Packaging Development Department to fabricate components accurate to 0.005″ within 24 hours of receipt of CATIA™ data. This enables us to completely mock up all prototype parts as soon as they are designed.

StereoLithography, in concert with "Process Driven Design," can improve quality by:

- Allowing us to quickly produce prototype parts (within 24 hours of the receipt of CAD data) directly from three-dimensional CAD models, into a finished part, bypassing the traditional methods where print errors, interpretations, machining, and inadequate cut sections delay the build of a correctly toleranced prototype.
- Allowing us to find our mistakes faster.
- Allowing us to evaluate assembly techniques with "process driven design" parts relating to fit/finish issues in less time.
- Allowing us to make patterns and die set combinations for sand casting and molding operations.
- Allowing us to build accurate plastic prototype parts as required.
- Allowing us to make complex prototype rubber parts by generating the tooling for a rubber mold.
- Reducing design cycle time as we integrate this process with simultaneous engineering through our management tracking system (MTS).

The two SLA-250s were installed in early January of 1990, and *we have been using the equipment nonstop, 24 hours a day, 7 days a week for two years, building well over 1500 parts of 500 different geometries*. Chrysler is predominately a CATIA™ (by Dassault) CAD user with over 600 seats of CATIA™ computer terminals. To date, 78% of our models have been built using CATIA™ data with ProEngineer™ and AutoCAD™ providing the remainder. This chapter will highlight how RP&M has been used by Chrysler in its automotive applications.

13.3 Automotive Applications

Automotive applications have centered around design aid models. These models have been used for proof of packaging, design verification, marketing presentations, component optimization, and vehicle development testing. Also, we have used secondary tooling from that model (or master pattern) to provide a fast means of prototyping components using traditional mold making or foundry techniques. Secondary tooling processes at Chrysler have included room temperature vulcanizing (RTV) molding, grey iron foundry/sand casting, resin transfer molding (RTM), vacuum forming and squeeze molding techniques (see Reference 5).

Proof of Packaging

Proof of Packaging for CAD designs was initially what sold Chrysler on StereoLithography. Our first example was for a new, more reliable starter for a turbo diesel Jeep Cherokee, which is exported to Europe. The starter was redesigned in France, as they did not have a physical model for us. The French group had designed the starter in CAD to fit their engine, and we were not sure if it would fit in thc Jeep.

JTE personnel loaded the CAD file on our SLA and built the part shown in *Figure 13-1*. We then installed it in our Jeep vehicle, as seen in *Figure 13-2*. With the model in place, we completed battery cable and wire harness routing development.

Design Verification: Distributor Cap and Body

Design verification now means building an SLA model of your "CAD" data to see if its "form & fit" are what the designer envisioned. The first example of design verification was on an engine upgrade program.

Two different engineers had designed the mating of a distributor cap and body. They were looking from different views. When the two parts were put together, the cap would not fit on the body.

The distributor cap engineer had an easier time changing his plastic part, so he revised the guides by 10°, and we rebuilt the part. Within 24 hours we had a second part (*Figure 13-3*) and proved that the new cap fit. This error would not have been realized by traditional methods until the first batch of prototypes arrived weeks later.

Design Verification: Shift Handle

An automatic transmission shift handle design was based on styling, and we did not feel we had to produce a tool of any sort to prove the concept. A manually

Figure 13-1. SL starter built with French CAD data.

Figure 13-2. SL starter installed Jeep vehicle.

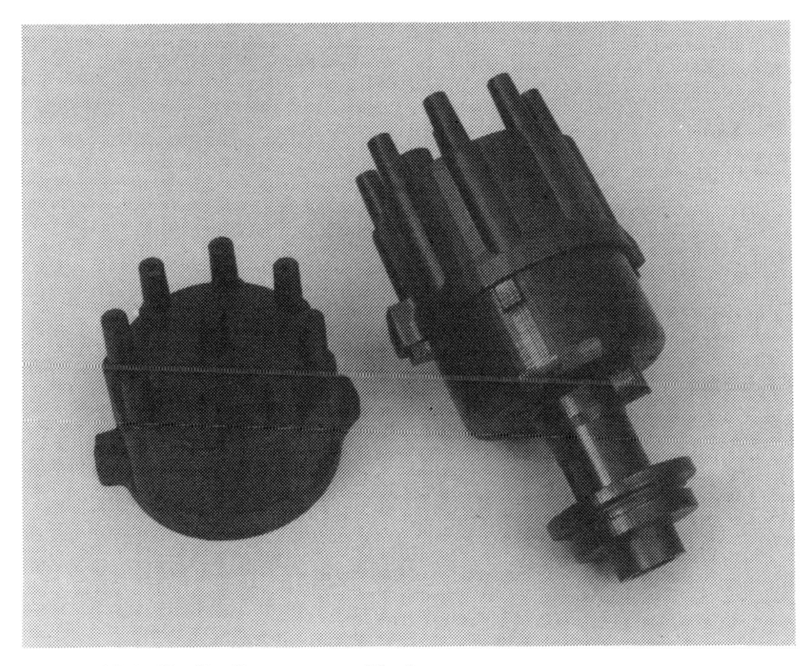

Figure 13-3. SL distributor cap and body.

built model and a cross-sectional drawing was provided to show styling intent. The hand-built model had angular irregularities that had to be corrected in the design process.

From the styling specifications and the drawing, a CATIA™ wireframe model was created. The wireframe was transformed to a solid via surfacing. The solid model provides a higher quality part with greater conformance to the complex external shape of the intended design.

The final solid model was transferred to the SLA, and an RP&M model was built. From the size and feel of the SLA part, we discovered that several revisions were required. The handle was too large, and a person with smaller hands would have difficulty grasping it and feeling comfortable shifting gears in the vehicle.

StereoLithography easily enabled us to scale down the model. We built three parts involving 8%, 10%, and 12% size reductions. The design team then chose the size they wanted, revised their CAD drawings accordingly, and went into production. This example proves the value of verifying these details prior to receiving the final prototype. *Figure 13-4* shows two SL models and a final shift handle.

Without an SL model, a functional prototype shift knob would have been manufactured before the desired changes could have been discovered. Due to the changes noted above, a second prototype would have been required. *Based on outsourcing estimates, this single SL model saved over $40,000 and 18 weeks of design time.*

This opened our eyes to possibilities we were missing. We had been neglecting much of the cost and time savings accrued with SL on prototype development.

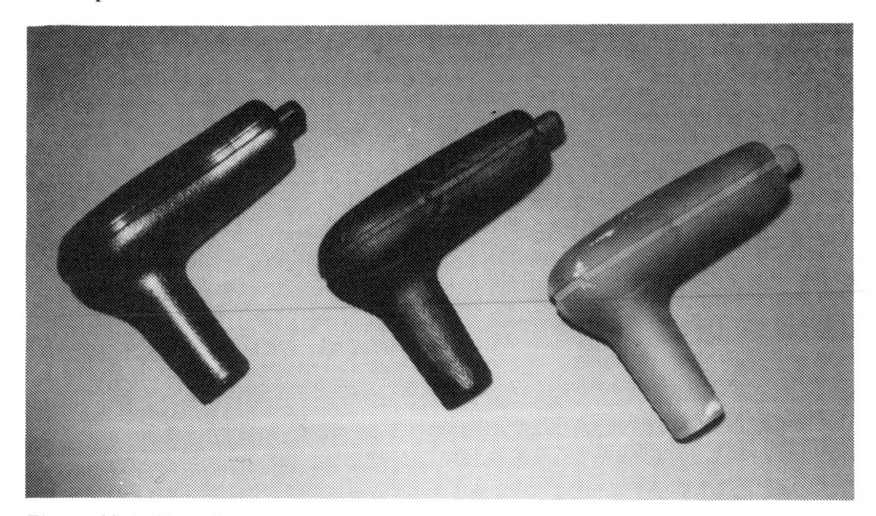

Figure 13-4. SL and prototype automatic transmission shift handle.

Marketing Presentations: Transfer Box Design

New Venture Gear™ (NVG) was asked to quote a transfer box design for a European customer. After the layout was modified to NVG's design and manufacturing methods, a quote was prepared for presentation at the customer's location.

Due to compressed timing, NVG would need to identify unique methods to reduce the design and development cycle. Customer supplied case casting drawings were used to make CAD solid models using ProEngineer. The CAD models were then used to make SL parts available for customer demonstration when presenting the quote, as seen in *Figure 13-5*.

The SL part had a significant impact on the customer. They were impressed that NVG was able to make a physical model for visual effects and packaging studies from rough casting drawings at the proposal stage.

The use of this model achieved several objectives:

1. Use for visual observation and packaging studies.
2. Ability to use a solid modeler with JTE's SL equipment.
3. Ability to reduce the design cycle by making physical parts available early in the program.
4. Time saving of at least six weeks versus conventional casting.
5. Cost savings estimated at $12,000.

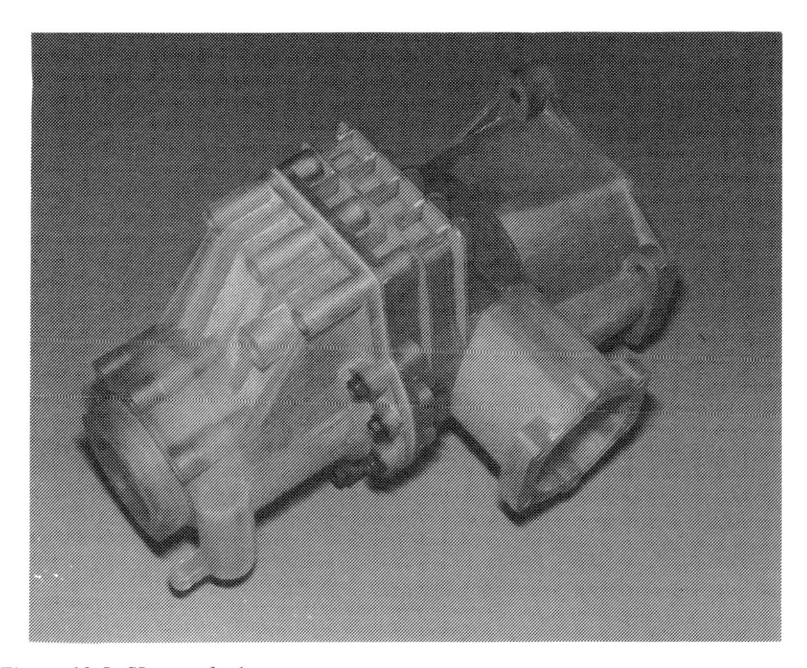

Figure 13-5. SL transfer box.

Airflow Testing: Ambient Air Duct and Cylinder Head

Engine Engineering used SL to assist in determining the optimum size for an ambient air duct. Three entries with varying diameters were made in the SLA lab and then flow tested.

An 80 mm diameter was determined to be the best choice. The part was further improved by varying the radius at the large diameter end. Another SL part was made and flow tested in the new configuration. The optimum size has now been selected, based on the test data.

Using SL, the parts were available in less time (*one day versus four weeks*). Aluminum stock is usually machined to the characteristics of a cylinder head for flow benchwork. The SL part shown in *Figure 13-6* took about two weeks.

We installed steel valves and springs and ran it on the flow bench, as shown in *Figure 13-7*, before making a functional prototype part. Initial air flow evaluations showed a *38% improvement in the air flow through the cylinder head!*

Once this change has been incorporated, we intend to generate a design in solid CAD to enable an SL master of the entire cylinder head. Using sand casting to build the functional prototype from the SL master should save many weeks and thousands of dollars compared to traditional methods.

Figure 13-6. SL cylinder head model.

Figure 13-7. SL cylinder head (with steel valves installed) on the flow bench.

13.4 StereoLithography and Secondary Tooling Applications

StereoLithography and associated conventional tooling applications used at Chrysler Corporation, Jeep and Truck Engineering Division have increased quality in product development activities. Conventional prototype tooling applications using StereoLithographic master models include the five techniques described in this section.

Vacuum Form Tooling

Vacuum form tooling produced from SL master patterns has proven to be of great advantage to Chrysler. Circumventing traditional wood pattern making has helped reduce design verification time significantly. Also, the learning curve using SL master patterns for vacuum forming proved to be relatively short. When building conventional wood patterns for vacuum forming, the pattern maker rotates the part section for optimum draw condition. The normal configuration is usually in the male or positive form because the vacuum chamber is located at the base of the male pattern and the heated plastic is pulled down over the male form.

When using an SL master pattern, this process is reversed during benching. The mold maker sets the pattern for optimum draw condition by temporarily rotating the SL part into position. Then runoff is added to the part with clay or wood and a female cast is produced. The female mold is taken using plaster or epoxy, as shown in *Figure 13-8*. In some cases, a combination of plaster with an epoxy face may be chosen when extended tool life is required.

The preferred method for any vacuum forming operation has always been to draw into the female side of the tool. SL master patterns force the use of female molds and excellent results are achieved.

Because many durable plastic parts are formed from ABS, vacuum form patterns are expanded 0.006″ per inch to compensate for shrinkage when the part cools. When the build parameters are set, we simply expand the model by 1%. *Figure 13-9* shows an SL master, the female cavity tool, and the vacuum formed ABS part.

Resin Transfer Molding (RTM)

Using SL master patterns to produce RTM injection molds has also proven successful at Chrysler. Complex part definition on the core side of molds is easily attainable using SL technology. These parts can be reproduced in durable materials such as polyester fiber filled resins and then mass produced. Some examples of automotive applications are: throttle body injection bonnets, center consoles, instrument panel components, and interior trim panels.

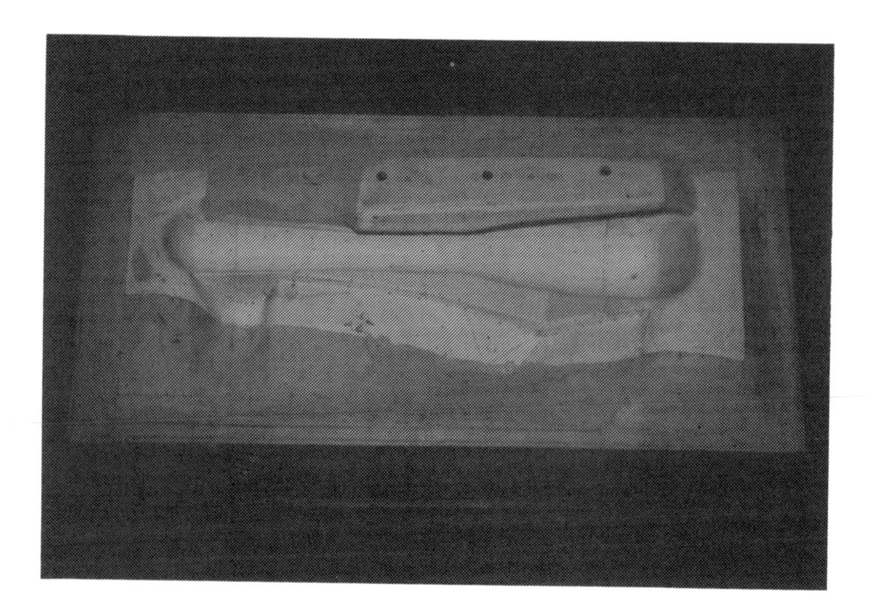

Figure 13-8. Epoxy cavity mold created from SLA master.

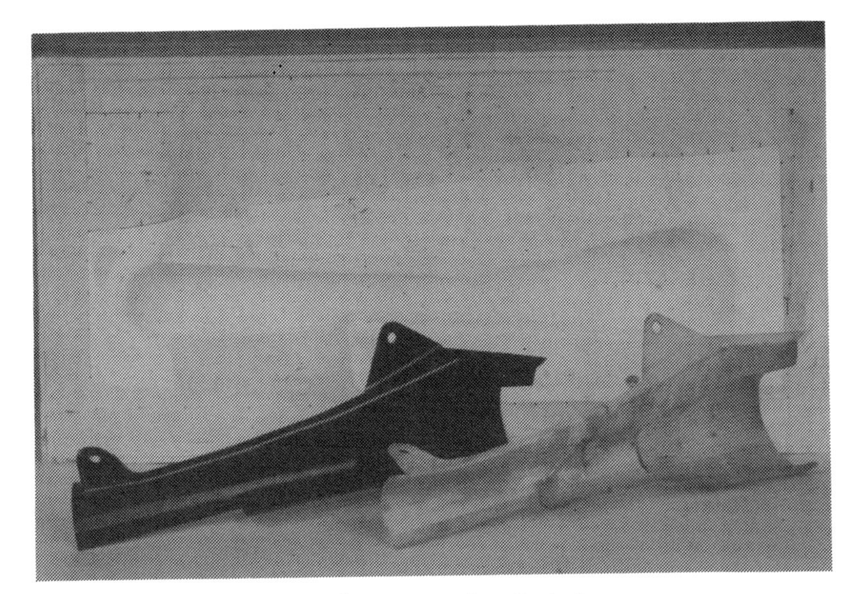

Figure 13-9. SL master, vacuum form part, and cavity tool.

The shop benching technique for setup of the SL master is very similar to that used in vacuum form mold production. An SL master pattern (*Figure 13-10*) is

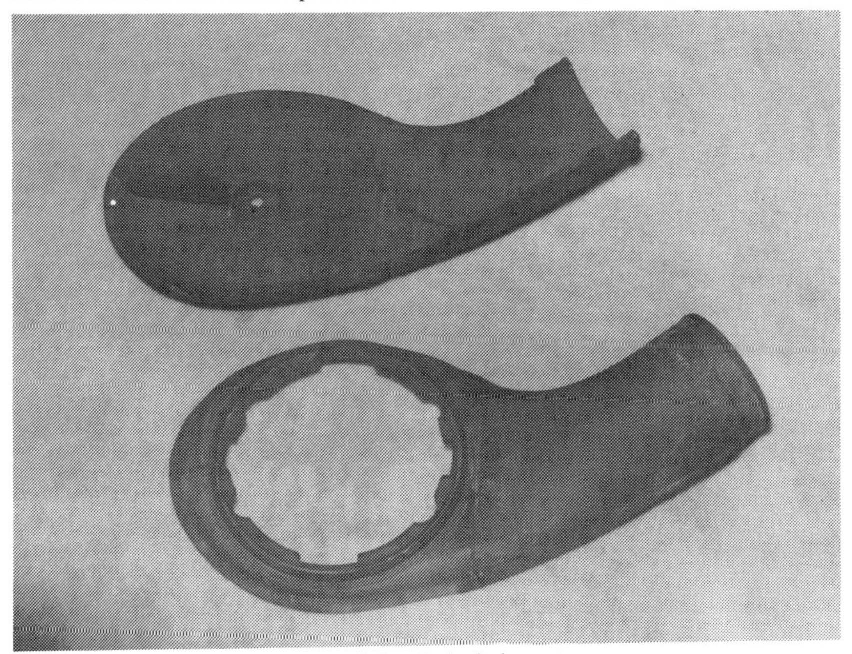

Figure 13-10. SLA master pattern of throttle body bonnet.

suspended temporarily and adjusted for best mold position. As in vacuum forming, a female mold is produced (*Figure 13-11*) with the parting line set at the trimline of the part. When the reinforced plaster mold is solidified, the SL master is *not* removed. A second mold is then produced in combination with the first half mold and, when solidified, the upper and lower tools are separated and the SL master is removed. The cavity form is now complete and the injection tool set is ready to be used, as shown in *Figure 13-12*. The final ABS part is seen in *Figure 13-13*.

Gray Iron Foundry Applications

Cope and drag patterns for gray iron foundry casting may be created by using SL master patterns to replace wood patterns. During the generation of the CAD solid model, it is necessary to utilize the expertise of a person familiar with foundry technique.

The CAD designer, with the assistance of the foundryman (molder) or patternmaker, should design the CAD model paying special attention to casting requirements such as draft angle, shrink rate, core making, potential back-draft problems and determination of parting lines.

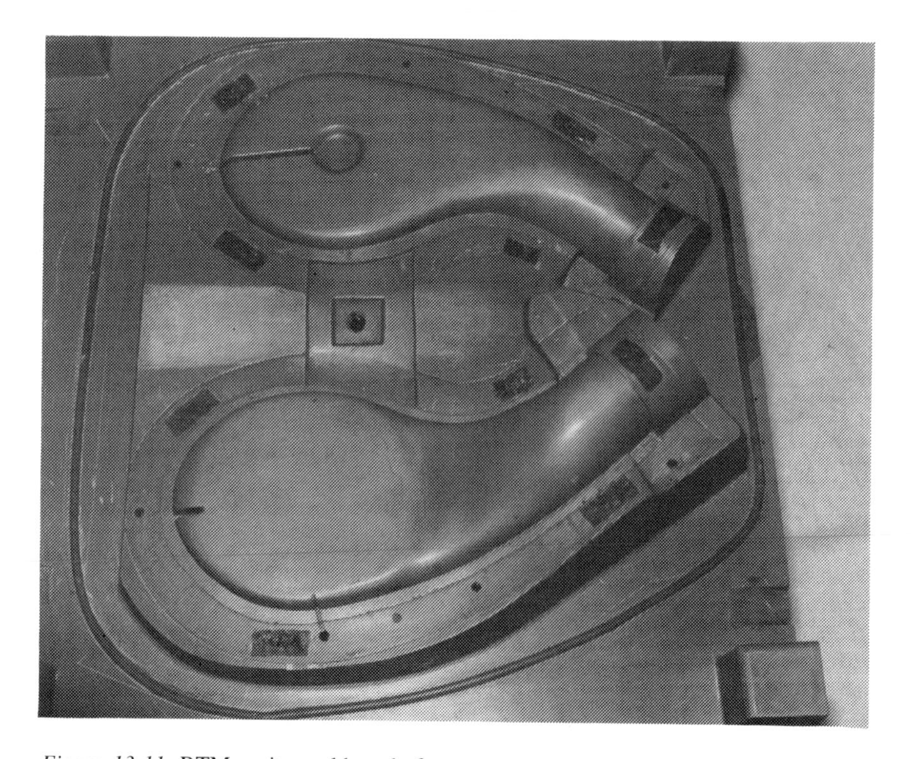

Figure 13-11. RTM cavity mold made from master pattern.

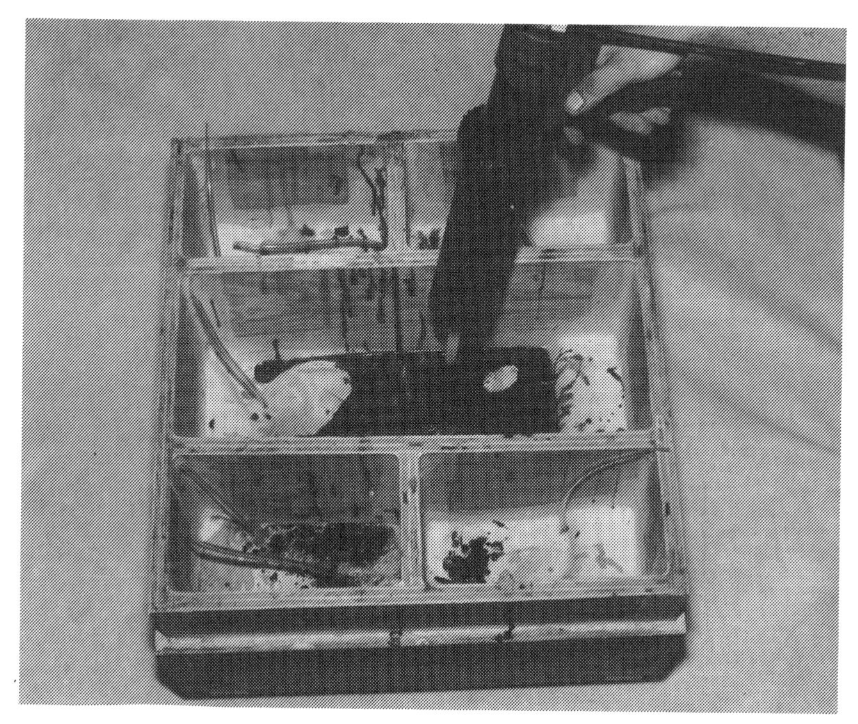

Figure 13-12. RTM cavity mold being filled from the backside.

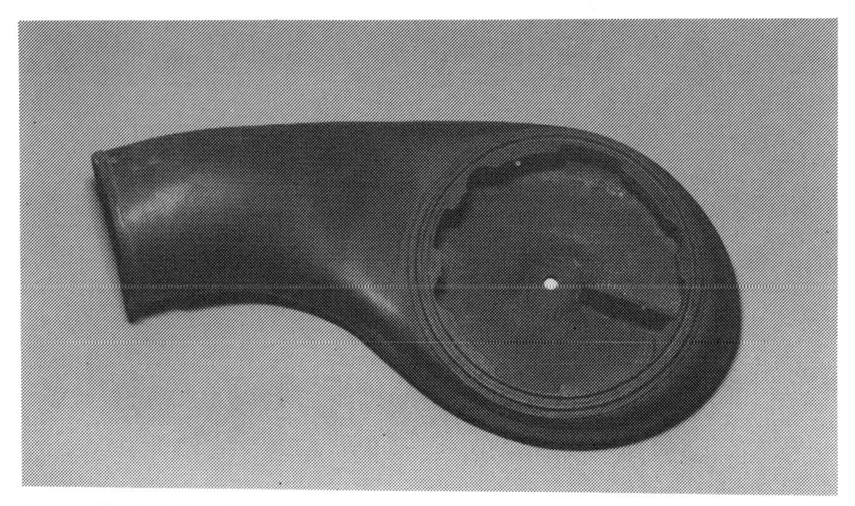

Figure 13-13. RTM throttle body ready for vehicle testing.

When the RP&M master pattern is completed, one may prepare the pattern using traditional bench methods. However, time is wasted if this is done.

The SL master pattern may be developed more extensively at the CAD design level by including some of the secondary foundry requirements directly. These include a runoff surface added to the parting line, and the match plate joining the cope to the drag (upper and lower elements of the foundry tool). Preparation within the CAD model to circumvent hand benching reduces the total time required to complete the casting process.

StereoLithography has been used by the Viper program to reduce costs for the V10 exhaust manifold and increase accuracy in the development of prototype parts. For example, to create a mold for the manifold, the pattern maker had to determine the outer and inner surfaces of the part. Using CATIA™, a solid model representing the inner core of the manifold, was designed and built on the SLA (*Figure 13-14*). The pattern maker applied a layer of wax over the core to define the outer surface of the manifold (*Figure 13-15*). A mold was taken of this surface, as seen in *Figure 13-16*. A second mold was taken directly from the core to define the inner surface. These molds were then used to create a metal casting (*Figure 13-17*) for final part production, as shown in *Figure 13-18*.

One advantage that SL provided was accuracy to design intent. Conventional drawings offer the pattern maker only cross-sectional representations of the part. These cross sections are "blended" using best estimates of design requirements. This can cause variations in the cross-sectional area, resulting in variable air velocity in the core. By using an SL model, we were able to create a prototype approaching the accuracy provided by CATIA™.

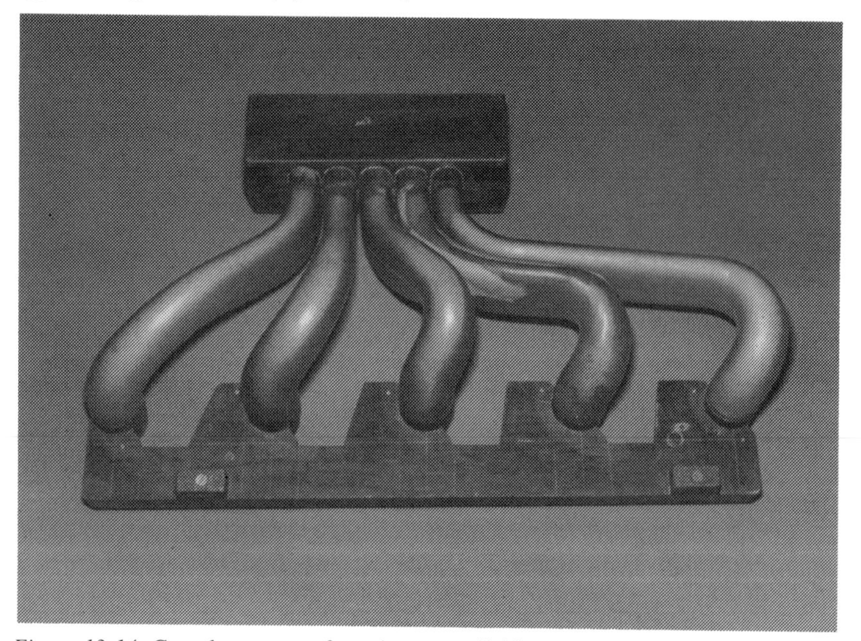

Figure 13-14. Complete pattern for exhaust manifold.

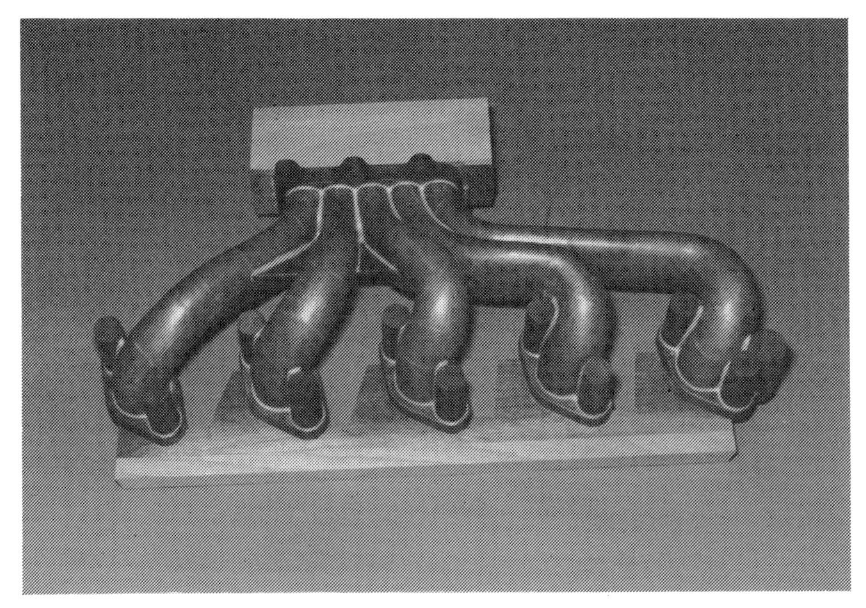

Figure 13-15. Pattern waxed off with stud mounting bosses and fillets added.

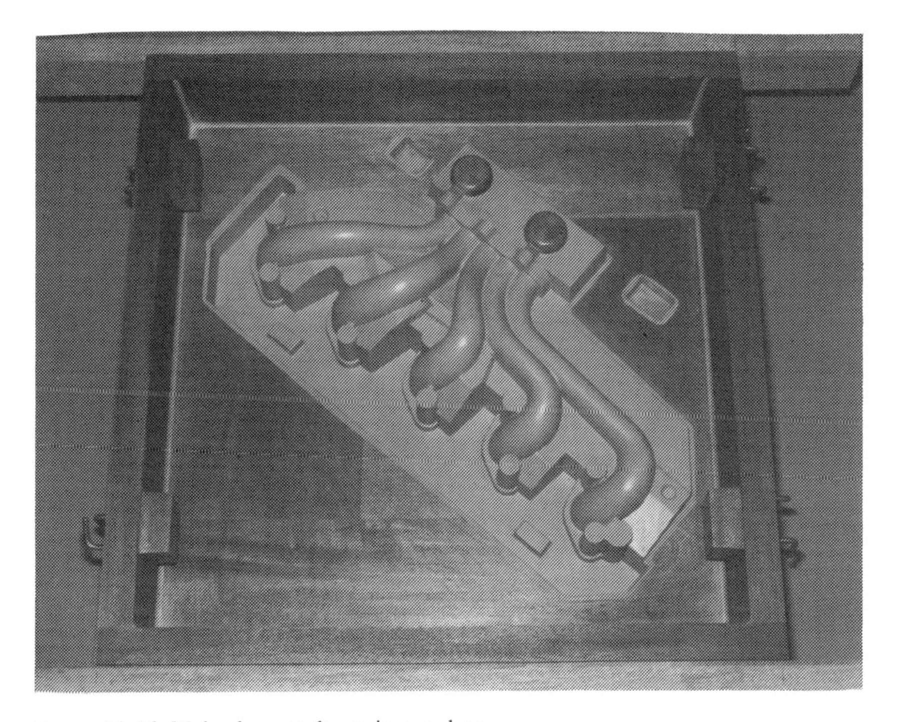

Figure 13-16. Male element shown in core box.

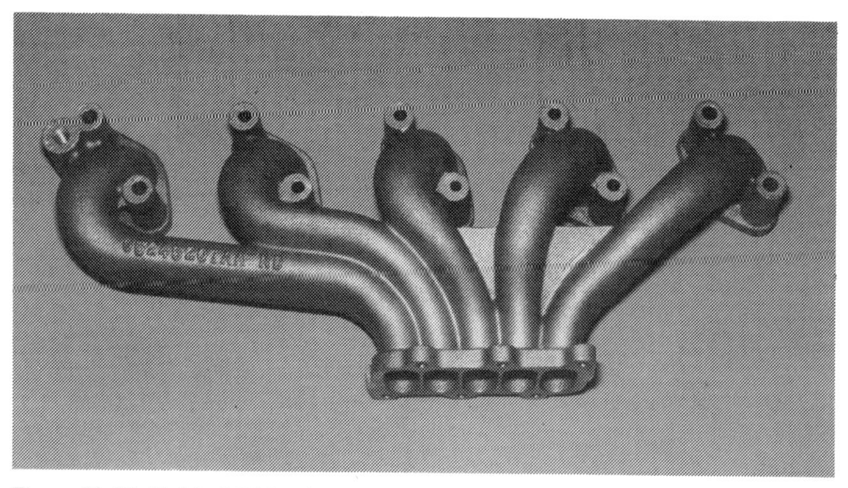

Figure 13-17. Finished V10 exhaust manifold casting.

Cost saving was the second benefit in this process. Pattern making costs were cut in half, saving approximately $6,000 over conventional methods. Due to the high accuracy (0.0025″), the SL master was used for *production tooling* (all 5200 vehicles), *saving over 18 weeks and $50,000 in tooling.*

Figure 13-18. Finished casting installed on Viper V10 engine.

Squeeze Molding

Squeeze molding is a relatively rapid and well established technique to reproduce parts in materials that do not lend themselves to photopolymer curing as required in SL technology.

An example of such a material is urethane. It is available in various durometers (hardness levels) and, as a result, closely approximates the material properties required by OEMs in their mass-produced plastic components, normally made by injection molding.

Squeeze molding is well suited to prototyping applications because it is a pourable, cold casting process and does not require heated molds or high nozzle velocities as in production injection molding. The benching technique for squeeze molding from an SL master pattern is a follows:

- The SL model (*Figure 13-19*) is set on a steel surface plate at the parting or trim line of the part.
- Polyvinyl acetate (PVA) parting agent is sprayed on the SLA model.
- A temporary plywood box or case is built around the SLA part.
- A plaster mixture (Hydrocal™) is poured over the SL part up to the top of the plywood case and allowed to solidify.
- The SL part is then removed from the plaster cavity.
- Sheet wax is applied to the cavity shape to simulate part thickness.
- A temporary plywood box is built on top of the cavity mold at the parting line between the upper and lower elements of the tool.

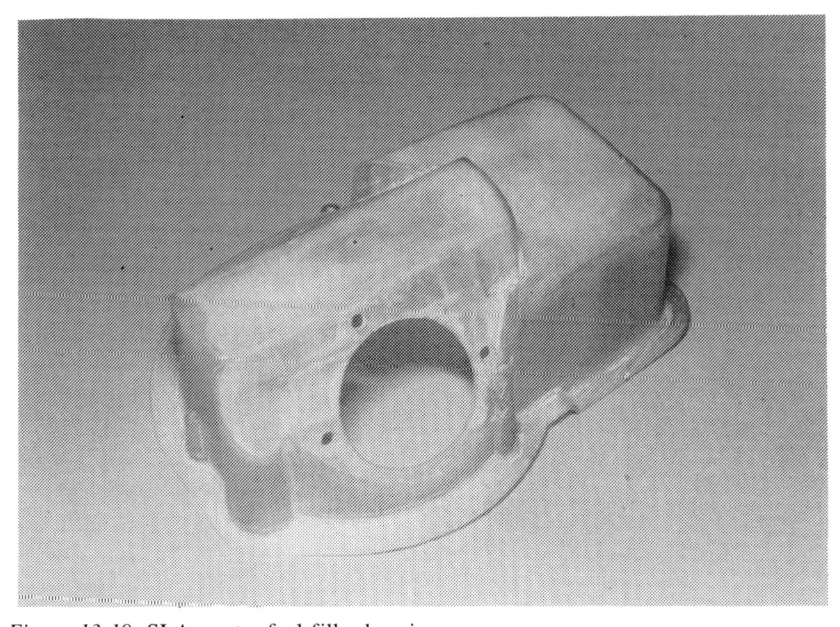

Figure 13-19. SLA master fuel filler housing.

- Plaster is poured into the waxed cavity to the top of the box and the male or punch side of the mold is created.
- The upper and lower elements of the squeeze molding tool set (*Figure 13-20*) are separated and the wax that was added to the cavity is removed, creating a material thickness void.
- The urethane liquid compound is mixed and a vacuum is drawn (to expel trapped air bubbles) and poured into the cavity.
- The punch element is forced into the cavity to displace the volume that the wax formerly occupied, and then clamped closed and allowed to solidify (approximately two hours).
- The mold is separated, the part is removed, and the flash is cut off. The part is now complete, as shown in *Figure 13-21*.
- Subsequent parts can be produced and demolded every three hours.

Note that sheet wax may be added directly to the SL pattern depending on part geometry.

As with any prototype production, the choice of appropriate technique is based on experience. There are numerous variables to consider, including part size, part geometry, cycle time required, number of parts required, etc. Part size is the primary consideration for squeeze molding. The maximum size successfully produced in this manner is approximately 6″ in depth and 24″ in length and width. Parts exceeding these limits would probably be produced more efficiently using resin transfer molding.

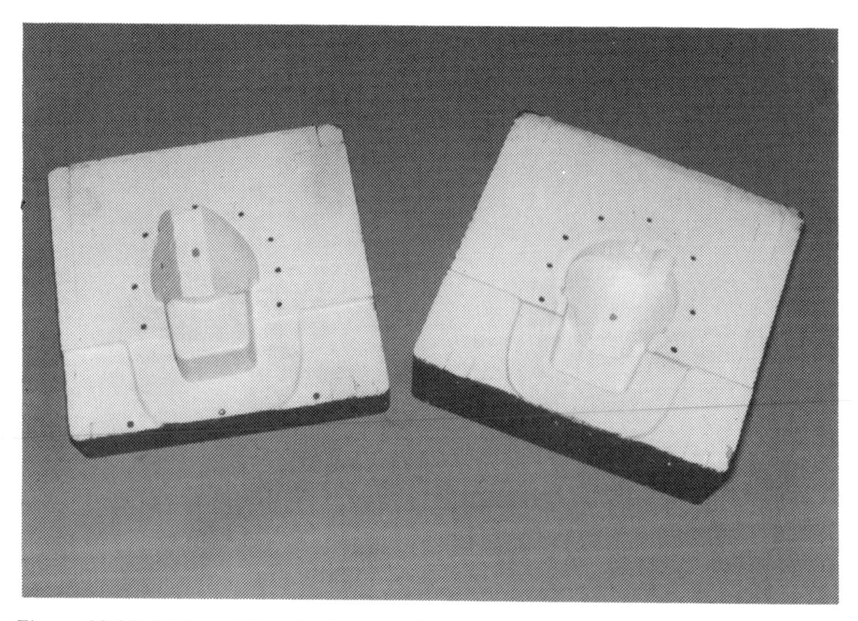

Figure 13-20. Male and female element of squeeze mold.

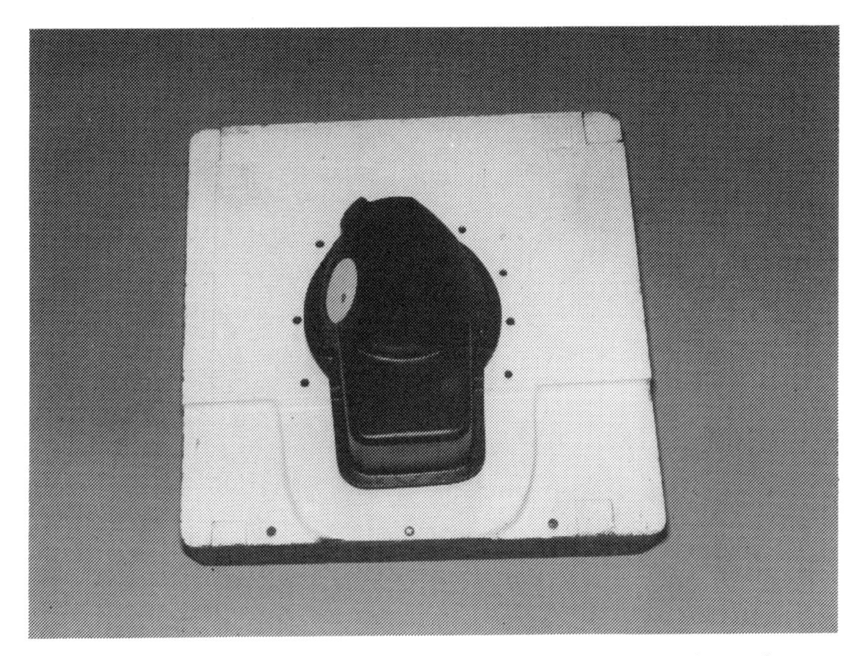

Figure 13-21. Finished squeeze molded part sitting on male element after curing.

Silicone Molding

Silicone molding compounds have been available to the prototype moldmaker for many years. The recent advent of RP&M systems to create accurate master models increases the use for these materials.

There are many advantages realized with silicone molding compound. First, extremely high resolution of master model detail can be copied to the cavity mold. Secondly, backdraft problems (die lock, or the inability to release the part from the mold cavity due to part geometry) are reduced. Such problems are commonly encountered using hard molds requiring loose pieces. These are virtually eliminated when using the silicone molding process, as shown in *Figure 13-22*.

Silicone molding compounds offer superior ability to recreate master model detail compared to traditional cold synthetic rubber molding material. Also, there is another advantage over cold molding rubber material, since about 15% more copies may be accurately produced from a single mold without degradation of the part tolerances.

When creating molds using silicone molding material, there is no need to use parting agents or materials like polyvinyl acetate (PVA) because the silicone material itself has *superior release characteristics*. Thus, it is not necessary to spray PVA onto the master model and wait for it to dry between applications (two are usually required). An additional 10% time advantage is realized as compared to traditional hard molding techniques.

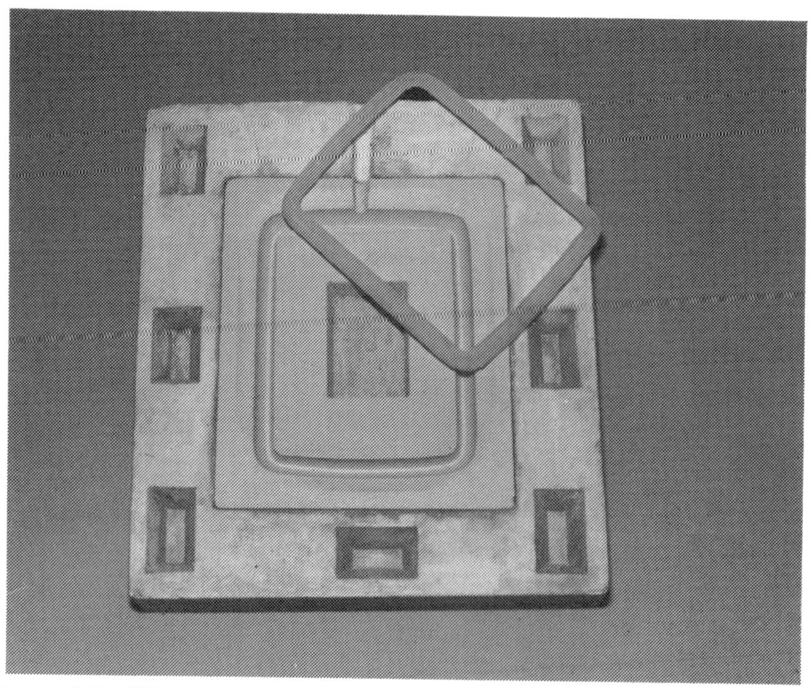

Figure 13-22. Silicone mold with SLA master shift boot retainer.

13.5 Problems

Brittleness

Initial SL resins were brittle and had a tendency to break when stressed. If dropped, they routinely shattered into hundreds of pieces. Here we were trying to build a door handle.

Someone obviously broke the SL part, as shown in *Figure 13-23*. We now have new SL resins that are much less brittle.

Supports

With StereoLithography, supports must be added to your CAD file to keep your model in place as it is building in the vat. Do not position supports significantly outside the part as this wastes time and material.

We now have an automatic support generation system call Bridgeworks™. Set the support parameters for the part and it automatically adds supports. This software is effective except in complex geometries, small overhangs, or thin sections. In such cases, separate supports must be added manually. Bridgeworks™ has cut our support generation efforts by about 80%.

Resin Swelling

We found that some resins swell with an increase in temperature. In one case, the vat temperature was in excess of 34°C and rising, due to an air conditioning failure in the facility. The recommended vat temperature is roughly 28°C.

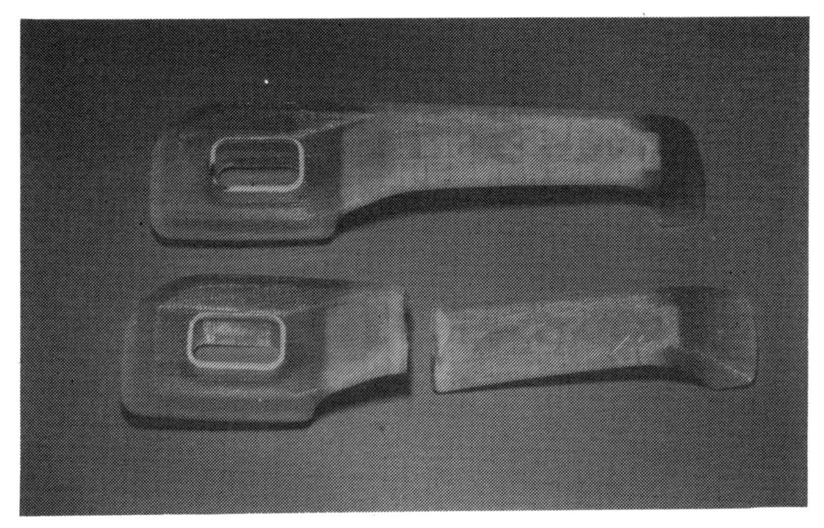

Figure 13-23. Broken SL door handle built with brittle resin.

Changes in temperature cause changes in laser curing since build parameters are entered at the start of the build. We were not building with good parameters.

Operator Error

The part shown in *Figure 13-24* was designed in millimeters, but the operator set the build parameters in inches, and a less than successful build occurred. The machines are not infallible. They require maintenance and monthly blade adjustments. The lasers do burn out. We are getting about 4000 hours on our lasers. This is our fourth laser in two years. Also, a power supply has failed. However SL should be viewed no differently than the standard machine tool. Things do go wrong with CNC and other machine tools. It is important, though, that you know what failures are likely. With RP&M, we are still learning what these will be and how much they will cost.

Many times bad files are transferred to us with incomplete sections, holes, and nontangent surfaces. The machinery builds exactly what is in the CAD file. The causes of problems that used to be blamed on ''interpretation'' differences between model maker and designer are now readily apparent. As noted in Reference 6, the machine only builds what you design!

13.6 Productivity Improvement

As we became more proficient in our use of the SLA units, *our productivity has more than doubled*. We are currently staffed by three people on one shift, viewing files, adding supports, inserting build parameters, and cleaning finished parts.

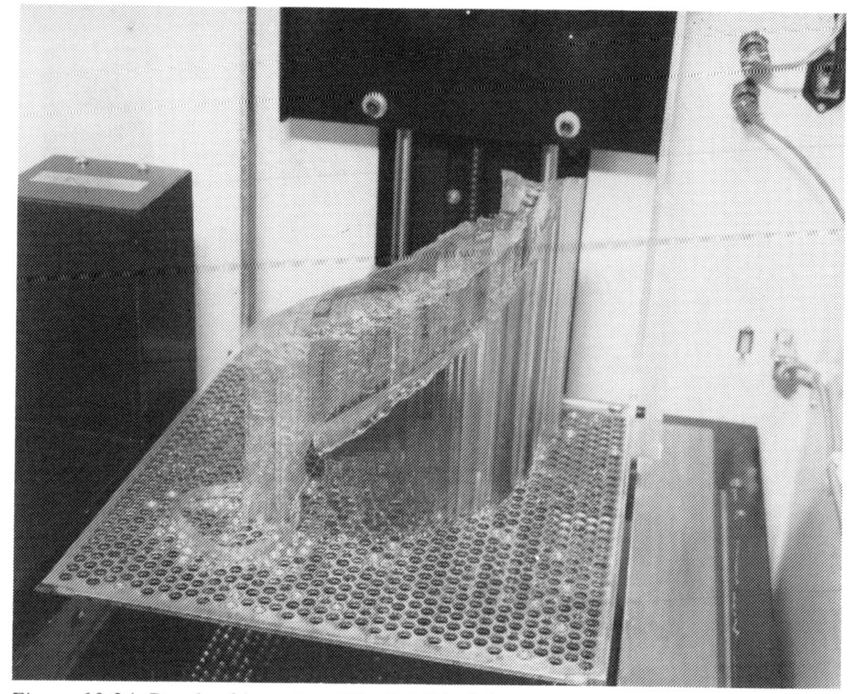

Figure 13-24. Result of inserting "English" build parameters on a "metric" part.

Automatic Support Generation

To improve our productivity standards, we are continually searching for software or hardware improvements as they are offered in the market.

One example was the testing and subsequent purchase of a software program called Bridgeworks™. This is an automatic support generation program designed to interface with our Silicon Graphics system. The program eliminates the need for designers to create supports on solid models and frees them to continue critical design endeavors. It can be used with any STL file from a CAD system.

Bridgeworks™ automatically creates support structures for CAD models submitted to the SLA department. Bridgeworks™ also eliminates the learning curve associated with SLA support structures. Nonetheless, designers still need to submit their model through the CATSLA macro for conversion to an STL format.

Efficiency and Productivity

Developing meaningful measurements of productivity and efficiency were vexing because the machines were initially building parts more slowly than we were preparing files or cleaning parts. In October 1990, we replaced the 286 microprocessor with 386 units. This *doubled* our productivity. We then had the equivalent of four SLA-250s with the former 286 MPUs. This implementation quickly reduced our backlog.

It was finally decided that the best measurement of how well our team was performing would be efficiency, production, throughput, and backlog, as illustrated in *Figures 13-25* through *13-28*.

Generally, we want production and efficiency to be as high as possible. Backlog and cycle time should be as low as possible (to keep the "rapid" in RP&M).

The data presented in *Figures 13-25* and *13-26* shows system efficiency and production comparisons through two years of constant operation. These figures summarize data supporting continual advances due to improved levels of operator and systems management of the SL equipment.

The data presented in *Figure 13-27* shows the effect of increasing and decreasing input data. The throughput tracking of SLA parts directly relates to the cyclical business levels within the automotive industry.

The backlog data in *Figure 13-28* shows the number of geometries waiting to be built at the start of each biweekly period. Obviously, as productivity and efficiency have increased, the backlog has decreased, so we must be doing something right.

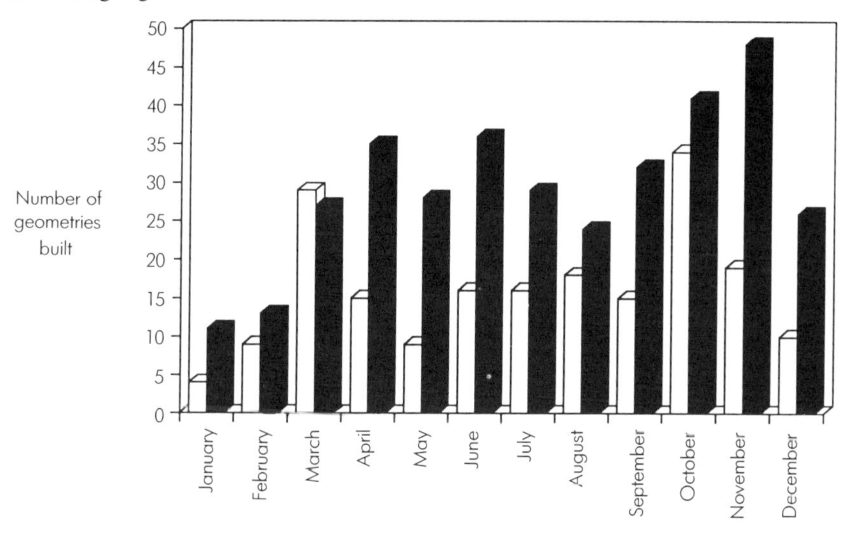

Figure 13-25. 1990-1991 SL efficiency graph.

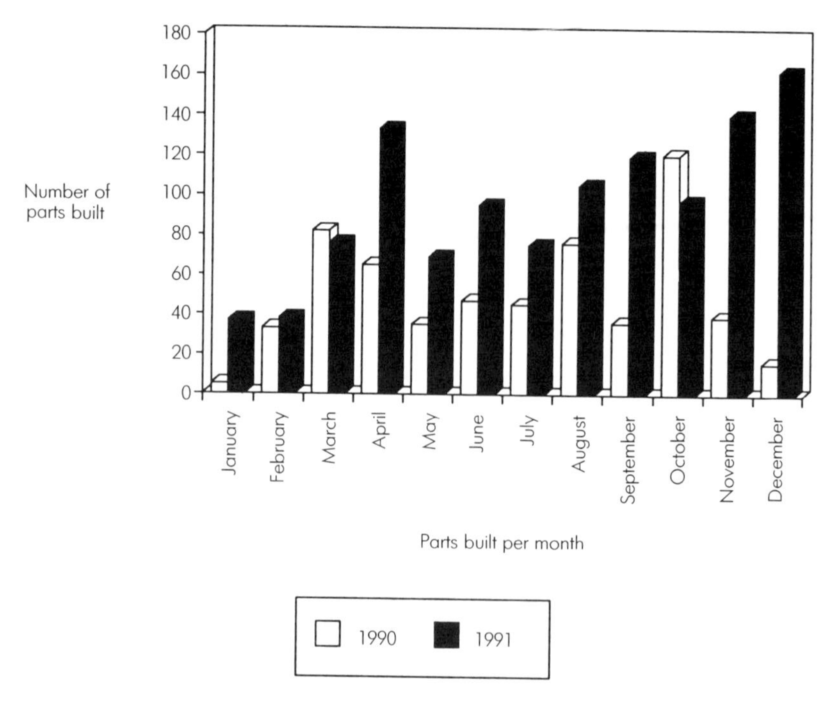

Figure 13-26. 1990-1991 SL production graph.

Research and Development

We are averaging 25% of our time on research and development in the SLA. This R&D consists of the following:

New Resin Trials. The StereoLithography department has been using a new tough and flexible resin in its SL equipment. The resin, by DSM Desotech, produces a part that can be bolted down, bent, twisted, flexed, and most importantly dropped without breaking. In testing the resin, we produced a model of the Chrysler pentastar (4″ x 4″ x 0.75″) and proceeded to drive over it with a truck. The part survived without a scratch. Next, the same model was dropped three floors from the roof of the building to the pavement below. We dented a corner with that test. Finally, the model was placed in a hydraulic press and subjected to a 3000 psi stress for several minutes. The part survived and quickly returned to its original shape. This material is not as strong when used in thin walled parts and may actually be too flexible for some applications. This resin is currently being used successfully in both our SLAs. In 1992, the resin was updated again with a slightly harder version, which improved rigidity.

Slice Software Parameter Comparisons. R&D testing of StereoLithographic parts involved the slicing of a standard SL model at three different layer thickness values, using WEAVE™. The layers were sliced at 0.005″, 0.010″, and

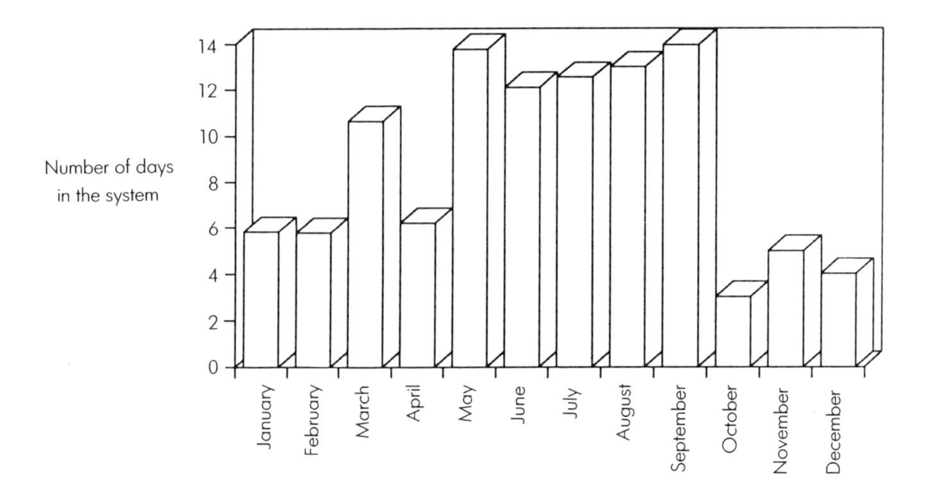

Figure 13-27. 1991 SL cycle time graph.

0.020″ with the same slice and build parameters. The SL part selected was the speed adapter used for other tests. The results were as follows: The test of the 0.005″ slice parameter was very successful, producing the best looking part with the fewest flaws. The actual build time to produce this part was 9.0 hours. The test of the 0.010″ slice parameter was also very successful, producing a fine part with very little damage due to holes in the skin fills. The actual build time to produce this part was 5.0 hours.

The test of the 0.020″ slice parameter was not very successful, the part had excessive layer delamination and skin fill damage. The part also appeared to have a course or rough texture. In building this part, the first two attempts failed. The actual build time to produce this part was 6.0 hours. The 0.005″ build proved to be the highest quality part, but the increase in the build time reduces the overall attractiveness of the part. The 0.010″ build was the best of both worlds with a good part produced in minimal time.

Accuracy Checks. As part of ongoing research and development within the StereoLithography department at JTE, an accuracy study was conducted on a part produced by one of our SLA systems.

A test part has been designed to check the x and y dimensions of the actual part versus the design dimensions. This test part was designed in ProEngineer

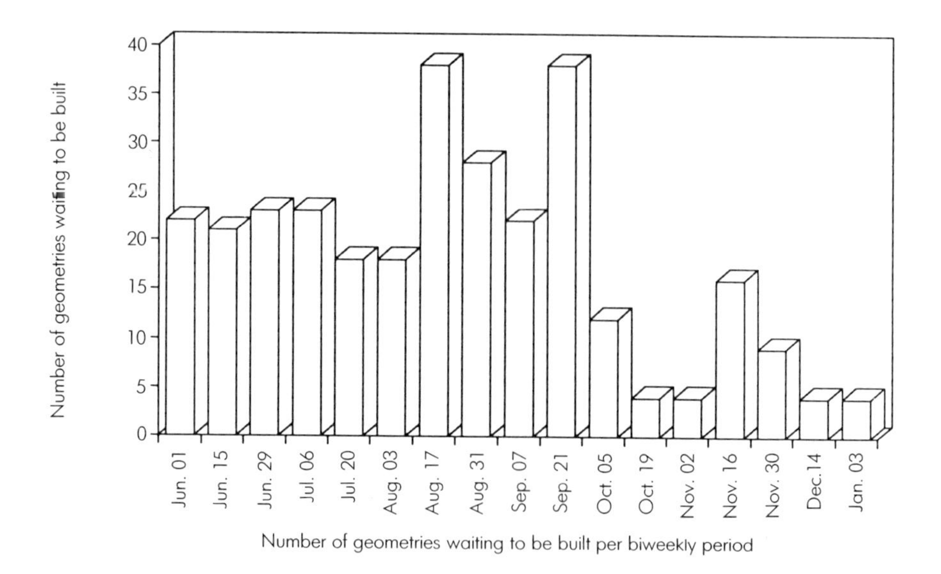

Figure 13-28. 1991 SL backlog graph.

and consists of a simple cross (200 mm x 200 mm) that will be built using a 0.5% shrinkage factor in the *x* and *y* directions. The part and support have been sliced and merged and are currently stored in the PC as file name: CROSS1. The test part was built with an overall accuracy of 0.1 mm for the *x* and *y* directions. It will continue to be built monthly to check the machine accuracy.

Translator Trials. We are currently testing two translators for solid modeling and StereoLithography. The first will produce a CATIA™ solid model from surfaces and faces. This translator will transfer to layer 254 any surface that is discontiguous or nontangent. Illegal faces (less than two or more than four sides) will also be transferred to layer 254, allowing the designer to easily repair the questionable areas. The second translator will create STL files (currently required input for the SLA) from skins, closed or unclosed volumes, and solids.

Time Comparisons: SLA-250 versus SLA-500

A future Chrysler product, a lower control arm, was built on both of JTE's SLA-250s on July 19, 1991. The part was cut into three sections, which were stacked on top of each other to save building time. The left and right parts were created, one in each machine. Upon completion, the models were assembled and bonded together during the postcuring process. Assembly took an additional working day to complete.

The build time results for this test were as follows:

SLA-250 (A): 78 hours (Right) SLA-250 (B): 62 hours (Left)

The variation in build times was due to the difference in the power of the lasers in each machine. The lower control arm was also built for us on an SLA-500 the following month. Both the complete left and right parts were built in the same vat at the same time, moving the parts into a vertical position. By standing the parts on end, the finished models had a smoother surface finish. Since the parts were already built in one piece, no additional assembly time was required.

The build time results for the SLA-500 were as follows:

SLA-500: 36 hours (Right & Left)

As you can see, for these parts, *the larger, second generation SLA-500 was about four times faster (36 hours versus 140 hours) than the SLA-250s.*

Software Upgrades An alternative method of building SL models has been tested at JTE with limited success. The new method, referred to as WEAVE™, produces a model with increased accuracy, reduced warpage, and a smooth surface without air bubbles, adding to the overall part appearance. The WEAVE™ method has produced several exceptional models. However, along with this success have been a few failures. When experimenting with new processes, setbacks are common until experience can replace the trial and error process.

Our findings thus far show that the WEAVE™ has some problems for models with thin wall sections on our machine. The WEAVE™process works very well for models with cross sections greater than 3/8″. The build method is thus determined upon receiving the SLA model.

This new part building method shows promise, and we will continue to develop this and other methods to constantly improve. Remember, our mission is to provide the best possible turnaround of CAD data to an SLA model for our customer's evaluation and use.

13.7 Economic Considerations

Getting people to try a new approach or use "state of the art" technology is difficult, but even more so if you charge them for it. At JTE, we took away the biggest hurdle to trying this technology by building any and all parts for free! No department has ever been charged to build a part at Chrysler. This has enabled us to debug the system and procedure by having a steady flow of CAD models to build. Obviously, we realize it is not truly free because we are actually subsidizing the other departments model costs. However, as noted in Reference 7, *we are saving enough with the process to justify our generosity!*

Part Costs

What does it cost to run the equipment and build a part? An analysis of our internal costs to build SL parts of various sizes is shown in Tables 13-1 and 13-2.

Table 13-1
SLA Part Analysis for SLA-250 with 386 Computer
Size/Time/Sunk Costs

Part size	Weight	Cost/ weight	Build time	Pre & Post processing	Depreciation— Cost of machines & computer equipment	Total cost
Small (2″ x 1″ x 1″)	0.04 kg	$4.00	5 hrs.	$51.12	(5 x $20.39) = $101.95	$157.07
Medium (4″ x 4″ x 3″)	0.18 kg	$18.00	10 hrs.	$51.12	(10 x $20.39) = $203.90	$273.12
Large (9.5″ x 9.5″ x 10″)	0.78 kg	$78.00	20 hrs.	$51.12	(20 x $20.39) = $407.80	$536.92

Table 13-2
Assumptions of Variable Costs Per Hour

Depreciation and cost of machines and computer equipment:	
Hourly depreciation of machines:	$4.81/hour
Cost per hour computer equipment:	$15.58/hour
Total:	$20.39/hour
Pre- and postprocessing cost (averages 2 hours/job):	
Operator cost per hour:	$25.56/hour ($51.12/job)

To give the reader a point of reference, consider that this part is a prototype of a complicated and unique centrifugal compressor wheel (the cold side of a turbocharger), as shown in *Figure 13-29*.

Typical prototype practice for this component is to mill one or two master parts out of a solid billet of aluminum for test purposes, a very timely and expensive process. Once an acceptable wheel is developed, which may take a few iterations, another NC milled part is used to make aluminum castings by a double reverse rubber/plaster casting method.

This is an acceptable way to do a development program, except for the cost and time. Each of these milled parts will cost about $11,000 and take 11 weeks.

Figure 13-29. Centrifugal compressor wheel SLA master and aluminum casting.

Through SL, the master part was created at about $500 and within 25 hours of SLA build time. With the SL part, the double reverse/plaster casting method is still used to make the functional part for testing and application. *Clearly, savings from $10,000 to $30,000 (three iterations) are possible on this part alone!*

Aggregate Costs

An SLA-250 system costs roughly $185,000, with $119,000 in additional costs required for start up activities. These costs were well documented in the site preparation guides written by 3D Systems.

Aggregate costs include all items required to operate on a day-to-day basis, and we found them to be substantial. Table 13-3 lists these items and costs. Total aggregate costs were $349,190 to be completely operational on a continuing basis.

13.8 Education

Our involvement in rapid prototyping does not stop with merely running the equipment. We have found that we must also serve to educate our engineers, designers, and managers regarding the time and cost savings that can be realized by using RP&M. We must also educate the designers as to the proper input required, so we published a technical manual, since interfacing Chrysler's CAD software with the SLA has been a greater task than first imagined. In that publication, topics such as wireframe to solid model conversion, support design, and file translation are addressed. The manual has been written for designers trained in CATIA™ and solid modeling.

Table 13-3
StereoLithography Aggregate Costs

Equipment	Quantity	Cost per unit	Total cost
SLA-250	1	$185,000	$185,000
SLA-250 maintenance agreement	1	36,000	36,000
SLA-250 upgrade	1	10,000	10,000
Personal IRIS	1	35,000	35,000
Personal IRIS upgrade	1	24,000	24,000
Personal IRIS maintenance agreement	1	4,100	4,100
UV curing oven	1	10,000	10,000
*Ultrasonic cleaner	1	2,000	2,000
*PDU power converter	1	8,000	8,000
*Personal computer	1	4,500	4,500
*Ethernet LAN system	2	1,200	2,400
*Ethernet bridge	1	2,400	2,400
*Transceivers, cables, misc.	1	900	900
*Air conditioning	1	3,500	3,500
*Post cure venting	1	2,500	2,500
*Cibatool resin	48	350	16,800
*Denatured alcohol	20	5	100
*Protective equipment	1	440	440
*Clean up materials	1	400	400
*Part finishing tools	1	350	350
*Hand-held UV gun	1	800	800
		Total cost	$349,190

*Aggregate costs totalling $45,900

The same holds true for educating Chrysler's suppliers. Video tapes have been prepared showing how the system works as well as application tapes that describe the ways "real" customers are using the systems to make parts. Slide presentations are geared toward each group's needs, whether they are suppliers or Chrysler engineers. We currently have five slide shows: Automotive Applications, Steps in the SLA Process, Secondary Tooling Applications, Chassis Prototype Tooling with SL, and Rapid Prototyping Technology Overview.

Furthermore, we give presentations to local universities to inform students and professors alike about this exciting technology. We also sponsor an in-house workshop to fully familiarize engineers, designers, and managers with the RP&M process as well as secondary tooling applications available to help reduce lead time and prototype tooling costs.

13.9 Prognosis for the Future

The long-range plan we are developing at Chrysler JTE includes the purchase of additional equipment as our backlog increases beyond a reasonable waiting period. However, we continue to research other alternative systems and conduct ongoing tests of parts built by the competitors of StereoLithography.

We can envision an RP&M division at Chrysler that includes machine tools of diverse capabilities with specific applications that will help us speed our products to market. Our benchmarking of the five major RP&M equipment manufacturers, shown in Table 13-4, lists our projected internal costs and the time to build a part if we purchased these systems and used them in-house. These considerations, as well as each system's unique capabilities, will guide our future investment in this technology.

REFERENCES

1. *SLA Today* No. 1, February 5, 1990.
2. *SLA Today* No. 2, September 14, 1990.
3. *SLA Today* No. 3, January 31, 1991.
4. *SLA Today* No. 4, June 4, 1991.
5. Schmidt, L.D., *Applications of StereoLithography in the Automotive Industry*, Successful Applications of Rapid Prototyping Technologies Conference, Society of Manufacturing Engineers, Troy, MI, April 23-24, 1991.
6. Schmidt, L.D., *StereoLithography: The Good, The Bad, and the Ugly!!!*, Second International Conference on Rapid Prototyping, University of Dayton, Dayton, OH, Conference Proceedings, pp. 184-192, June 23-26, 1991.
7. Phillips, W.L. *StereoLithography and Conventional Tooling Applications*, Second International Conference on Rapid Prototyping, University of Dayton, Dayton, OH, Conference Proceedings, pp. 163-164, June 23-26, 1991.

Table 13-4
Benchmarking of Major RP&M Equipment Manufacturers

Variable	3D Systems baseline SLA-250 with 286 MPU	3D Systems SLA-250 with 386 MPU	3D Systems SLA-500	Cubital Solider 5600	DTM Sinter Station 2000	Stratasys 3D Modeler	Helisys LOM 1015
Cost of equipment	$195,000.00	$195,000.00	$395,000.00	$490,000.00	$397,000.00	$182,000.00	$95,000.00
Depreciation cost/hour[1]	$2.02	$2.02	$4.10	$5.09	$4.12	$1.89	$0.99
Service contract expense	$36,000.00	$36,000.00	$85,000.00	$49,000.00	$68,000.00	$7,000.00	$17,000.00
Preprocessing time	2:55	:34	:34	:21	:55	4:20	:46
Machine build time	10:00	5:06	4:44	10:00	3:00	8:00	9:51
Postprocessing time	1:45	1:45	1:45	1:00	1:20	:15	:25
Total process time	14:40	7:25	7:03	11:21	5:15	12:39	11:02
Total material cost/part	$4.00[3]	$4.00[3]	$4.00[3]	$5.96[2,7]	$5.89[2,6,8]	$4.00[3]	$3.82[3]
Maintenance cost/part	$41.10	$20.96	$45.90	$55.94	$23.29	$6.39	$19.12
Preprocessing cost[4]	$194.76	$38.02	$38.02	$23.35	$61.36	$288.81	$51.36
Machine build cost (less attendant)	$20.24	$10.32	$19.39	$50.85	$12.36	$15.11	$9.71
Cost of attended operation[5]	$0.00	$0.00	$0.00	$220.00	$66.00	$0.00	$0.00
Postprocessing cost[5]	$38.50	$38.50	$38.50	$22.00	$29.26	$5.50	$9.24
Total part cost	$298.60	$111.80	$145.80	$378.09	$198.16	$319.81	$93.25

Total part cost formula: preprocessing cost + postprocessing cost + build cost (machine + attended operation) + material cost (part + incidental) + maintenance costs.

1. Eleven year straight line depreciation - Salvage value $0.
2. No supports required.
3. Assumes supports are negligible relative to part cost.
4. Combined hourly rates of individual personnel is assumed to be $51.12/hour.
5. Hourly rate of post processing and attended operation personnel is assumed to be $22.00/hour.
6. Ignores 10% loss of powder material lost in reprocessing material.
7. Includes cost of non-reusable wax and resin.
8. Nitrogen cost is assumed to be $15/24 hours - the part assumed to use 4.5 hours worth or $2.81.

chapter 14

AMP Incorporated: A Simultaneous Engineering Case Study

All I could see from where I stood was three long mountains and a wood; I turned and looked the other way, and saw three islands in a bay.

—Edna St. Vincent Millay
Rockland, Maine
1911

14.1 Introduction

This chapter may well be titled "Teamwork," as it discusses the relationship of rapid prototyping and simultaneous engineering. A case study is presented from AMP Incorporated of Winston-Salem, NC.

AMP is a leading supplier of connector and interconnection devices for the electrical and electronic equipment industries, with annual sales in excess of $3 billion. The corporation produces approximately 100,000 different types and styles of connectors in 160 facilities in the United States and 29 other countries, employing over 24,000 people.

At any one time, the domestic organization alone has over 100 active new product programs; and when international organizations are considered, this figure more than doubles. Over the 50 years of its existence, AMP has enjoyed

---

*By **Thomas A. Kerschensteiner**, Manager, Product and Process Development, AMP Incorporated, Winston-Salem, NC.*

remarkable success attributed largely to its technical and innovative capabilities. AMP has reached the enviable position of being nearly five times as large as its nearest competitor. This success has led to a high level of confidence with respect to the corporation's new product development function, indicated by the fact that the corporation ranks 13th in United States patent issuance and 34th worldwide.

AMP's Director of Engineering, James Coller, notes that, "If a picture is worth a thousand words, then a sample is worth a thousand pictures."

14.2 Simultaneous Engineering

The benefits of implementing and utilizing the principles of simultaneous engineering have recently been the subject of much discussion and activity.

For our purposes, simultaneous engineering is defined as *the current or simultaneous development of the product, the process, and the related tooling and equipment*. Simultaneous engineering also emphasizes strong leadership and the existence of multidisciplinary teams whose members continuously interact and focus on program objectives.

Team members include engineers, designers, and managers. Each of these individuals comes from a different division or department within the company since simultaneous engineering must cross departmental lines. Some companies have located all team members in a single office area to enhance communication.

Likewise, suppliers or service bureaus must be considered a part of the team. Clearly, rapid prototyping systems (and other areas as well) must work within the time and budget restriction of product development.

While product design and manufacture is technical in nature, the reader should note that rapid production of a prototype part has advantages for a marketing and sales force. Their participation in the team is an important consideration.

The factors necessary for success in the simultaneous engineering effort are shown in *Figure 14-1*. Graphically, the distinction between traditional (sequential) and simultaneous engineering can be seen in *Figure 14-2*.

Implementation and utilization of tools such as SL have a major impact on this dramatic compression of time. The equation of time-to-market versus cost is borne out by the following economic model put forth by McKinsey & Co: "High-tech products that come to market six months late but on budget will earn 33% less profit over five years. In contrast, coming out on time and 50% over budget cuts profit only 4%."

Evidence to support that simultaneous engineering effort relates to the cost of change is seen in *Figure 14-3*. This is, at times, referenced as the "Rule of Ten" since the cost of change increases by a factor of 10 during a product's advancement through its development.

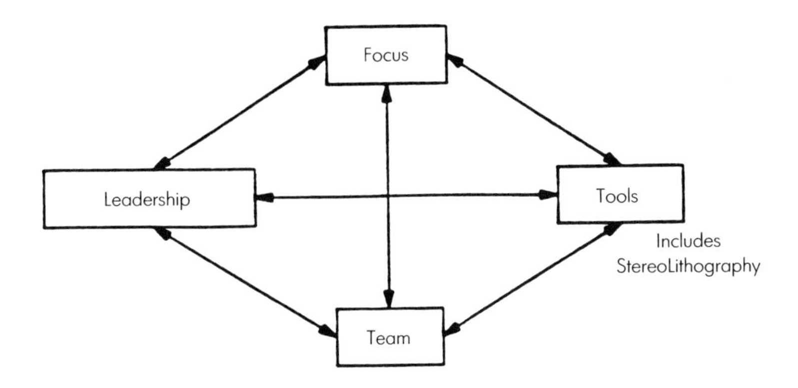

Figure 14-1. Success factors.

14.3 Starting Simultaneous Engineering at AMP Incorporated

Early in the simultaneous engineering implementation process, a specific effort was initiated to develop a set of recommendations for management. Outside experts were consulted, and other organizations that had successfully implemented the concept were studied.

The committee conducting this study felt that to set goals for improvement, they must map the process and establish a benchmark of typical project completion times. Since the available information was not consistent among the various organizations, a decision was made to construct a computer model of the new product development function. Typical tasks were identified and, using the judgements of experienced development engineers, times were assigned to each.

Because projects vary in complexity and levels of innovation required, the committee decided to generate three sets of data for low, moderate, and high-complexity programs. Arbitrary distributions were established around each of the three mean times for completion. The model was validated by empirical information gathered during the 1970s. The results of this modeling are illustrated in *Figure 14-4*. Actual times are not shown. Rather, the information obtained by modeling the process is related to the time commonly believed to represent the average new product development and introduction cycle required to compete in a global market in the next decade. This time, identified by "X" in the figure, demonstrated the need for significant cycle time improvement. A goal of 50% reduction was set.

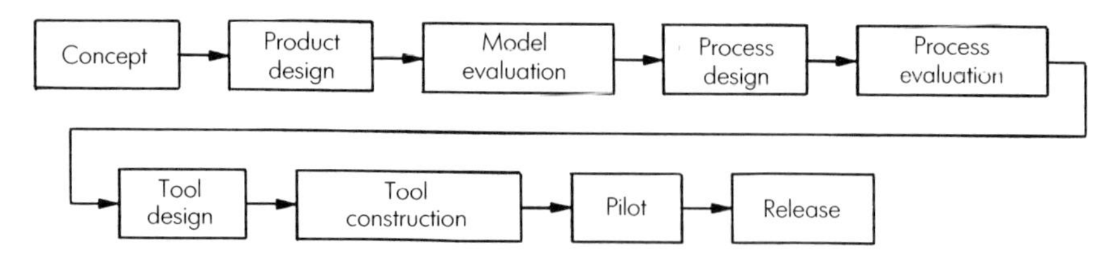

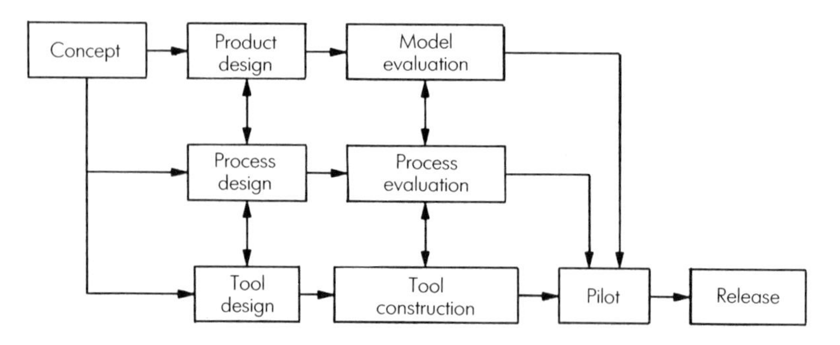

Figure 14-2. Traditional engineering versus simultaneous engineering.

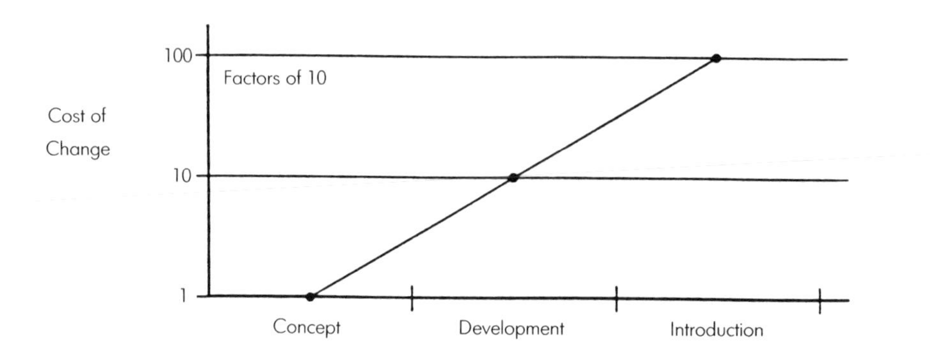

Figure 14-3. Cost of change.

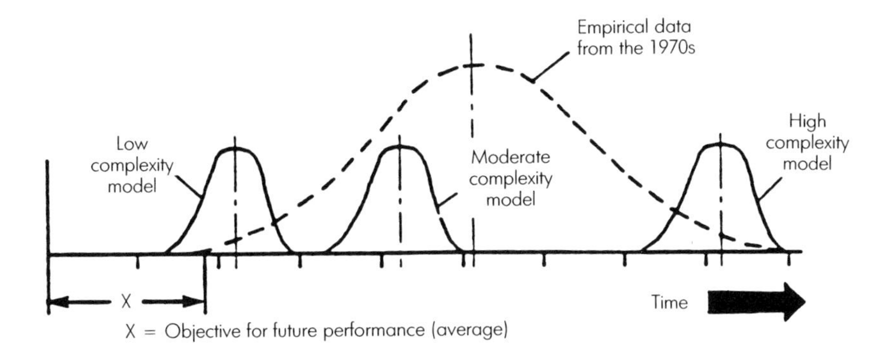

Figure 14-4. Historical projects' time-to-market.

14.4 History and Implementation

The Automotive/Consumer Business Group of AMP was chosen in late 1987 as one of five companies to participate as a Beta or test site for the newly introduced technology of StereoLithography.

Since its inception, visionary management viewed SL as a key strategic tool in its efforts to implement the simultaneous engineering principles.

Two weeks of intensive training of AMP employees from five disciplines (model making, CAD/CAM, product design, process development, and business planning) took place in early January, 1988. Installation of the SL unit occurred on February 12, 1988.

The machine was installed into the existing model shop organizational structure, which supported development engineering with traditional machining techniques in plastics and metal. The intent was to use the organization's existing resources, *not* create an aura of exotic technology, as is often the case.

Initial requirements for the SL output were kept simple and clear. This was done in the hope that its apparent "user friendliness" would create demand among engineers or customers. Only two requirements were necessary to receive a model:

1. Supply a three-dimensional CAD model.
2. Process a typical model shop request.

Also, an attempt was made not to oversell the technology. Early resins were brittle and accuracy was not as good as it is today, so "form and fit," not function, became the rallying call.

The early multidisciplinary training team helped resolve any initial misunderstandings relating to use and implementation. Further, it served as a basis for future team use of this technology.

The group chosen at AMP for the initial use of rapid prototyping technology possessed a combination of several key factors which aided the integration of SL into the business.

A high degree of CAD/CAM expertise. Nearly all engineers and designers had three-dimensional modeling capability and expertise.

A new and challenging market to enter. As part of a strategic thrust, a focus on the automotive marketplace was to be the key. This implied a higher level of development activity than was typical; notwithstanding the traditionally high levels already existing within the corporation.

Computer-aided engineering and analysis training. As part of the development engineering activity, large capital investment and training for engineering analysis had already taken place. While the direct correlation between simultaneous engineering and SL integration is difficult to see at first, the fact is that engineering analysis and verification must take place within the product and process cycle. Product that meets specified and anticipated values the first time shortens the time-to-market cycle. Other activities, such as those supported by SL, become even more critical in the time continuum, creating a higher need and value for them.

Familiarity with model shop. Rapid prototyping technology became friendly rather than foreign. There were no new organizational structures, procedures, or bureaucracies to deal with. Additionally, and very key, the model shop had a reputation of quality and product timeliness that needed to continue. The quality of the output of an SLA machine greatly depends on the individual establishing the building parameters, and the person postprocessing the models. By placing the SL unit in the shop, not only was the need familiar, but the customary quality level of finish, detail, and care was preserved and in some instances amplified.

Not to be overlooked are the financial aspects of a model shop implementation. The charge-out rate for the shop did not greatly increase with the introduction of the SL machine because the asset base only increased due to the machine. Thus, the cost for the product was quite attractive, especially in light of prior time-consuming manual model shop construction techniques.

This favorable cost factor, coupled with its timeliness, no doubt helped initially to win supporters for the technology even before they realized the true positive impact and implications it would have on interactions with customers or other team members.

A tool for the organizational change to simultaneous engineering. As management espoused the benefits of the simultaneous engineering concept and organized in that fashion, the use of a three-dimensional model of a product produced early in the development cycle created a hub around which members of a multidisciplinary team could develop much faster and more meaningful feedback of the product and process.

14.5 Use of RP&M in the Automotive/Consumer Business Group

StereoLithography is an important tool in achieving the goal of simultaneous engineering. Such leverage results in:

- Faster time to bring a product to market;
- Higher confidence in design acceptance by the customer;
- Greater levels of design for manufacture;
- Increased levels of business, and
- Potentially reduced cost.

Cited next are several examples indicating these results.

Example 1

AMP had embarked on a significant development program involving several multinational automotive companies. The product itself was intended to meet a recently released standard.

The engineering group at AMP immediately began developing concepts using SL. The initial design and prototype model were completed in less than three weeks. Process engineers were called in to review the product design for its ability to be easily manufactured. Elements such as moldability and robotic exchange were reviewed. Consequently, an integrated manufacturing plan and proforma cost model were developed for the product in *two additional weeks*.

Concurrently, the SL prototype model was shipped overseas, where AMP sales engineers explained the product to the customer's engineers. Together, they requested further product features and capabilities.

SL was used four additional times to iterate and optimize the product in the eyes of the customer. StereoLithography showed the ability to verify, iterate, and improve a design.

As this iterative process was occurring, assigned manufacturing people were involved. As a result, the manufacturing process actually became simpler and less capital investment was required.

The result was that within *four months* the product could be manufactured and was exactly what the customer expected.

Example 2

A major U.S. company initiated a design competition for a next-generation interconnection system. This procurement solicitation involved five competitors. Each company was given four hours to make a presentation of its proposed solution to the customer's need.

StereoLithography samples of proposed product were shown by only one company, along with a complete and comprehensive engineering analysis of the contact physics and a manufacturing plan. As a result, AMP Incorporated was awarded the business. To date, there have been no fundamental or major changes to the product, its design, or its manufacturing plan. The product line is on schedule and on budget, and the customer is delighted.

As part of this program, a limited number of molded shields of different configurations were produced from silicone rubber molds. These were done in two weeks, versus 16 weeks for conventional methods. Inputs from five submanufacturers regarding their design, fit, and performance were included, and the necessary modifications were made for final tooling.

Using SL early in this program helped many of the following key elements succeed:

- Customer acceptance,
- Mold engineering quotes, and
- Ergonomic considerations of the system such as interferences.

Based on this experience, *AMP will not issue a major customer quotation package without a StereoLithography model.*

Example 3

While there have been many major successes and applications of StereoLithography within the company, perhaps the most important involve preventing errors. Error detection is made easier with the product in hand as opposed to prints or CAD files. This also is true for validation when the product is finally correct.

Some specific examples are:

- A 16-position, two-part connector with nine circuits erroneously lining up with seven circuits. The cost and time savings were estimated to be *$80,000 and four months.*
- Circuit configuration angles mismatched. Cost and time savings were *$40,000 and three months.*
- Design with too much interference or improper polarization of mating halves.
- Improvements to product packaging.
- The application of KOJURI principles for global products.

While it is sometimes difficult to admit one's mistakes, one engineer upon finding an error said: ''*[The SLA-250] is the best thing we've ever purchased.*''

14.6 SLA Use, Efficiency and Key Success Factors

The SLA machine itself has been kept busy and has run virtually without flaw for almost four years. For example, during 1,029 working days, it produced the following results:

- 543 models–nearly one every two days,
- 3,794 pieces–6.9 pieces/model; 3.7 pieces/day,
- 9,757 machine hours to produce these pieces,
- Model shop equivalent estimated to be 30,885 man hours or 15 years of one staff person.

Key factors to successful RP&M integration are:

1. High level of three-dimensional modeling capability.
2. Rapid new product development.
3. High perceived internal value.
4. High perceived external value.
5. High quality of finish and detail.
6. Requirement for use on all programs.

Integrate as a basis of business with:

1. Business planning.
2. Finite element analysis.
3. Design for manufacture.
4. Manufacturing/capacity planning.

14.7 Outcomes Due to StereoLithography

Simultaneous engineering represents change, as does SL. Together, these have lead to major advances within the organization.

1. AMP has multiple SLA machines throughout the company—both domestically and internationally. These systems serve as simultaneous engineering cornerstones.
2. The model shop does nearly no plastic modeling with what was formerly considered conventional technology.
3. The new product development organization gives tours to customers and suppliers showing the flow of information from analysis to concept to production prototype.
4. Engineering is now product and process driven. This represents a departure from the design only perspective of the past.
5. The RP&M technology is pulling the engineering and design organization toward higher levels of three-dimensional modeling.
6. The technology is also allowing more interactive solutions to design problems—resulting in better designs.

At this point it is not clear whether SL supports the simultaneous engineering activity or whether simultaneous engineering supports StereoLithography. What is clear, however, is that together they are very complementary, and they create a strong combination of application and technology for corporate strategic competitiveness in the 1990s.

ACKNOWLEDGMENTS

The author wishes to acknowledge the efforts and contributions of all product and process development engineers, as well as the timeless commitment of Mr. Timothy Newman for his assistance in the implementation of StereoLithography technology at AMP Incorporated.

Further, Mr. Donald Ratledge and Mr. Delmont Bates must be recognized for their commitment to quality of product through the postprocessing operation.

Lastly, Mr. James Coller, Mr. Linn Lightner, and Mr. James Wise are especially cited for their assistance, guidance, leadership, and implementation of simultaneous engineering principles.

chapter 15

DePuy, Incorporated: A Medical Case Study

Successful innovation is a feat... Not of intellect, but of will.

—Joseph Schumpeter
Economist
Early 20th century

15.1 Introduction

The Medical industry is facing a dichotomy. The goal of providing the best patient care possible is contrasted with rising cost. Health care providers are forced to limit available funds and related services or share the expenses with patients, employers, government, or private sources. We are not simply competing for market share and position, but rather have at the very heart of our existence the desire to offer the latest and most technically capable devices to our customer. Competing in this global environment, where nationality and surgical expectations may dictate product requirements, is only possible as we embrace new technology and apply it to gain competitive advantage.

The medical industry has seen great advancements in the quality of life offered to patient health care recipients. Many of these are related to various technologies such as imaging systems, laser scanning, robotics and rapid prototyping that are either coming of age or are now affordable for implementation. The medical industry and particularly orthopedics has certain applications, which, while not necessarily unique, are extremely well suited to these technologies and may foster their integration and expansion. This chapter will

---

*By **David G. Trimmer**, Director of Engineering Services, DePuy, Incorporated, Warsaw, IN.*

develop the relationship among these technologies and present case studies from the orthopedic field to show the economic and social issues involved in the utilization of rapid prototyping.

DePuy Incorporated, of Warsaw, Indiana, a leading orthopedic manufacturer, designs and manufactures replacement joints and implants for the musculoskeletal system of the human body as well as related instrumentation. Degenerative and arthritic problems (joint diseases) result in very painful or nonfunctional joint movement. Most commonly utilized are devices for the hip and knee, where relief of pain and improved mobility are of primary concern. However, products for other areas such as the shoulder, ankle, elbow, wrist, etc. are also available as well as trauma and sports medicine devices. They consist of metal components made from a chrome-cobalt alloy, titanium, or stainless steel that are attached to prepared bony structures and surfaces. Ultra high molecular weight polyethylene bearings are used between the mating joint surfaces. The components are often anatomically shaped or contoured designs versus basic geometric shapes, as shown in *Figures 15-1* through *15-5*. They are produced as a family in a range of sizes that can be selected at surgery to match the patient requirements, or, if desired, can be provided on a custom prescription basis.

The orthopedic industry consults with leading surgeons in the development of new implant systems and related instrumentation. Their operating room expertise and medical training coupled with our bioengineering and manufacturing capability provides the synergy to advance the state of the art of these implants, while developing instrumentation for precise, repeatable surgical techniques. It should be noted here that this is a very complex design environment; soft tissue

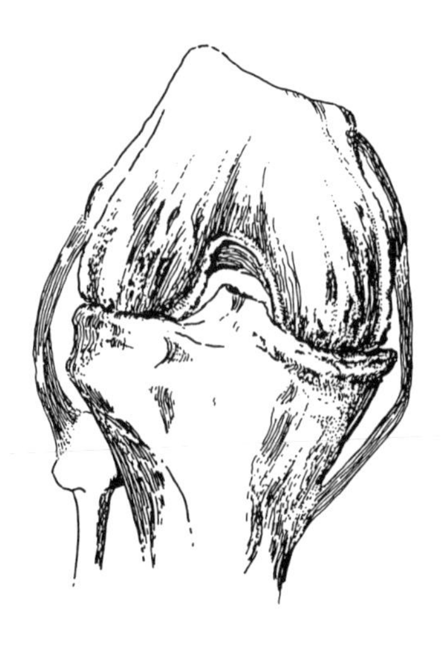

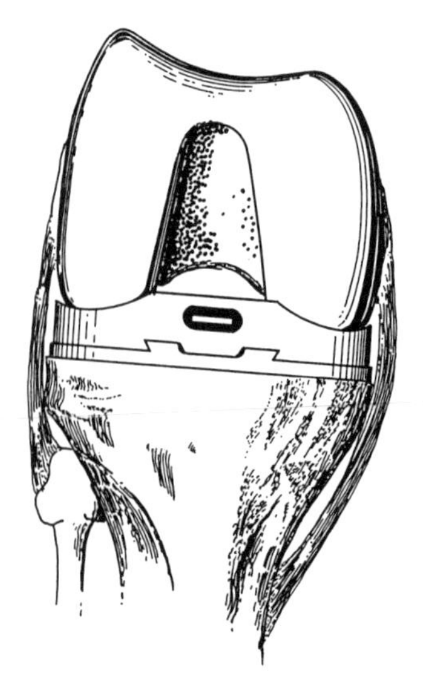

Figure 15-1. Knee before and after implants.

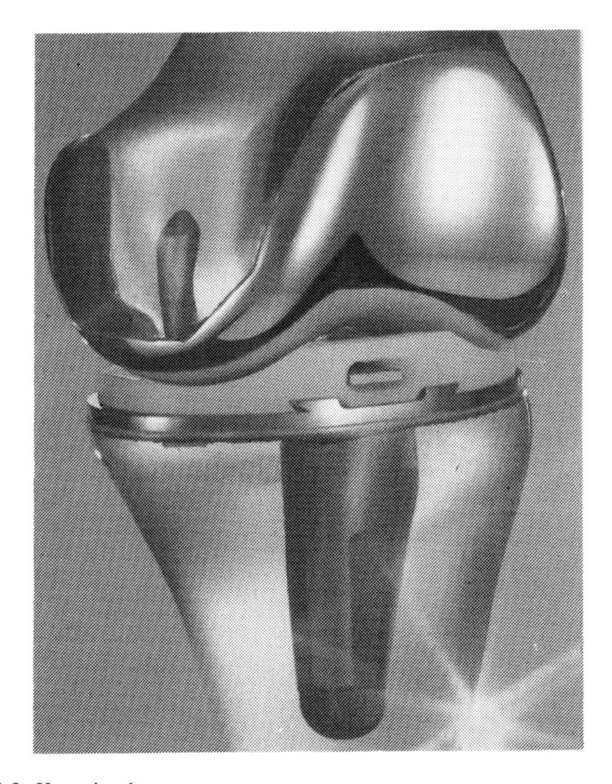

Figure 15-2. Knee implant.

integrity and bone remodeling all relate to the success of the implant. This interdependent relationship among experts in different specializations makes rapid prototypes essential to concurrent product and process development. *Figure 15-6* depicts the relationship of the development team consisting of a product engineer, marketing product manager, quality engineer, manufacturing engineer, purchasing manager, developing surgeons, and engineering support personnel. Each person on the team is responsible for making decisions related to their area of expertise and thus keeping the project advancing in a timely manner.

Ideas, as conceived by the engineer or surgeon, may start out as hand drawn sketches. The dilemma comes as individuals try to communicate a complex three-dimensional concept in this two-dimensional modality. The engineers understand how to read and interpret the engineering drawing with its geometric dimensions and tolerances. They may even be able to envision the actual items by imagining the orthogonal views in their minds. Other members of the team also create these mental images, but in actuality, some of their views may be quite different. Features that seem insignificant on a blueprint or sketch can

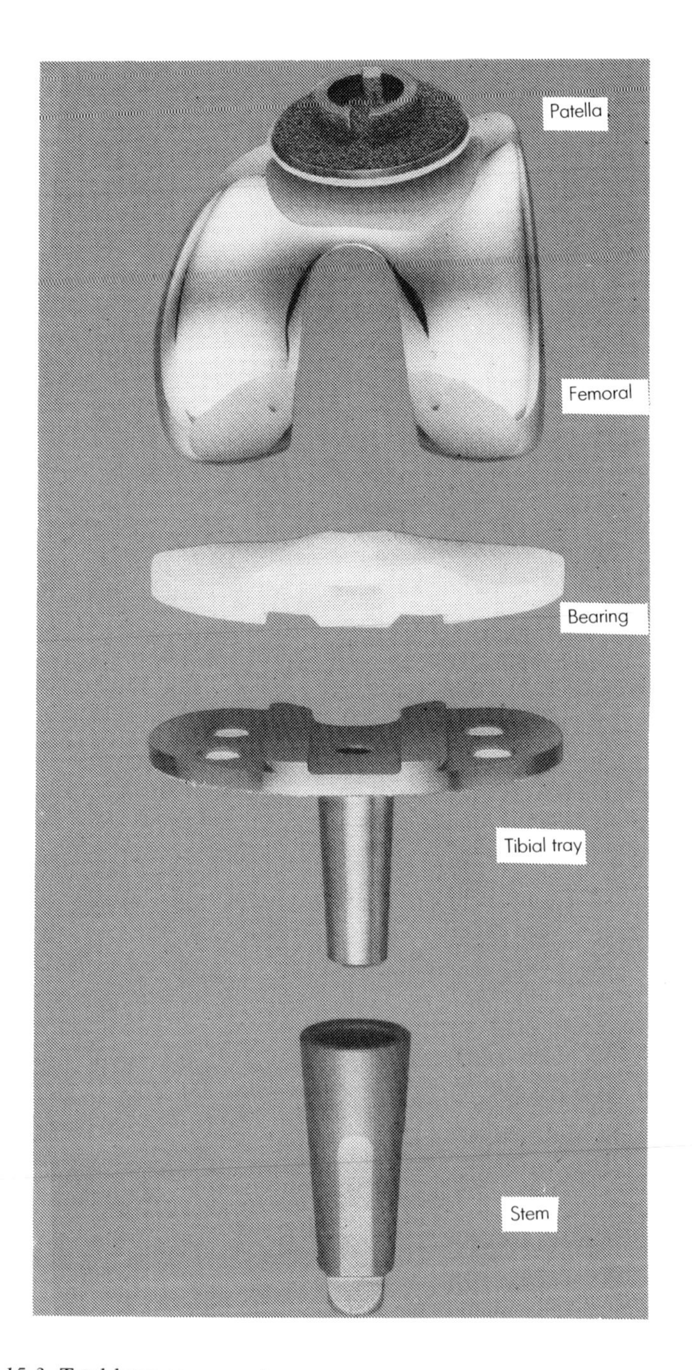

Figure 15-3. Total knee components.

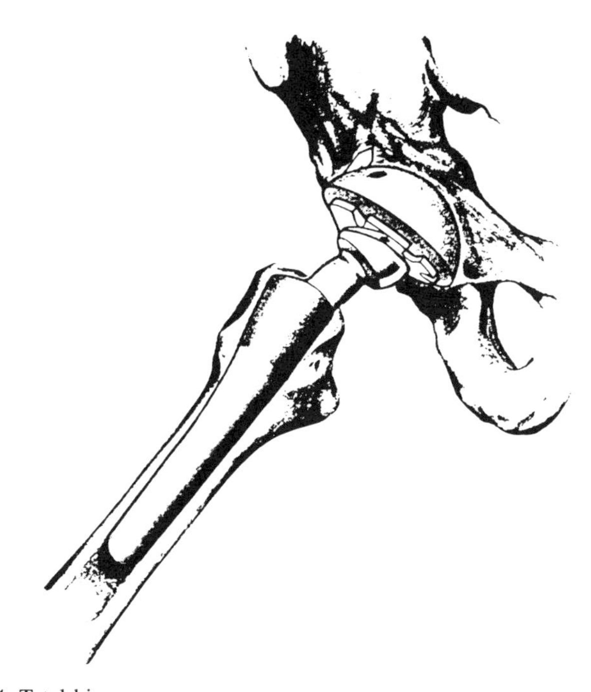

Figure 15-4. Total hip.

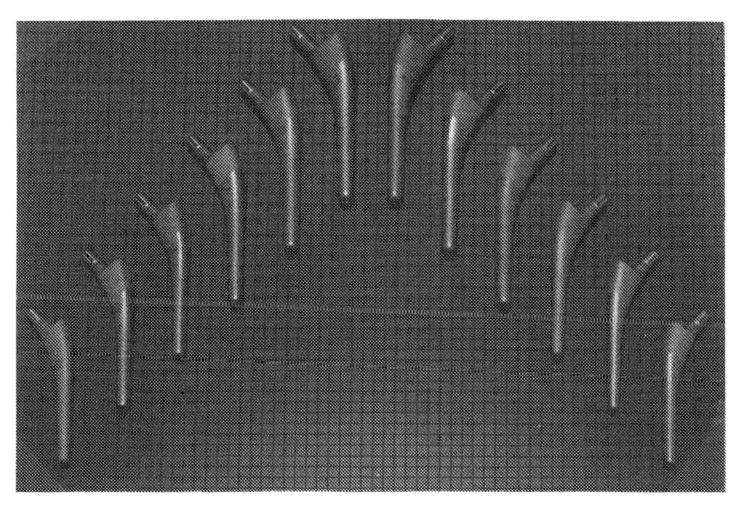

Figure 15-5. Range of hip stems.

suddenly become very expensive manufacturing challenges when visualized and understood. The goal of rapid prototyping is to bridge this gap by providing actual full-size physical models that each party can touch, analyze, and use for

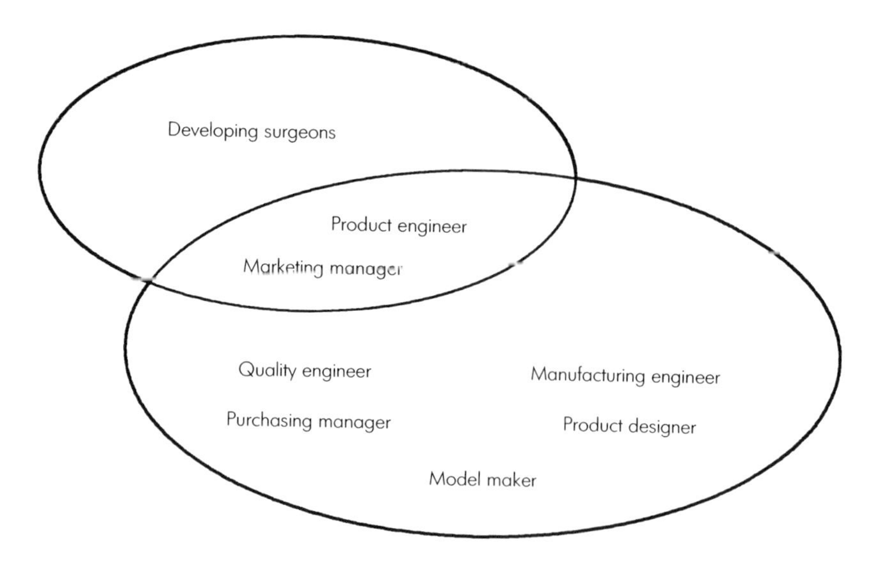

Figure 15-6. Development team.

further development. This capability has revolutionized the design review meeting. The constructive dialogue among the members, as they pass the models to one another, suddenly allows everyone to express concerns and suggestions in a manner based on a common level of visual understanding. This, coupled with the ability to redesign and present new versions within days, excites the team members and encourages them to commit their efforts and abilities into promoting the team goal of concurrent product and process development.

A conventional machining prototype shop staffed with model makers and state of the art CNC or "computer numerically controlled" equipment, along with conventional machining and welding capability, is highly utilized at DePuy. They make many parts, including functional instruments for use in surgery. The most difficult items to produce are those with sculptured surfaces that require surface machining. StereoLithography fills this need nicely, as the physical part configuration is not a limiting factor in part building. There are also many instances where a concept design does not need to be functional to convey the design intent and manufacturability. Rapid prototyping is desirable here, as time is very critical and several iterations may need to be reviewed prior to requiring a functional device. DePuy has two product designers assigned to the rapid prototype area. They each have a Sun Sparc II workstation and use Computervision 4X design software. Each of the two 3D Systems StereoLithography SLA-250 machines has a vat size of roughly 1000 cubic inches (10″ on a side), which allows for building multiple implants at one time. In addition, there is one postcure apparatus and temporary tooling equipment for urethane and spray metal tools.

15.2 Shoulder Design Case Study

The following case study for a new shoulder implant design will describe how rapid prototyping impacts concurrent product and process development. The major goals are decreasing the time to develop a product, while enabling manufacturing to launch the product quickly, resulting in a larger return on investment. The shoulder implant design consists of three basic components, with size variations allowed for each component. *Figure 15-7* diagrams the components. *Figures 15-8* through *15-10* show the size variations that were developed.

The development team for this project included two leading orthopedic surgeons who are both specialists in shoulder surgery. These experts from different regions of the country were joined by a multidisciplinary team within DePuy. Previously, the concept stage often involved four to six months of discussion and prototyping of one implant size, followed by another nine to 12 months of design and development where details and sizing were established. Therefore, over a year of product definition was then followed by another 12-18 months of manufacturing processes required to produce launch quantities and have the product "on the shelf" and ready for sale. During this time, Marketing was preparing the product literature, and the development team was securing the appropriate FDA labeling approvals necessary for the product.

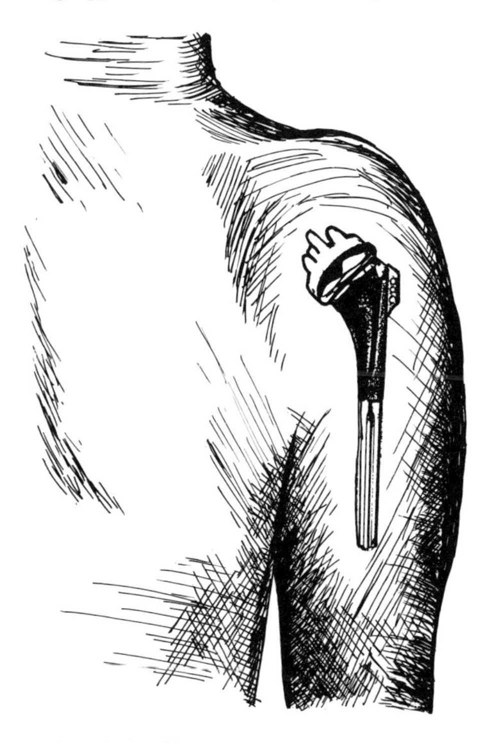

Figure 15-7. Illustration of total shoulder.

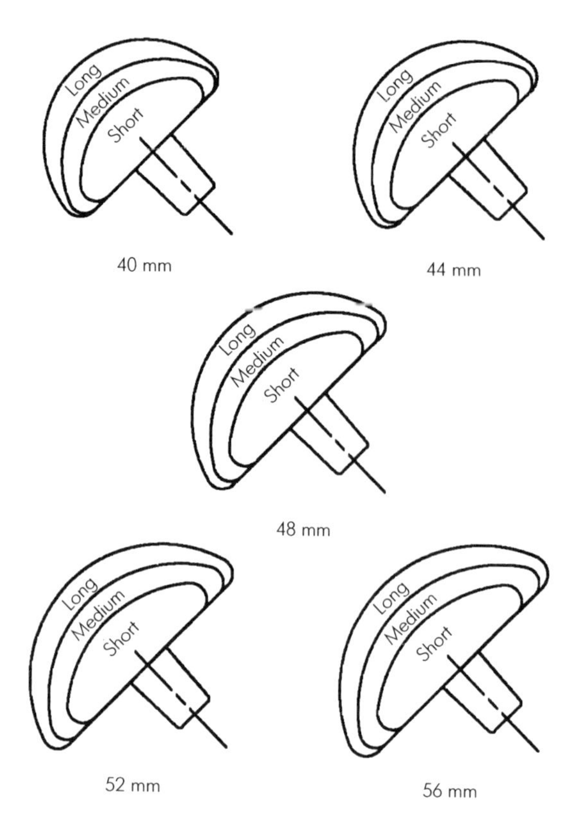

Figure 15-8. Humeral heads.

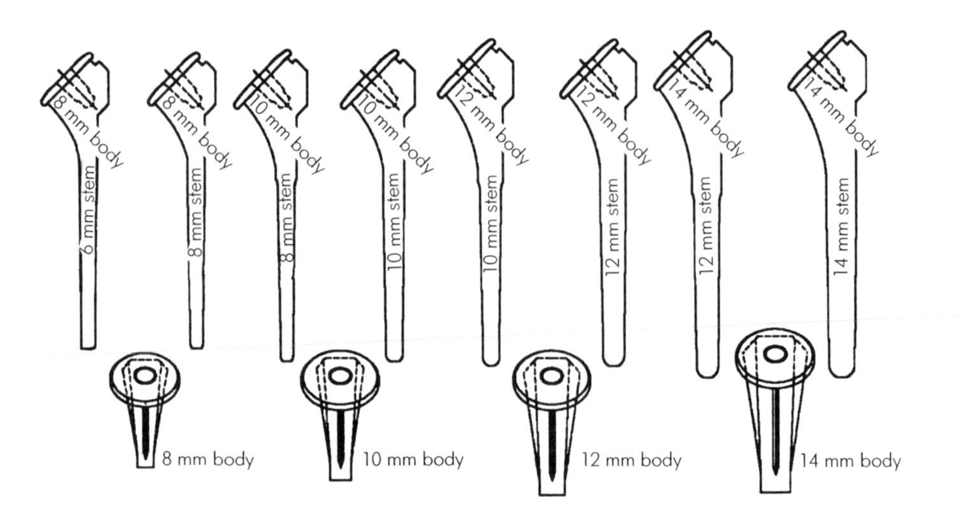

Figure 15-9. Humeral body.

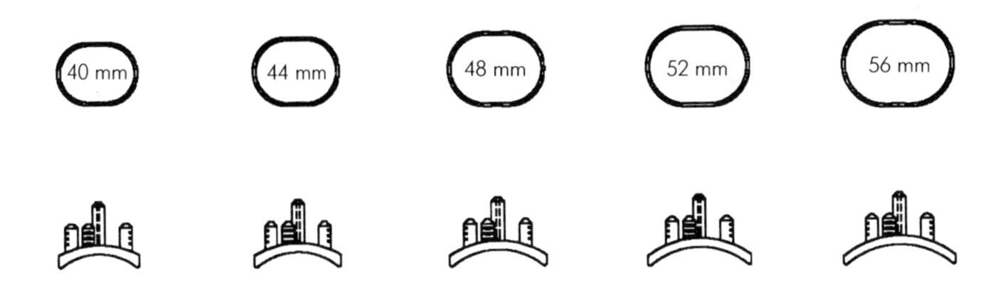

Figure 15-10. Glenoids.

At this point, the surgeon could finally start using these devices in surgery. Often a request for what seems like a minor change after using the product is, in reality, a major investment on the part of the manufacturer. The surgeon may request a small dimensional change. Hard tooling changes for investment castings, inspection gauges, process plans, and engineering drawings are all affected along with inventory considerations. Related matching instrumentation may need to be exchanged and coordinated with the release of the revised product. This is very expensive and limits the ability to react quickly.

The goal for the shoulder implant design team was to define, in the concept stage, a full range of products that met manufacturing requirements and surgical expectations. The surgeons met with the development engineer and marketing manager to discuss design features. They returned with hand-drawn sketches indicating basic requirements and dimensional criteria for the stems, heads, and bearings. A product designer, using Computervision CADDS 4X software on a SUN Sparc II workstation, was then assigned to the team and developed a three-dimensional computer model of the device. Together with the engineer, they blended the basic dimensional features into a fully surfaced or ''solid'' model that represented their interpretation of the design intent. This was done for one size of each component and then given to the rapid prototype area to make plastic models. The design took five days to develop, and the rapid prototype group required only two days to build and finish the models. The parts were built using .005″ layer thickness, as mating taper features were included and limited post finishing was desired (*Figure 15-11*). We made three sets of models so that one could be sent to each surgeon and one set used in-house. Within two weeks of the original meeting, a conference call verified the design features as each party simultaneously reviewed the plastic RP models.

This discussion provided information for iteration of the design. The rapid prototyping group produced a second set of models for verification. Multiple models were again produced and not only went to the surgeons, but also were used at the first design review meeting. Team members were initially skeptical of these early meetings. Since time is a precious commodity, members would historically prefer to wait for engineering blueprints to be released to have

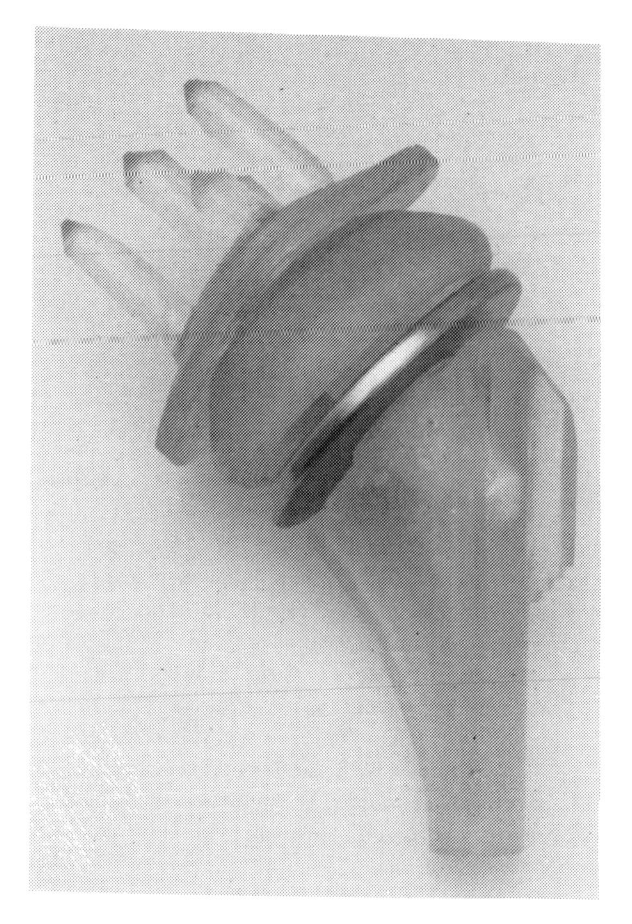

Figure 15-11. Shoulder prototypes.

something concrete to work against. Prior to StereoLithography, these early meetings were held around a conference table covered by graphic plots and computer output, with various people trying to convey to each other the changes necessary to make the product usable in surgery, manufacturable, and marketable.

This is where StereoLithography has a major impact. It is very rewarding and exciting for people to come to a concept meeting and have actual parts to review. There is no substitute for having physical models to stimulate and convey ideas. The models can be produced quickly and efficiently, since no tooling or fixtures are required. The shoulder heads were initially conceived as solid metal components, as shown in *Figure 15-12*. Weight is an important factor, and much of the assembled weight was in the head. By reviewing rapid prototype models, it was suggested that a two-piece cast head could be made that would be hollow. The design questions centered on how to design the interface for maximum weld integrity. Since different junction designs and tolerances were being studied, the parts were modeled on the computer and a series of SL prototypes made for review. The welding engineer could then make recommendations from studying the physical models. Next, the selected design was sent to the casting vendor for confirmation regarding his process. *Figure 15-13* shows the resulting design. This is just one of many iterations that occur in a timely manner when concurrent development and rapid prototyping technology is used.

The early discussions are not only desirable for design and manufacturability reasons, but for costing as well. Management strives to predict the return on investment for a new product so that business decisions and priorities can be established. Purchasing of investment castings and tooling as well as cost estimating are much easier when you have a sample part to review with potential vendors. The purchasing department had previously been reluctant to approach a casting vendor until the engineering drawings were completed. This precluded the potential vendor from having any significant input during the concept and development stage. The vendors were also reluctant to estimate against a project

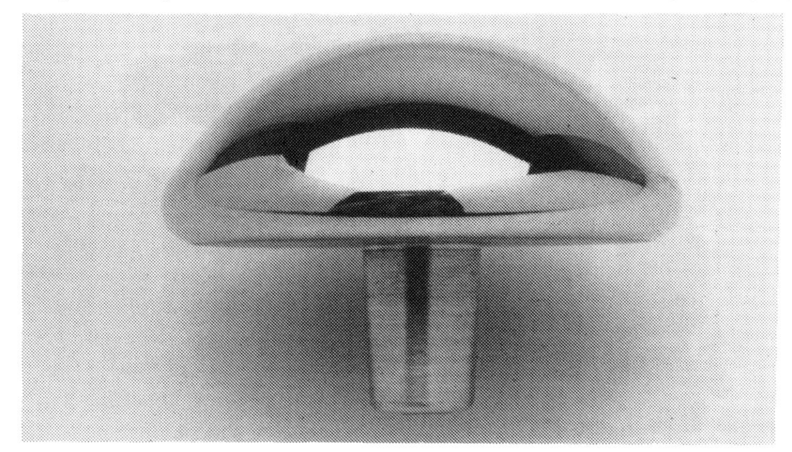

Figure 15-12. One-piece humeral head prototype.

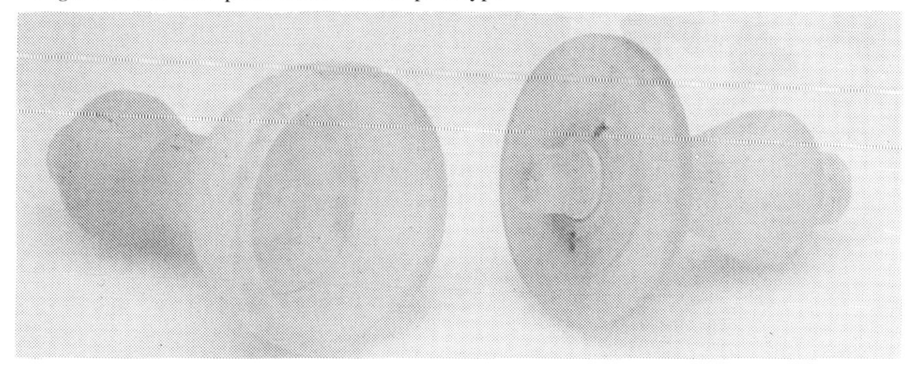

Figure 15-13. Two-piece humeral head prototype.

without a detailed review of engineering prints. The purchasing department can now approach the potential vendors with a set of plastic RP models that are very usable for costing purposes. From these improved estimates, purchasers can select the best vendor for the project. The early identification of the project vendor establishes a partnership relationship and is critical to the next stage of the development process. The vendor will subsequently provide process input for the project. The design of temporary tooling can then be specific to the wax and process shrinkage details desired by that particular casting vendor.

At this point, the product designers were striving to develop the appropriate range of sizes. There are some proportional features, but they are not always equally proportional in diameter versus length. Simply applying a scaling factor to the CAD database will not produce the desired results. In addition, the mating instrumentation includes reaming tools that may or may not be available in standard millimeter sizing. The combination of all these criteria results in modeling each component in all possible sizes, to work out blending features. The designs are then modified to allow for the specified process material needed for casting and finishing. StereoLithography models for masters in the temporary tooling process are then created. PUR-FECT Tool, a Polyurethane tooling compound from CIBA-GEIGY, was originally used for the shoulder implant project, and nine molds were created. The quality assurance department at DePuy has a laser scanning coordinate measuring machine (CMM), which allows us to scan and measure the master SL parts for qualification prior to making the molds, and then allows for inspection of the resulting castings. The process shrinkage is verified in this manner and compared to that expected by the casting vendor. The combining of rapid prototyping and casting technologies helps us maintain the swift flow of information and decisions necessary for the project. It also helps in the production of hard tooling, which will ultimately provide a near net shape casting as the process variables are better known.

The intent of this epoxy tooling was to expedite the investment casting process, and also to secure castings to enable manufacturing to process orders for early clinical surgery on a prescription basis. The manufacturing and quality assurance people can also evaluate their process at this time. The surgeon can now give final design approval, or request further modifications. We have not yet produced expensive permanent tooling, and the cost for the temporary tooling is reasonable. At this stage, we are still receptive to any additional iteration that may be required, and in fact welcome modifications that will lead to producing a superior product. Everyone wins in this new development protocol, especially the patients, as we can now provide the most effective device without incurring excessive cost.

At this point, we have the design requirements to start mass production of the implants. The hard tooling process requires the surface machining of elaborate multipiece aluminum tools for wax injection. The tools must be produced and qualified prior to the purchasing department authorizing quantity orders of castings. CAD data is used in this process when interacting with the tooling vendors to ensure that the design criteria is maintained. Since the hard tooling process takes several months, we use temporary tooling to fill this void. If the earlier tools that were qualified are suitable, they will be used. Product or process changes that affect the design dimensionally may require a new set of temporary tools. This will be accomplished by using StereoLithography to make a set of plastic masters for the process. These models will be built, very carefully, with a 0.003″ (0.076 mm) layer thickness for maximum resolution and minimal post processing. Spray metal may be used, as it is not susceptible to moisture and will provide a stable tool that can be employed by the casting vendor to make several thousand wax patterns. The shoulder implant project used urethane tooling to provide wax patterns for the first 500 pieces of each size (*Figure 15-14*). This allowed us to proceed with the product launch as an extension of the development process. It should be noted that this approach also requires the acceleration of all the other activities normally related to launching a product. The completion of literature, approvals, and case and tray designs for presenting the instrumentation also had to be accomplished concurrently.

We no longer have the luxury of waiting more than two years to get new ideas to the marketplace. If market preferences change, or competitive products get established, the product has a very small chance of providing the expected return on investment. The shoulder implant project was the first large-scale product to fully utilize all stages of Rapid Prototyping & Manufacturing. *The return from launching a single product several months early pays for the entire technology investment.*

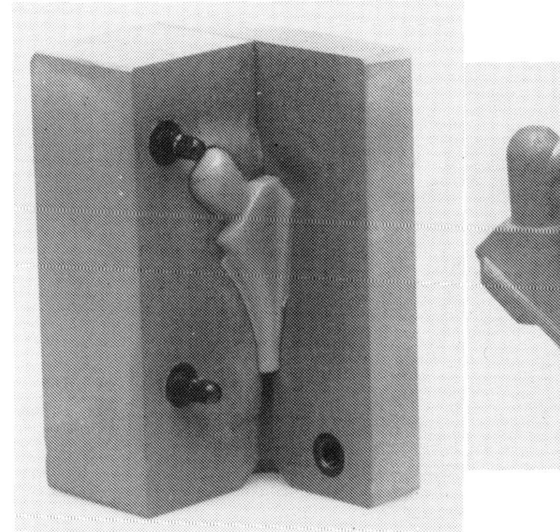

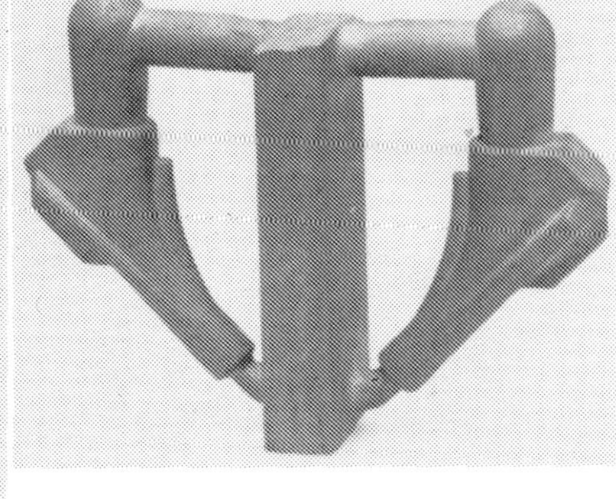

Figure 15-14. Shoulder mold.

15.3 Custom Implants

Custom implants are necessary for those situations when an off-the-shelf standard size implant is not suitable. These are usually complex cases involving trauma or disease resulting in bone deformity or loss. A custom implant is produced on a prescription basis and is unique for each patient. The surgeon and the custom engineer work together using either patient X-rays or CAT scans to develop the implant design. When it is desirable to make a custom implant, the shape is often of major concern. Free-form or anatomical geometry is often desirable to provide maximum fit and fill for the implant. These are usually the most complex implants to manufacture and would require extensive surface machining. Custom implants are, by their very nature, "one of a kind" designs and, therefore, time and cost must be controlled. Since these are elective surgeries, they can be scheduled to allow for manufacturing time, but the patients are usually experiencing discomfort and may be quite anxious to have the surgery.

The first step for the development of a custom implant is also computer aided design. The difference here is that the design criteria revolve around patient data supplied as X-rays or CAT scan information, as shown in *Figure 15-15*. The

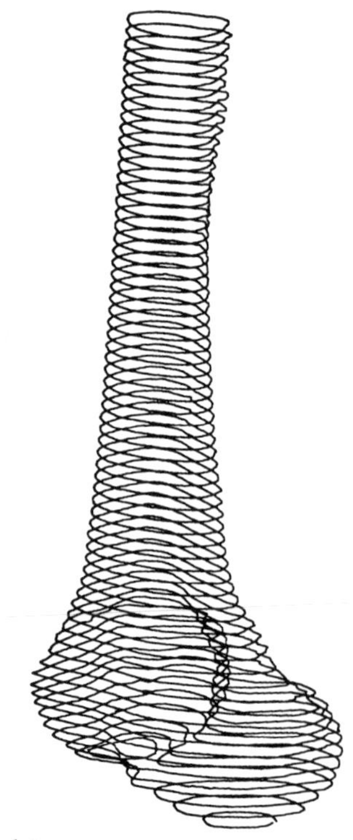

Figure 15-15. CAT scan data.

actual bone geometry may also be modeled to serve as a reference, as shown in *Figure 15-16*. The engineers on these projects use Computervision 4X software on a Sun Sparc II workstation. They have direct access to the StereoLithography units and prepare their models to be run on these systems. We often merge items with those from other runs to maximize machine utilization.

The RP prototypes can then be sent to the surgeon for review and approval prior to fabricating the actual implant. Investment casting is used for many of these items, and since the tooling is only needed for one part, the urethane material is used. This is where a resin that can be directly investment cast is ideal. New resin development is providing avenues for accelerating the production cycle by producing RP&M models or masters that are suitable for use in the shell investment casting process. This eliminates the cost and time of soft tooling to produce a wax master. We have cast some parts this way, and the new resins are indeed more suitable to the process.

15.4 Conclusions

Rapid prototypes have allowed a paradigm shift in concurrent process and product development. Each person's job is more than just another step in the chain of events. Every individual can see the goal, the product, and can relate to the customer. They are a vital part of a team that is alive and functioning. The job is more challenging and rewarding, as individuals are empowered to make

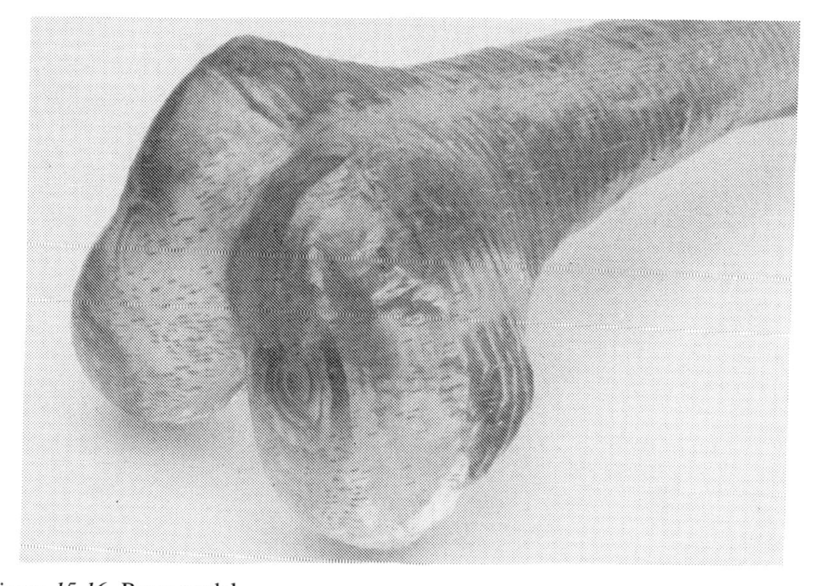

Figure 15-16. Bone model.

decisions and see the direct results of their collective efforts. If we are to continue as a society that manufactures products, then Rapid Prototyping & Manufacturing is essential for our survival.

chapter 16

Alternate Approaches to RP&M

I am never content until I have constructed a mechanical model of the subject I am studying. If I succeed in making one, I understand; otherwise I do not.

—William Thomson (Lord Kelvin)
Molecular Dynamics and
the Wave Theory of Light
(Lecture notes)

16.1 Introduction

The earlier chapters of this book have concentrated on descriptions of the StereoLithography apparatus (SLA) developed by 3D Systems, Inc. A variety of other rapid prototyping systems are in varying stages of development and commercialization. In this chapter, an attempt is made to provide a survey of those systems. The field is dynamic and rapidly expanding, so the reader is cautioned to further investigate new developments and techniques that arise after this book has been published.

The techniques and machines that will be described here can be divided into four broad categories. Almost every RP&M device operates by a layering process, so they are either layer-additive or layer-subtractive. Layer-additive techniques form new material as they go, attaching it to previously formed layers. Layer-subtractive techniques start with a full area layer of material, bonded to the previous layer, and then cut away unwanted material. The various RP&M machines operate on either a point-by-point basis or an entire layer at a time, and they are either laser based or not. At least one system has the capability of adding material in an arbitrary fashion. However, even this device is typically operated in a layering manner.

By ***Allan Lightman****, University of Dayton Research Institute, Dayton, OH and* ***Adam Cohen****, F Cubed, Los Angeles, CA*

The 3D Systems' SLA uses a layer-additive laser point-by-point technique (Chapters 1, 7, and 8). There are several other systems which are of this class. They will be described insofar as they have implemented innovations not used on the SLA or vice versa. The reader will have to ascertain the relative benefits from these innovations for their own applications.

16.2 Layer-Additive Laser Point-by-Point Fabrication

3D Systems

As a starting point, the operation of 3D Systems' SLA machines will be reviewed. The object is generated one layer at a time, starting from the bottom, using a UV laser to "draw" the layer structure on the surface of a photocurable resin (*Figure 16-1*). Where the laser exposure exceeds a threshold value, the liquid resin polymerizes and solidifies. The laser drawing speed is automatically adjusted so that the energy deposition is sufficient to polymerize the resin to the desired cure depth. The first layer formed is adhered to a platform positioned slightly beneath the surface of the liquid resin. After each layer is completed, the platform is indexed down into the liquid-resin bath by an additional depth of the new layer to be formed. Fresh liquid resin coats over the last solidified layer and the process is repeated, with the new layer adhering to the previous one. When the final (top) layer has been formed, the object is complete. The platform then

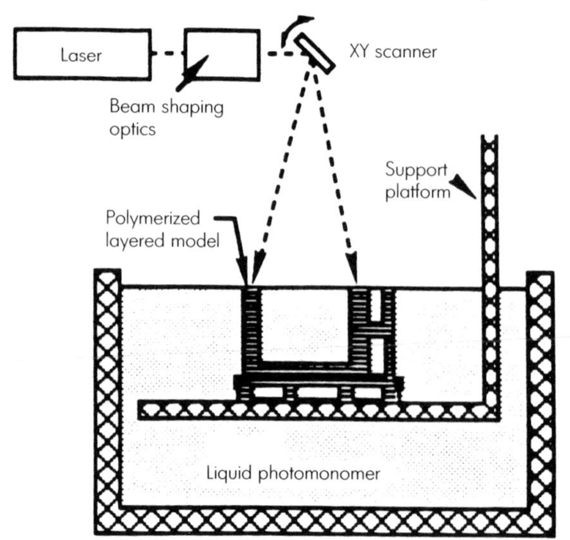

Figure 16-1. Laser photopolymerization build-by-layer RP&M system.

rises up out of the pool of liquid resin and the necessary cleanup and finishing stages are performed (as discussed in Chapter 9).

Consistently producing dimensionally accurate parts with the desired mechanical properties and surface finishes requires a much more sophisticated understanding of all the parameters and their interactions. These are detailed in the earlier chapters of this book. In this chapter, systems will be compared according to their fundamental mode of operation. There have not been enough, if any, case studies to gauge the impact of these variations. At this juncture, only the differences will be illustrated.

The laser and the laser pointing system also require some discussion. In 3D Systems' SLA, the UV radiation is generated by either a UV HeCd laser (325 nm) or a UV argon-ion laser (351 and 364 nm). In either case, the beam is positioned using reflecting mirrors mounted on a pair of orthogonally scanning galvanometers, whose position is directed by computer. The lasers are focussed to provide an illumination spot between 200 and 300 μm (0.008″ to 0.012″) in diameter on the liquid resin surface. The laser intensity footprints are different in the two cases. The HeCd laser operates in a multimode manner as discussed in Chapter 3, and is only quasi-Gaussian. The argon-ion laser, however, generally produces predominantly a Gaussian fundamental mode structure. The laser scan speed is automatically adjusted by the computer to provide the required exposure (mJ/cm^2) necessary to polymerize the material to a depth of from 100 μm to 500 μm (0.004″ to 0.020″), as chosen by the operator. Other requirements for the laser are discussed in Chapter 3.

The galvanometers provide vector-scanning mode operation and the exposure patterns have evolved considerably during the development of the technology. The galvanometers are located in an axially symmetric position and the resulting "best surface of laser focus" is that of a sphere centered at the galvanometers. This spherical surface is designed to intersect the flat liquid-resin surface in a circle whose radius is determined by the focusing lens. The laser's depth of field at focus is inversely proportional to wavelength and directly proportional to the focus-spot diameter. The short wavelength of the UV radiation and the weak focussing employed result in sufficient focus depth to provide acceptable beam diameter over the entire liquid resin surface.

The need to flat-field-correct the scan pattern becomes more of an issue when either the laser wavelength increases (visible or infrared), tighter focussing is used, such as a beam diameter of 50 μm (0.002″), or a larger surface needs to be scanned. Often two or more variables are changed at the same time, exacerbating the issue. These items will arise in the discussion of other RP&M devices.

DTM[1]

DTM Corporation's selective laser sintering (SLS) system is an RP&M technology which generates three-dimensional parts by fusing powdered-thermoplastic materials with the heat from an infrared laser beam. A thin layer of powdered thermoplastic material is evenly spread, by a roller, over the build region. Then, the pattern of the corresponding part cross section is ''drawn'' by the laser on the powder surface (*Figure 16-2*). With amorphous materials, the laser heat causes powder particles to soften and bind to one another at their points of contact, forming a solid mass. This process is called fusing or sintering. With crystalline materials, the heat causes the powder layer to melt, forming a liquid, which hardens to a solid upon cooling. Each layer formed resides within the remaining powder of that layer. After all layers have been formed, the finished part is imbedded within a cake of loose powder, which is later removed.

The part is formed in a cylinder of powder supported by a moveable piston (*Figure 16-3*). The cylinder is flanked on both sides by removable cartridges containing feed powder. A counter-rotating/leveling roller spreads the powder in either direction as it moves from one cartridge to the other. A 50-watt carbon dioxide (CO_2) laser produces infrared radiation at a wavelength (10,600 nm) that is absorbed by many materials, resulting in a rise of their temperature. The laser radiation is focused onto the surface of the powder in the part cylinder. Galvanometer scanning mirrors, operating under computer control, direct the beam's position.

The part building chamber is purged with inert gas and heated, raising the temperature of the uppermost layers of powder to nearly the fusing or melting temperature of the material. Maintaining the hot environment reduces the additional laser energy needed to heat the powder to a fusing temperature. It also reduces the thermal shrinkage of the layers during fabrication, which should help reduce part distortion.

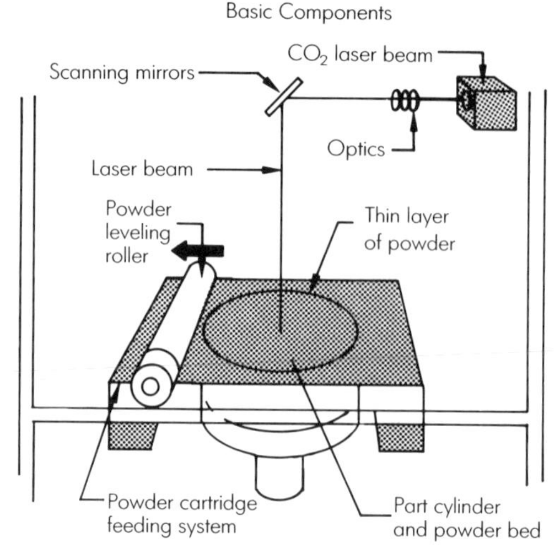

Figure 16-2. Sintering of powder using a CO_2 laser to write the layer pattern and deposit required energy.

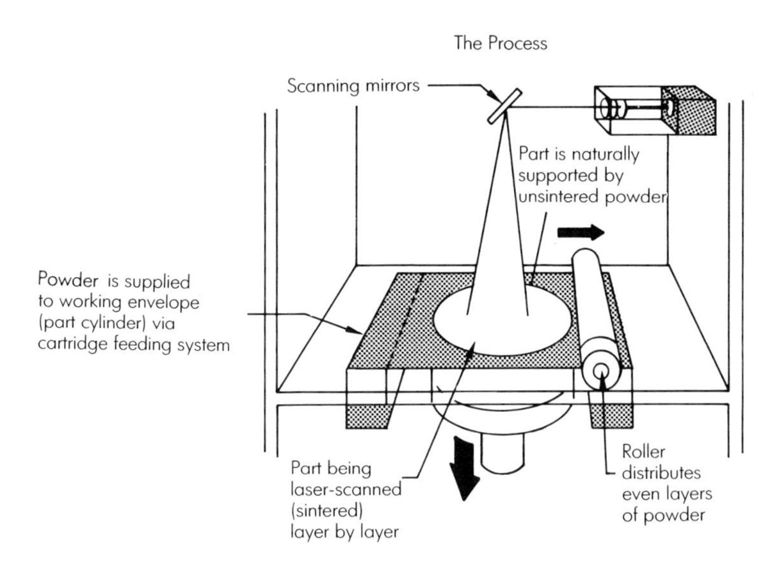

Figure 16-3. Powder supply system using a feed cartridge and a delivery and levelling roller.

The nitrogen purge reduces residual oxygen to 1% or 2%. This decreases the explosion hazard when using certain powders and avoids oxygen contamination of the bonding surfaces. The cooled nitrogen is also used to balance the heat loading at the powder surface maintaining it at the required temperature.

The scanning mirrors direct the laser beam across the powder surface, heating the material in the pattern of the sliced cross section. Normally the area of the cross section is filled in a raster pattern, a series of parallel, closely spaced lines. To obtain smoother surfaces, or when building thin walls, the outline(s) of each cross section may be drawn as well.

By generating the part within a form-fitted "cake" of supporting powder, the part is fairly well stabilized during building. Furthermore, as with all other RP&M techniques that build parts within a solid host, the environment keeps unsecured "islands" precisely located in space until bridges to the remainder of the part are formed later in the building process. The relative freedom from designing and removing supports facilitates building many parts simultaneously because they can be suspended anywhere within the powder cake without special provision.

Extracting the part from the powder host is a quite different procedure from the postfinishing process for parts fabricated from liquid resins, so it will be described in detail. After all layers are built, the part piston is raised. Some of the excess powder will fall off the part and a spatula or knife may be used to remove additional powder. The powder can be pushed back into the cartridge(s) for reuse. Except for parts made from wax, a layer of powder about 2.5 cm (1″) thick is left covering the part. This serves as insulation while the part cools, since rapid cooling can cause distortions. The warm part is removed and transferred to a cooling rack, where it remains for several hours.

After cooling, the part is taken to a "rough breakout station" where the powder is removed by hand using various brushes. Because powder (particularly wax) tends to adhere to the part surfaces, low-pressure air and dental tools are sometimes required to remove it, especially from inside holes, grooves, and other internal features. Loosened powder is sieved to strain out any large clumps and contaminants, such as brush bristles. The sieved powder is then collected in a canister and returned to the cartridge(s) for reuse.

All rapid prototyping systems produce a vertical "stairstep" surface finish because they build with discrete layers. However, the surfaces of SLS parts are particularly rough due to other factors. The most significant of these is that the material is a powder and not continuous. The particles of powder are relatively large and are not completely transformed by the building process, yielding part surfaces with a granular texture. The particles used in SLS are roughly spherical and range in diameter from 80 μm to 120 μm (0.003″ to 0.005″). Amorphous powders tend to retain their shape during the process and, therefore, are more likely to produce rough surfaces. Crystalline powders liquify and are smoothed by surface tension. However, these surfaces have exhibited some pitting. Surface finish varies with the orientation of the surface.

Another contributor to roughness is the raster-scan laser drawing technique which results in a horizontal stairstep effect on perimeters of the cross section that are not exactly parallel or perpendicular to the raster lines. DTM rotates the orientation of the raster by 90° on alternate layers, distributing the roughness more equally on all surfaces. Surface finish may be further improved by outlining each cross section prior to drawing the raster. This process eliminates most of the roughness caused by raster scanning, but it also reduces the part building speed.

Porosity is another consequence of building with powdered materials. Many SLS parts are not solid, but contain small voids between neighboring powder particles. While crystalline materials such as wax and nylon are melted during the process and can produce fully dense parts, sintered amorphous parts can be quite porous. For example, sintered polyvinyl chloride (PVC) has a density of only 60% (it is 40% air), whereas polycarbonate reaches a maximum of 85% density. The main drawback associated with porous parts is their lower strength compared with solid parts.

Shrinkage occurs in transforming the material from a loose powder to a solid mass. While it is highly desirable to eliminate porosity, this characteristic must be traded off against increased shrinkage and the potential for distortion. The crystalline materials exhibit the largest volumetric change going from a loose powder to a fully dense solid. Most of this shrinkage occurs along the vertical axis and is compensated for in the building process. Amorphous materials shrink considerably less because they are only sintered. Shrinkage may lead to defects and distortions such as curl. These are difficult to anticipate before actually building and measuring the part. If a grid support structure is required to anchor the part (as with the SLAs), it is built onto a slab of solid material placed over the piston before part building begins. Production equipment has not been available long enough for end users to develop statistical data on part accuracy and error distributions.

Materials used for SLS include PVC, polycarbonate, investment wax, nylon, and ABS (under development). The variety of materials which may be adapted for SLS use is considerably greater than that of RP&M processes which require a specific class of chemicals such as photopolymers. With the large number of candidate materials for SLS, it is likely that reasonably close approximations to final production materials will be available. Also, given their dry, solid nature, changing from one material to another should be reasonably straightforward and less messy than liquid-based rapid prototyping systems. Furthermore, the materials are inert and nontoxic, which simplifies handling and disposal issues.

The DTM sintering system was developed at the University of Texas. Work is continuing at the university to extend the concept to powdered metals and ceramics. In the case of metals, the energy requirements increase substantially, presenting modified process control requirements. Researchers are attempting to determine which materials and procedures can be implemented.

Du Pont/Teijin-Seiki

E.I. Du Pont de Nemours & Co. designed and built the SOMOS machine. This machine operated in a manner similar to the 3D Systems SLA, using a UV argon-ion laser source. While in operation, it appeared that the system exposed the resin layer using a raster-scanning mode, rather than a vector scan, with a fast shutter used to define the boundaries. The pattern fills in the area between the part borders while curing a large fraction of the resin. The remaining postcure is accomplished under ambient fluorescent lighting. The machine used proprietary Du Pont resins. Full operating details have not been released. Du Pont licensed Teijin Seiki Co. of Japan to commercialize the SOMOS technology in Asia (October 1991) with the option of extending the license worldwide. It is anticipated that operation specifications will become available as Teijin Seiki brings the machine to market. Under the agreement, Du Pont will restrict its involvement in the RP&M field to the supply of UV curable photoresins for use in SLA-type machines. To date, the only SOMOS machines operating are for internal Du Pont use.

Sony

The Sony solid creation system (SCS) is marketed by D-MEC, Ltd., a subsidiary of Japan Synthetic Rubber (JSR). The system's basic operation is very similar to 3D Systems' SLA units. Three models are available. The small unit uses a UV HeCd laser, the large unit uses a UV argon-ion laser, and the middle unit can use either. All units use galvanometer scanners to direct the laser over the resin surface.

Noteworthy is the size of the resin vat in the largest model, 1000 mm x 800 mm x 500 mm (40″ x 32″ x 20″). These systems also implement focus adjustment permitting variation in the size of the laser spot. The maximum spot dimension is listed for the three models at 100 μm, 150 μm, and 300 μm (0.004″, 0.006″, and 0.012″), but can be made somewhat smaller. Also implemented is a high-speed acousto-optic modulator to shutter the beam. Currently, marketing is restricted to Asia and Japan.

Mitsubishi

CMET, Inc. markets the Mitsubishi solid object ultra-violet laser plotter (SOUP) line of laser-lithography machines. There are four models in the SOUP line. All models build objects in the same manner as the SLA machines. The significant differences arise in the laser scanning mechanism and in the implementation of variable laser spot size.

The model 400GA/GH is their only unit to use galvanometer scanners. It can use either UV laser, HeCd, or argon. The laser spot diameter is fixed in this unit at 100 μm (0.004″). The slice thickness can be varied from 50 μm to 300 μm (0.002″ to 0.012″). The smaller dimension is substantially less than claimed by any other manufacturer for this class of machine.

The other three models of the SOUP line use *XY* translation stages, with fiber-optic light-conduit laser beam delivery, for scanning the UV beam. This is distinct from rotating mirror scanners. The smallest unit uses the HeCd laser and the two larger units use argon lasers. All units incorporate variable scanner spot dimension, 0.13 mm to 2.0 mm (0.005″ to 0.080″).

The manufacturer claims that parts fabricated from the SOUP resin do not require postcuring, a step required by almost every other manufacturer of this class of rapid prototyping machine. Detailed data regarding the extent of cure, modulus, strength, etc. have not been made available for parts without postcure. Data is not available on part accuracy.

Mitsui

The Mitsui Engineering & Shipbuilding Co. (MES) has produced the computer operated laser active modeling machine (COLAMM). This machine is another in the SLA class with the distinction that it literally builds the object upside down. Rather than building the part from the bottom to the top, as in the SLA, COLAMM builds from the top to the bottom. It forms the material additive layers from underneath, and the platform moves up after each layer is completed (*Figure 16-4*).

As in most of the machines in this class, the part is fabricated from a UV photocurable resin. The initial layer is attached to a platform which then indexes

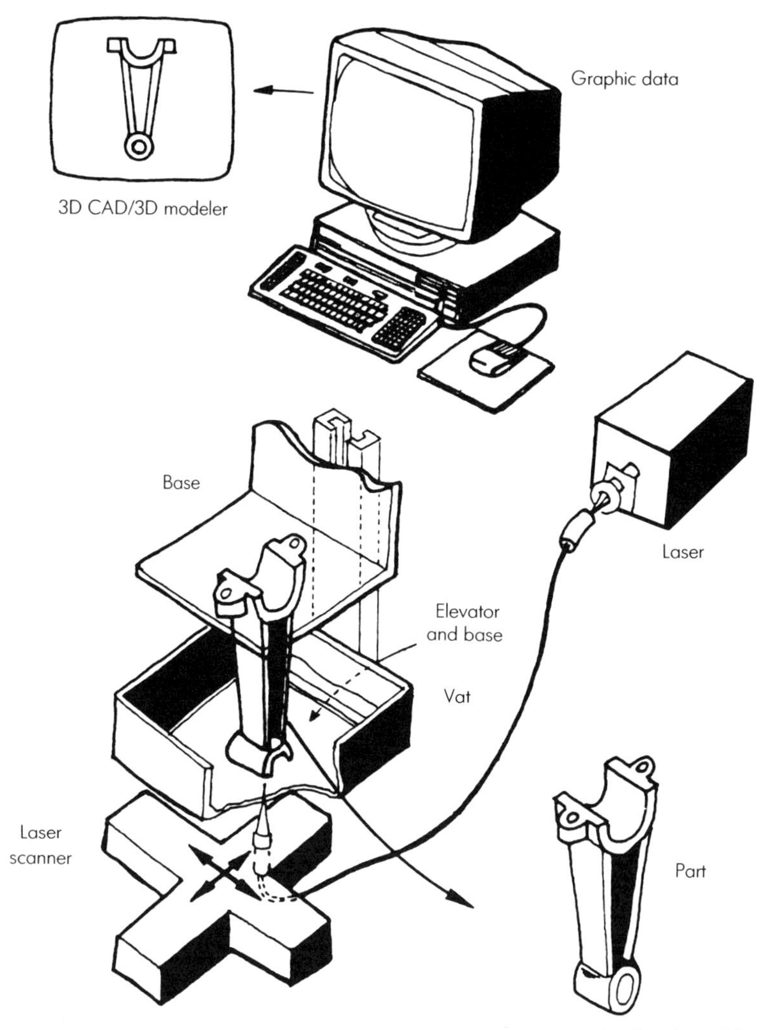

Figure 16-4. COLAMM system which builds parts from underneath, drawing object up from resin pool.

away from the fabrication layer, providing room and new material for each successive layer. In this system, the platform moves up, out of the shallow pool of resin. This eliminates one potential difficulty, parts swelling from residing in the resin too long. Other problems may arise due to the tension on the green part as it hangs from the platform, but no data is available to assess these issues.

The liquid resin to be exposed exists in a layer between the previous layer (or initially the platform) and a transparent vat floor. The UV HeCd radiation is scanned using an *XY* plotter mechanism. The beam spot diameter can be varied from 100 μm to 500 μm (0.004″ to 0.020″). After a layer is finished, the platform is indexed upward one layer thickness. The manufacturer claims that the adhesion to the window is weak and that the bond easily separates when the platform rises. There may be restrictions on the cross section of the part because sufficient structural strength is needed to insure the parts integrity. The liquid resin flows into the void created as the model rises. The surface does not require leveling since the window boundary is flat, so there is no delay in drawing the next layer.

Another aspect that might impact the part fabrication and material characteristics involves polymerization occurring at the bottom of the bath, which is away from the surface exposed to the oxygen in the air. Mitsui has not expressly commented on any differences in the material. Measurements at Ciba-Geigy indicate that oxygen concentration in the resin is reasonably homogeneous throughout the vat.

EOS

Electro-Optical Systems GmbH (EOS), of Germany, is marketing two models of their Stereos 400, an SLA-type machine. The smaller version has a maximum model size of 0.4 m x 0.4 m x 0.25 m (16″ x 16″ x 10″), and the larger version extends the depth to 0.6 m (24″). They operate with the typical UV lasers, HeCd or argon-ion. A noteworthy point is that EOS offers interchangeable vats, allowing easy exchange of the resin, as with the SLA-500.

16.3 Layer-Additive Nonlaser Point-by-Point Fabrication

Stratasys

The Stratasys, Inc. process, fused deposition modeling (FDM), uses thermoplastic wire-like filaments which are melted in the delivery head. The material is then extruded from the head and deposited on a layer-by-layer basis. The layering lamination technique is based upon the rapid solidification (approximately 1/10 second) of the molten laminate material from the modeling filament. The semiliquid thermoplastic material is deposited into thin layers, building the model upwards off a fixtureless base (*Figure 16-5*). The plastic or wax material solidifies in place, positioned by the *XY* controlled extrusion head. It is essential that the thermoplastic material temperature be maintained just above the solidification temperature. Also, the model temperature must be maintained just below the solidification temperature to ensure proper adhesion between the layers. A precision volumetric pump is used to control the material passing through the extrusion orifice. The extrusion process shears the material and it quickly solidifies while bonding to the previous layer by heating it and then fusing. The model is fabricated upon a piston which is lowered, between layers to make room for the next layer.

There are some features of the FDM technology worth noting. Layer thickness is claimed to be adjustable from 25 μm to 1.25 mm (0.001″ to 0.050″). The layer thickness control is achieved by varying the speed of motion of the delivery head, with the top speed being 380 mm/sec (15 ips). The width of the extrusion can be varied from 230 μm to 6.25 mm (0.009″ to 0.25″). Repeatability and position accuracy are listed as ±25 μm (±0.001″) with overall tolerance of ±125 μm (0.005″) within its working volume of 300 mm x 300 mm x 300 mm (12″ x 12″ x 12″). The repeatability and position accuracy claims are exceptional for RP&M machines and should be confirmed independently. A larger working volume machine has been shown. The basic machine has outside dimensions of 0.9 m x 0.9 m x 1.8 m (3′ x 3′ x 6′) making it one of the smallest machines commercially available.

The FDM material is delivered by the extrusion head tracing out the CAD layer information. At the layer end, the machine pauses while the platform indexes downward to make room for the next layer. This produces a seam at the pause location. Plans are underway to provide a quick downward index of the platform while the FDM head is still in motion. When implemented, it is anticipated that this will create a small upward ramp to the next layer and should eliminate the seam. It is important that the head be kept in continuous uniform motion. If the head pauses, material melts near the tip and forms little bumps that are visible on the surface. This can alter the surface finish if a new spool of material is required. An automatic changeover technique is under development but is not currently available. Temperature control of the FDM head and part are critical to success. If the door is opened too long, the part will cool and problems with adhesion to the next layer will result.

Flat or near-flat surfaces require a support structure. This must be cut away when the part is finished and should not be broken off. The creation of skins may lead to surface quality versus size limitations from surface undulations due to the support structure. Also, the support structure must be built up to secure flat surfaces at the top of objects, such as for closing the skull of the model head that is often displayed.

An advantage of the FDM process is that the material delivery process is on-demand and does not require a large reservoir of expensive material at the onset. Also, there is very little waste upon material changeover. It should be noted that the current cost of the material is $350 for a 1.6 km (1 mile) long spool of 1.3 mm (0.050″) diameter, making it almost twice the cost of liquid photopolymer on a volume basis. Consequently, the material costs for manufacturing prototypes using this process will exceed those for SLA RP&M systems.

Stratasys has used a variety of materials. Any material exhibiting thermoplastic behavior is a likely candidate for this process. These include a tough nylon-like material, a machinable wax, and an investment casting wax. The

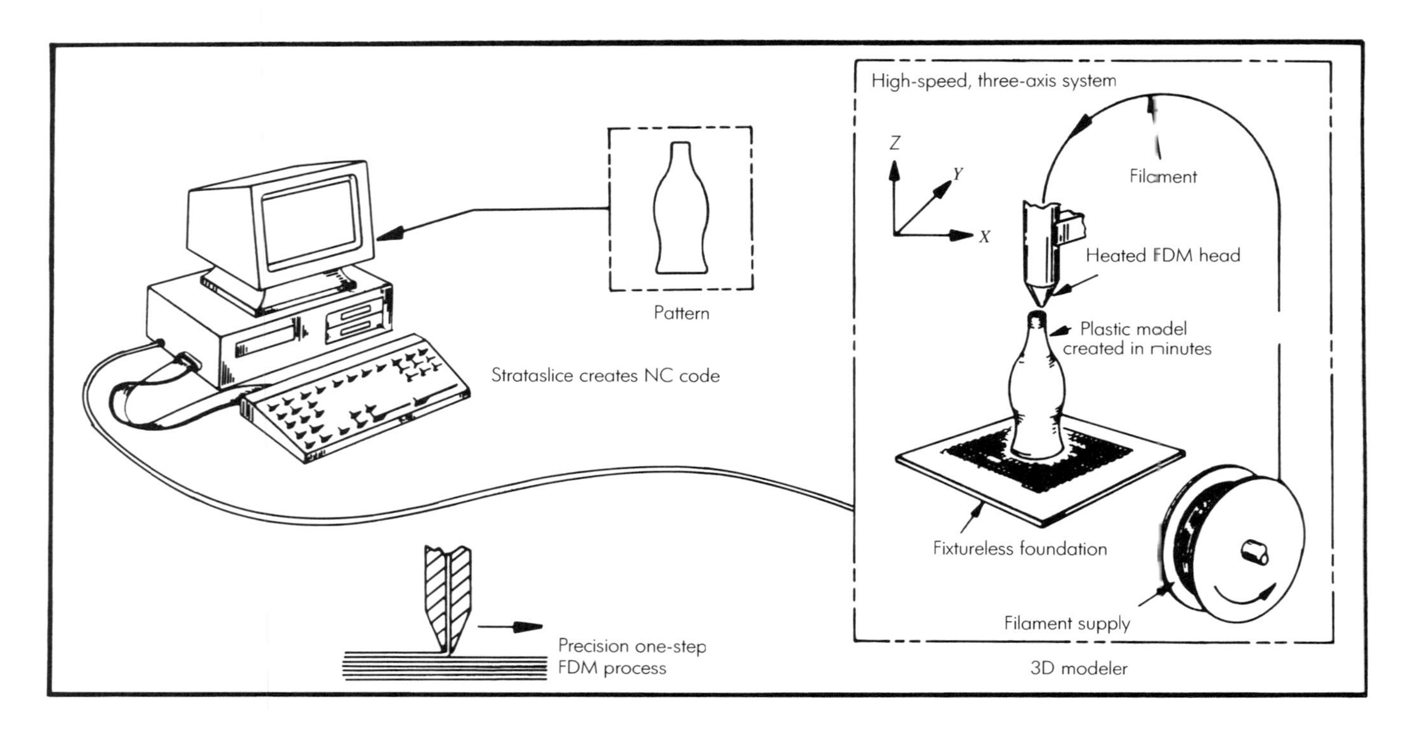

Figure 16-5. Fused deposition modeler producing objects by delivering manufacturing material from extrusion head positioned by *XY* translation stage.

latter provides models that can be directly invested and cast. The nylon-like material has properties resembling conventional nylon while providing a melt temperature of only 130°C (266°F). A polymer material having good strength and flexibility has recently been introduced. (Specifications for this material have not been made available at this time.) Stratasys claims that it features improved bonding and surface finish compared to the nylon-like material. The investment wax offers some machining capability if required. Wax parts can be polished using a Q-tip dipped in alcohol and chucked in a Dremel Moto-tool.

MIT[2]

Massachusetts Institute of Technology (MIT) is developing a process called three-dimensional printing (3DP). The 3DP process can use a number of powdered materials including those well-known to the investment casting industry in the production of shells: refractory powder, such as silica or alumina, and a liquid colloidal silica binder. Three-dimensional parts and ceramic molds are fabricated by selectively applying binder to thin layers of powder, causing particles of powder to stick together. Each layer is formed by generating a thin coating of powder and then applying binder to it with the ink-jet-like mechanism. Layers are formed sequentially and adhere to one another to generate a three-dimensional object in the usual manner for most RP&M systems.

The 3DP apparatus includes a cylinder fitted with a piston which can be lowered in small increments under computer control. The piston is coated with a thin layer of powder supplied by a hopper. Above the powder is an ink-jet-like mechanism which is supplied with binder and can move along both horizontal axes. Small droplets of liquid are continuously ejected downward from the nozzle toward the powder surface (*Figure 16-6*). Unwanted droplets are skimmed before reaching the powder by electrically charging them at the nozzle and then deflecting them from the stream by applying a voltage to electrodes located below the nozzle. The nozzle is moved across the powder surface in a raster scan while the electrical signals control the deposit of binder onto the powder in specific locations.

When molds are fabricated from ceramic powder within the cylinder, they are placed in a furnace to cure the binder and strengthen the mold. Following this step, excess unbound powder is removed and the object is ready for use. A 3DP-produced ceramic mold is shown in *Figure 16-7*, both whole and in section. A number of molds, some including integral cores, have been made and castings produced from them. When metal powders are used, the excess powder is removed after all the layers are fabricated. Finally, the part is put into the furnace where the binder is burned off and the metal sinters. Measurements of shrinkage during the sintering process have not been published.

3DP has several advantages relating to cores. Since molds are fabricated as a single unit consisting of a shell and core, the registration of cores to shells is precise. With investment casting, cores sometimes shift from their intended

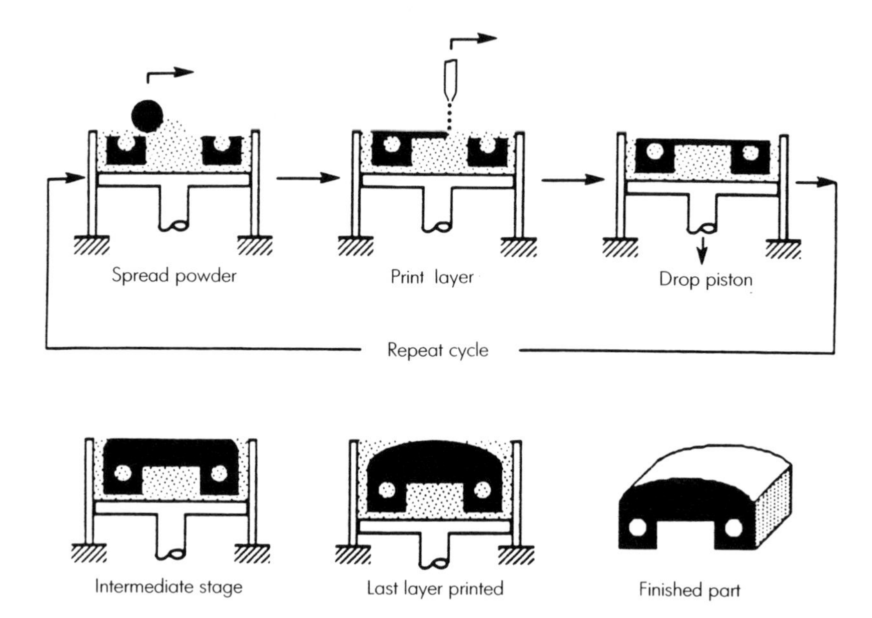

Figure 16-6. Three-dimensional printing writing the part geometry in powder by delivering binder from ink-jet-like printer head.

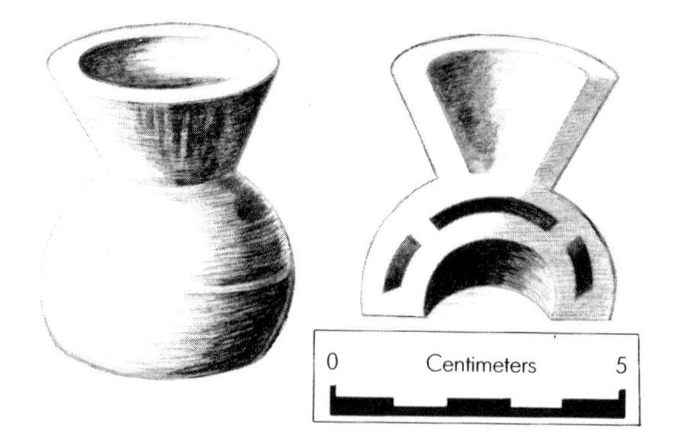

Figure 16-7. Ceramic shell fabricated using 3DP.

position, creating inaccuracies. In addition, cores can be made hollow exposing more of the core surface area to the caustic solution, and dissolving the core at a faster rate.

Currently, the major problem with 3DP-produced molds is inadequate surface finish. There is also some concern that unbound powder may be difficult to extract from long narrow passages, and will displace the poured metal in such regions. So far, the powder has not caused a problem, although the molds have been of simpler geometry than those in common use. The 3DP system is not yet commercially available.

Perception Systems

The ballistic particle manufacturing (BPM) technique, developed by Perception Systems, builds objects in a layering fashion, supplying material through a drop-on-demand ink-jet-like mechanism. Molten wax droplets, approximately 50 μm (0.002″) in diameter, are ejected at a rate up to 12,500 droplets per second. The droplet adherence to the object is explained by Perception as a momentary boundary melting of the previously deposited material upon impact of the hot droplet. The new structure then cools and solidifies. Multiple delivery systems can be operated in unison to increase the material deposition rate. Both the object material and any wax used to support the object are deposited on each layer. The wax is typically applied using a flooding procedure.

The delivery system is moved via an *XY* translator, precisely positioned above the model. Due to randomness in the droplet dispersion, accuracy in the model will depend upon the distance traveled by the particle before impacting the model. The thickness of the deposited layer is a function of the delivery rate of the material. Process control attempts to provide consistent operating performance. Test pieces with layer thicknesses of 90 μm (0.0035″) have been produced.

An interesting innovation (see also Cubital) is the use of a soluble material for the support structure. After the model is fabricated, the part is then placed in a warm water bath and the support material dissolves.

Incre

Incre, Inc. is developing a method of depositing molten metal droplets to build up structures. There is some similarity to the ballistic particle technique (see Perception Systems). The material used has been largely limited to tin but some work has been done with aluminum. The advantage to the technique is that it can directly fabricate metal prototypes and, the developer claims, it should be able to directly fabricate large structures.

16.4 Layer-Subtractive Laser Fabrication

Helisys[3]

The laminated object manufacturing (LOM) technique of Helisys, Inc. builds parts by laminating and laser-trimming the material delivered in sheet form. The sheets of material are laminated into a solid block using a thermal adhesive coating. Each sheet is attached to the block using heat and pressure to form a new layer. Next, the contours associated with the current cross section are cut into the layer using a 25 or 50 watt CO_2 laser. Areas of the layer outside these contours (outside the part volume) are then typically cut into small pieces, called "tiles," to facilitate their removal later.

After all layers have been laminated and cut, the result is a part imbedded within a block of supporting material. This material is then broken loose into chunks along the tile cuts. Parts may then be coated with a sealant to keep out moisture.

Figure 16-8 shows the configuration of an LOM machine. Sheet material is supplied from a continuous roll on one side of the machine and taken up on the opposite side. Beneath the sheet is a platform on which the block is fabricated and which can be moved vertically under computer control. The CO_2 laser beam is delivered via an *XY* scanner system that contains the final beam focussing optics. A heated roller is located above the paper on a linear stage.

The depth to which the material is cut increases with higher laser power and slower cutting speed. To achieve a uniform cutting depth while the optics head accelerates and decelerates, the laser power is varied in proportion to the speed of the head. To increase part building speed, more than one layer can be cut at a time. For example, two sheets of material may be cut together. Obviously, the cutting depth must then be doubled and the vertical resolution of the part is reduced. However, this option may be useful for part sections involving vertical borders.

The entire surface of the material is coated with adhesive, therefore each layer adheres to the previous layer at all points of contact (everywhere the previous layer has not been cut). *Figure 16-9* shows a cross-sectional view of a part as

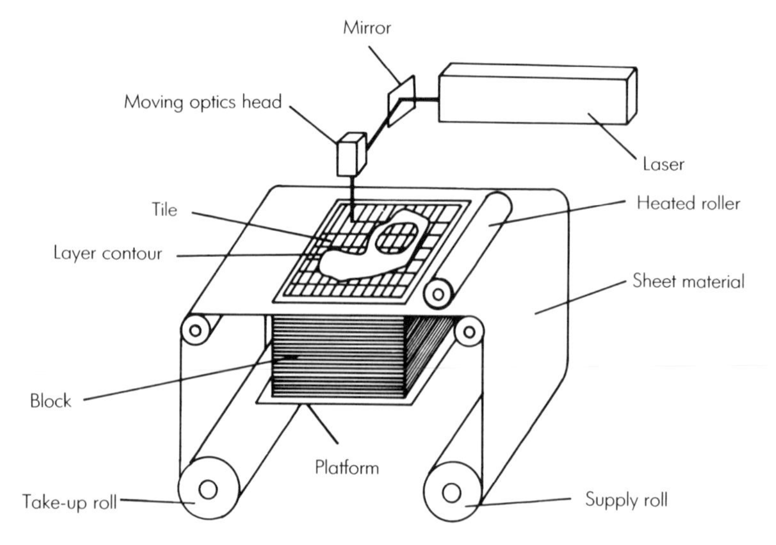

Figure 16-8. Laminated object manufacturing system showing material delivery system, laser cutter, and part support platform.

designed and built from layers. In *Figure 16-10*, the same part is shown within a block of support material. While the laser can cut through the layers to separate the part from the support, it cannot cut between layers. Thus, all up- and down-facing surfaces of the part adhere to the block. At present, the adhesion of as-designed horizontal surfaces is reduced by a method known as ''burnout.'' For each horizontal down-facing surface, the mating surface of the previous layer is cut with a tightly-spaced crosshatch pattern. The pattern is also cut into every horizontal up-facing surface of the part. Because cutting reduces the area of contact and weakens the material, the mating surfaces can be separated more easily. Currently, up- and down-facing regions of as-designed angled surfaces are not burned out.

Several problems are generated by the presence of support material surrounding the part. The adhesion of up- and down-facing part surfaces to this material means that cleanup is a manual process needing care to ensure that only waste material is removed and delicate sections of the part are not fractured. The burn out method is far from perfect, and relies on the use of a weak material,

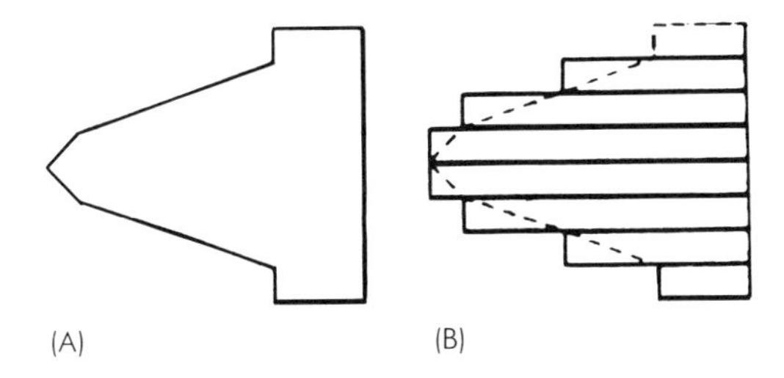

Figure 16-9. Cross-sectional view of part as designed and as fabricated by LOM.

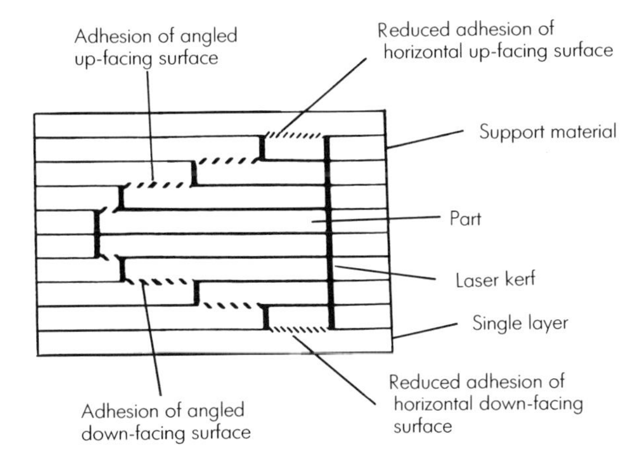

Figure 16-10. Interlayer adhesion issues in LOM illustrating potential complexities in removing excess material after construction.

since it is the material which yields, not the adhesive. Whether this or some other method of selective adhesion is employed, it must be applied not just to horizontal surfaces but also to as-designed angled surfaces, especially those which are nearly horizontal where the adhering areas can be quite large.

Because of excess material trapped within the walls, a hollow structure with closed surfaces cannot be fabricated as a single piece. While hollow parts are not always required, the difficulty of removing unwanted material extends to any part with narrow passages, internal cavities with restricted access, blind holes, etc. The scrap material is as strong as the part itself. Tiling precuts the scrap into smaller chunks, but these still adhere to all up- and down-facing surfaces and do not simply fall away. Another issue is whether the chunks are small enough to be manipulated through small openings.

One possible solution to these problems would be removal of excess material as it is produced on each layer. Helisys proposes using a vacuum to suck away loosened pieces before they fall inside the part; however, this is difficult to implement. Furthermore, removal of this material eliminates a major benefit of the current process: building the part within a solid block of material.

Typically, the majority of material consumed by LOM does not contribute to the part itself. Rather, it remains with the original continuous sheet or ends up as support material, to be scrapped after building. If more expensive materials than paper are used, the cost of such waste may be significant.

LOM parts are formed from alternating layers of material and adhesive. Thus, many of their physical properties are inhomogeneous and anisotropic. For example, strength and modulus can be different parallel to the layers than perpendicular to them, especially with respect to shearing forces. Another concern is delamination of the finished part under stress, because of failure of either the material or the adhesive.

LOM uses a subtractive method of layer formation. Material is removed to create a layer with the required cross section, whereas all other RP&M systems create layers by addition of material. This makes LOM potentially the fastest technology for building parts with a high ratio of volume to surface area (thick-walled or large parts).

By generating the part within a form-fitted block of support material, the entire geometry is stabilized during building and is prevented from distorting under its own weight. Furthermore, in slicing the CAD geometry into layers, isolated "islands" are sometimes produced. LOM avoids the need to design specialized supports to keep these precisely located in space until bridges to the remainder of the part are formed later in the building process.

Since only contours need to be drawn with LOM, an *XY* translator is practical and has several advantages compared with the alternative of directing the laser beam with rotating mirrors. One, there are no geometrical distortions resulting from the mapping of angular coordinates onto a plane, thus saving the computation needed to correct them.

Two, a laser beam can be precisely positioned more easily by a linear stage. Finally, the distance between the laser optics and the layer surface can be very small and independent of the part size, making possible a machine capable of building extremely large parts.

These benefits have been introduced in several of the machines that build models using UV photocurable resins.

Potential accuracy of LOM parts is high. The technology can build parts from very fine layers, since manufacturing thin, uniform sheet material is not very difficult. Because layer thickness can be determined before the contours are calculated, material thickness variations can be compensated for by adjusting the location of the slicing plane. Shrinkage is not an issue. If the layer happens to shrink horizontally during lamination, there is no loss of accuracy because the contours are cut afterwards. If the layer shrinks in thickness, the error is accommodated on the next layer when the accumulated height of the part is measured. Finally, if materials with low moisture content and expansion are used, the dimensional changes produced by the heat of lamination can be minimized. Thus, stresses which tend to produce layer curling while building or creep after building can be avoided. Data has not yet been made available from long-term multiple part measurements to quantify the accuracy of LOM fabricated parts.

Materials used in LOM tend to be inert and nontoxic. Waste is easily disposed of and there are no hazardous part-cleaning agents required. Given suitable adhesives, the process lends itself to a great variety of materials, from paper to plastics to metals, and with any desired coloring. Many materials useful for part building are not expensive. Since the layer is formed from a solid, dry sheet, different materials could, in principle, be used on different layers within the same part. This last capability would require the use of sheet material delivery rather than by roll, and a material carousel would need to be implemented.

Sparx

Sparx AB of Sweden has introduced the Hot Plot rapid prototyping system. It is the current low-cost leader in this technology area but requires considerable manual assistance to produce the object. The process operates by cutting layers of expanded polystyrene with a heated electrode, manually removing the excess material, and then stacking the layers with the aid of a mounting fixture. The material has a self-adhesive layer for bonding between layers. Alignment in the cutter and mounting fixture is assured by registration pins and corresponding holes in the layer. The cutter is moved on an *XY* translation stage under control of the CAD data files.

The currently available material is supplied in 1 mm (0.040″) thick layers. Once the object is built, any excess adhesive can be removed using either antistick powder or antiadhesive varnish. The company has indicated that it will provide thinner material if the demand materializes. As with all types of RP&M

systems, restrictions due to coarse layer thickness, weak material structural properties, and adhesion should be balanced against cost and speed when determining thc applicability of the technology.

16.5 Layer-Additive Nonlaser Fabrication

Cubital[4]

Cubital, Ltd., an Israeli company, has developed the Solider system, a rapid prototype fabrication process based upon solid ground curing (SGC). In SGC, each layer of the part is generated by a multistep process. A thin layer of liquid resin is prepared and then exposed to UV through a patterned mask having transparent areas corresponding to the cross section. UV radiation passing through the mask solidifies the exposed areas of the resin. The remaining uncured resin, while still a liquid, is then removed and replaced by wax. In the final step, both resin and wax are machined to a uniform thickness, forming a flat substrate on which the next layer is built. After all layers have been formed, the result is a part imbedded within a solid block of wax. This is then melted and dissolved away, leaving behind one or more parts composed of fully cured resin.

Figure 16-11 shows how the masks are created, describing each layer. Ionography technology is used to produce masks representing the required cross section. The layer is exposed through the masks, and then the masks are erased for reuse. The object processing is illustrated in *Figure 16-12*. A layer of photopolymer is spread on the support carriage. It is then moved to the exposure station, which includes a powerful flood UV radiation source and a shutter. The liquid resin is cured in the desired pattern by exposure through the masks created by the mask plotter. An air-knife wiper then removes unsolidified resin. This is followed by the wax applicator which coats the cured polymer with a layer of molten wax. A cooling plate then solidifies the wax. Finally, a fly cutter mills the polymer/wax layer to the correct thickness and provides the flat surface for the next layer. A vacuum system (not shown) sucks away chips as they are produced. The carriage supports the block of wax and polymer during the building process, moving it horizontally from station to station. The carriage also moves downward to accommodate new layers as they are added.

Typically, several parts are built simultaneously during a single building cycle. The machine operator selects the parts to build and translates and rotates the data files representing them so that the group fits within a block of the smallest possible height. A further goal is to nest parts as closely together as possible, maximizing the number that will be fabricated at a time. After the parts are arranged, their computer files are processed together to generate the cross sections, which control the mask plotter.

Generating the part within a form-fitted block of support wax, the entire geometry is stabilized during building and is prevented from collapsing under its

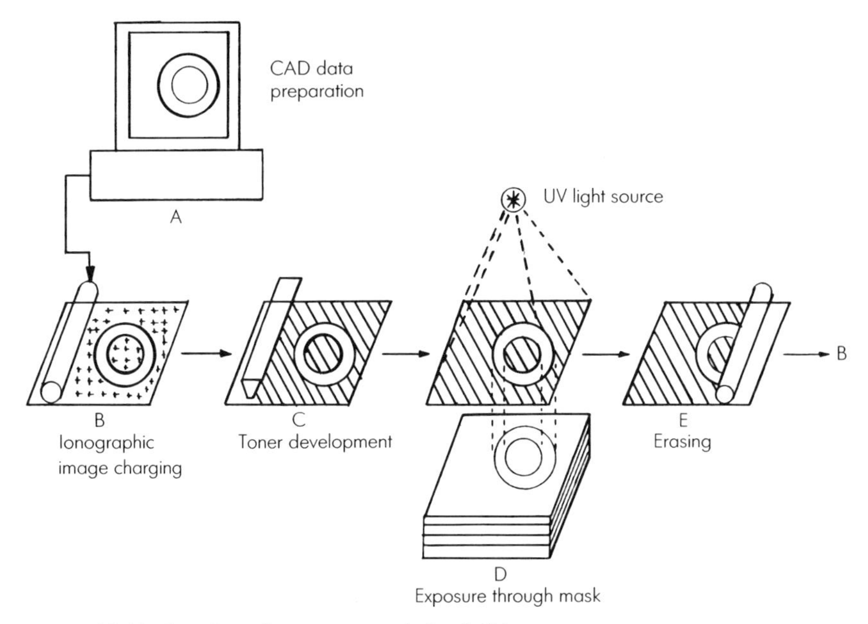

Figure 16-11. Creation of exposure mask for Solider system.

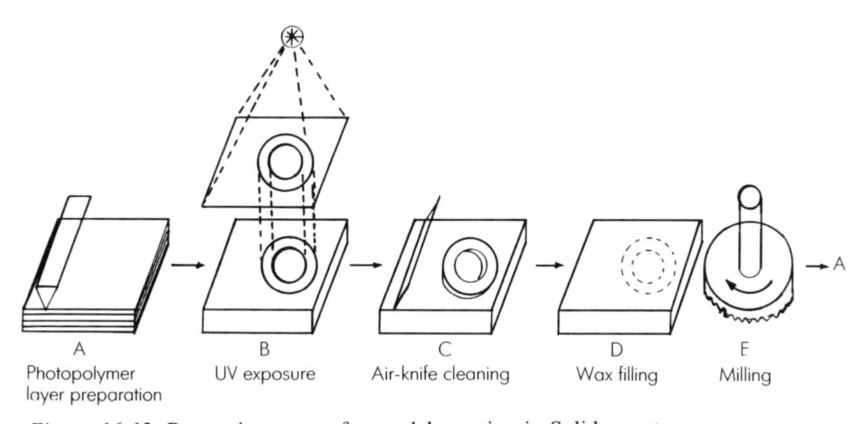

Figure 16-12. Processing stages for model creation in Solider system.

own weight. Furthermore, when isolated "islands" are produced during the slicing operation, SGC keeps these islands precisely located in space until bridges to the remainder of the part are formed later in the building process. The freedom from designing and removing supports facilitates building many parts simultaneously because they can be suspended anywhere within the block without special provision. It also makes feasible the one-piece fabrication of entire functioning assemblies, composed of multiple parts which move in relation to one another. For example, Cubital has built a set of meshed gears which rotate on their shafts when one is turned.

High-viscosity, low-shrinkage resins can be readily accommodated with SGC technique. Full curing of the part on a layer-by-layer basis is possible with SGC. This gives the part its full mechanical strength immediately, which can aid in reducing potential warpage afterwards. The even illumination of individual layers in the SGC system contrasts with the postcure illumination in systems that first build "green" parts. In the latter case, it is impossible to achieve uniform exposure throughout the volume of the part, and this may result in nonuniform shrinkage and subsequent warpage.

Dimensional accuracy along the building, or vertical axis, is determined by the thickness, flatness, and parallelism of the layers. Whatever shrinkage occurs in the resin layer during exposure, and whatever nonuniformity in thickness there may be in the cured resin, thickness, flatness, and parallelism are determined entirely by the fly cutter and other mechanical parts. The fly cutter's performance for surface flatness and for effects of blade wear should be studied.

Some practical problems are encountered. Ideally, the mask should prevent any UV radiation from reaching the resin directly beneath it during exposure. In practice, however, shadowed areas of resin do receive some exposure (probably because the toner layer is not sufficiently opaque at the UV exposure wavelength). The result is partial curing and an increase in viscosity. Therefore, resin that does not contribute to the part and is wiped off cannot be reused and must be discarded. Consequently, resin consumption per layer is constant, independent of the part geometry, resulting in the waste of expensive material.

SGC is most cost-effective when there are enough parts to build such that it is possible to make efficient use of the block by nesting many parts within it. On the other hand, if one needs to build a single part immediately and there are no other parts to share the block, this part can be very expensive. With a 50% part fill factor (quite optimistic) the waste material costs equal the used material costs. At current resin prices of about $100 per liter (quart) this cost may be significant.

Another factor that requires consideration is the proper disposal of the discarded resin. Uncured or partially cured resin is a hazardous material and proper disposal procedures must be followed. This is the case with the waste materials from all the processes using photopolymers compounded in the Cubital system by the large volume of both the waste resin and the wax chips/resin chips from the fly cutter. One technique is to fully cure the material, using sunlight, so that it is rendered nonhazardous and then dispose of it with the regular trash.

The ionography process used to create the masks deposits electric charge on the glass plate in raster lines, 300 per inch (11.8 per mm), typical of most laser printers. Therefore, each part cross section must be converted into a set of parallel lines when plotting the mask. The process introduces rough surfaces (stairsteps) on horizontal edges whose orientation is not perfectly parallel or perpendicular to the raster lines. The stairstepping effect occurs in the vertical direction in all RP&M machines that build by layering. In the SGC Solider machine, the surface finish also suffers from this irregularity in the horizontal plane. The technology for inexpensively producing finer raster pitch is being actively pursued, driven by the laser printer market. When implemented in the SGC, it will help reduce the stairstepping. At 1000 lines per 2.5 cm (1″) the stairstepping will not be a problem.

Another issue relating to the accuracy of the vertical surfaces results from the build procedure. The air-knife wiper does not remove every drop of unexposed liquid resin, particularly along the edges between exposed and unexposed areas. If this were left uncured, the resulting parts would have unacceptable wet, sticky surfaces. This residual resin is cured during a second UV exposure. The result is surface roughness and dimensional inaccuracy.

In the SGC system, the building time per layer is constant, independent of the part geometries. Using a 4 kW lamp, the exposure of the resin is much quicker than with laser-based systems, especially for large or multiple parts. But SGC's other steps do slow the process down. Building time is determined primarily by the number of layers which must be generated. Each layer represents a fixed overhead of building time, currently about 90 seconds.

The SGC technology is well-suited to building very large parts. Full curing to final strength and the use of supporting wax allows SGC to make large parts. Also, since resin is applied and removed on each layer, different resins could theoretically be used on different layers of the part. Indeed, several resins could even be used on the same layer, before application of the wax. Providing additional resin applicators further increases the complexity and size of the machine, but opens the door to constructing multicolor parts or assemblies containing parts made from different materials (rigid and elastic).

The SGC is a significantly larger machine than any of the others discussed in this chapter. It has a footprint of roughly 1.7 m x 4.1 m (6′ x 13.5′). It is massive and fairly noisy mostly due to the fly cutter and vacuum system. There are numerous moving parts and many subsystems. Production supervision is required on a regular basis probably including a full-time operator.

Light Sculpting

Light Sculpting, Inc.'s design-controlled automated fabrication (DesCAF) is another sequential building technique where the entire layer is exposed simultaneously. Masks are created and the photopolymer is then exposed to radiation through these masks. In one of the illustrations of this technology, the company positions a windowed mask carrier into the liquid resin bath.

The company has proposed to use a liquid crystal display (LCD) as an electronically rewritable mask. This would save the problem of hard generating new masks for each layer. This has been tried in Japan and to date the LCD material starts to fail after limited UV exposure (10 hours). Light Sculpting has also investigated the use of variable-density masks for building a certain class of geometries using a single mask.

The issue of the polymerized material sticking to the window is mentioned in the literature. Some coating release material is used to reduce the adhesion of the layer to the window, as with the MITSUI process. Generally, it is very difficult to obtain a full explanation of the process, much of it clouded in proprietary positions of the company. Finished parts have been shown without detailed explanation of how they were obtained. Detailed data is not available on part accuracy.

Carnegie Mellon University[5]

Researchers at Carnegie Mellon University's (CMU) Robotics Institute are developing a new RP&M technology which produces metal parts. The new technology is named MD* (pronounced MD star), a shorthand notation for repetitive masking and depositing. The process builds an object one layer at a time, using metal deposited from a thermal spray gun. Atomized droplets of the desired metal, generated within an electric arc or plasma, are propelled at high speed toward a substrate where they rapidly solidify to form a thin layer of metal. By placing a stencil, or mask, against the substrate, metal is deposited in the exposed areas, corresponding to a single cross section of the object. The mask is then replaced with a complementary mask, exposing the previously covered regions, and a low-melting-point alloy is sprayed. The result is a layer of relatively uniform thickness made up of both part and support metals. The process is repeated for the additional cross sections of the part, with each layer serving as a substrate for the next one. After all layers are generated, the support metal is removed by melting it at about 138°C (280°F), leaving behind a three-dimensional part composed of the desired metal.

The masks, which are disposable, can be laser cut from pressure-sensitive labeling paper. Each mask is then carefully aligned using pin registration. Features as narrow as 0.10 mm (0.004″) have been sprayed. The method should allow parts to be produced with very little porosity, achieving densities exceeding 99%.

The metal shrinks upon cooling, producing residual stresses that may be locked into the part, leading to delamination or distortion. Shot peening of layers during the building process may help control these stresses (annealing would probably liquify the low-melting-point alloy). Another area for development is in improving the layer thickness uniformity to avoid accumulation errors over many layers. Finally, there is a goal to operate the apparatus within a vacuum or inert gas atmosphere so that oxides, which can produce brittleness, are not formed. Working under vacuum is a technological challenge, but it has the advantage of avoiding entrapped air which decreases final density.

The MD* process should be capable of producing parts with layers 30 μm (0.0012″) thick. Any number of materials can be deposited within the same object, thereby allowing complete "assemblies" to be fabricated with no assembly labor. An example of this would be an integrated electronic/mechanical package in which shielding, heat sinks, heat pipes, and housing are fabricated as an integrated module by MD*. Integrated circuits could be inserted during the building process and all components would be encapsulated within the final module. This is possible since, by carefully controlling the deposition rate, sprayed materials can be prevented from heating the parts excessively. The process is under development and commercial systems are not available at this time.

U.S. Navy[6]

The U.S. Navy's David Taylor Research Center is studying a rapid prototyping process which uses an electric field to shape resins into three-dimensional objects. The process, called "programmable molding," uses a class of materials known as electrosetting liquids. In the presence of an electric field, these materials exhibit an accelerated rate of curing. Insulating material is deposited on thin sheets of conductive material (such as aluminum foil) to form electrodes whose shapes represent cross sections of the object to be generated. These are then assembled into a stack with spacers in between them, forming a mold. After connecting the electrodes to a power supply, the mold is immersed in a bath of resin so that the resin floods the spaces between the electrodes. Upon application of a high voltage, the resin begins to solidify more quickly between the electrodes than in the surrounding bath, eventually hardening into a solid object. The resins are electro-rheological fluids, which means that they undergo an instantaneous increase in viscosity when an electric field is applied. This allows the mold to be removed from the bath before full curing has been achieved, so that liquid outside the electrode regions can flow off. After the part has solidified, the foil remains imbedded within the resin. The protruding areas of each sheet of foil can be trimmed off, but the foil trapped in hollow volumes, such as the inside of a bottle, might be difficult to remove.

So far, a piping flange, a shoe sole, and several other parts have been built. The sole is of particular interest in that it exploits a unique characteristic of electrosetting: the ability to alter material properties such as density, hardness, and tensile strength as the material is curing. Properties can be modified on a local basis. Thus the shoe sole, composed of the same resin throughout, was fabricated as a sandwich of spongy rubber within two slabs of hard rubber. As a practical matter, most of the parts have been built with fairly thick layers, typically about 3 mm (0.125″). However, if demanded by part geometry, thinner layers are claimed possible.

The process can work with a diversity of resins, including epoxies, polyurethanes, and silicone rubbers, some of which are suitable for building functional parts. Electrodes are produced using a personal computer and a laser printer to print toner on the foil. Combined with a high-voltage power supply, it is possible to have the basic equipment for building parts for only about $5000. However, such a system would only be suitable for occasional use because the construction of the mold and other operations would be entirely manual. The Navy is trying to automate the process starting with a system to assemble electrodes and spacers into molds.

The main impetus for the research is the need for replacement ship parts that can be made while out at sea. Rather than carry a large inventory of spares, a ship could stock raw material, a simple apparatus, and data files for all parts. This is another process under development. Commercial systems are not available.

Formigraphic Engine/Battelle

One of the earlier concepts for free-form fabrication was patented by Formigraphic Engine. The idea was to produce a part inside a liquid vat by a process of selective polymerization requiring two lasers (two different wavelengths) to initiate the reaction. Placing windows on the sides of the vat permits the lasers to be aligned perpendicular to each other (*Figure 16-13*). This limits their overlap region to a small volume controllable by focusing lenses. "Writing" the part would be performed by moving the lasers so that the crossing region traced out the solid's volume.

In concept, the approach is elegant. In practice, there are many technical issues complicating the development. Battelle participated in the development. At this time, nothing has been released and all development effort has ceased.

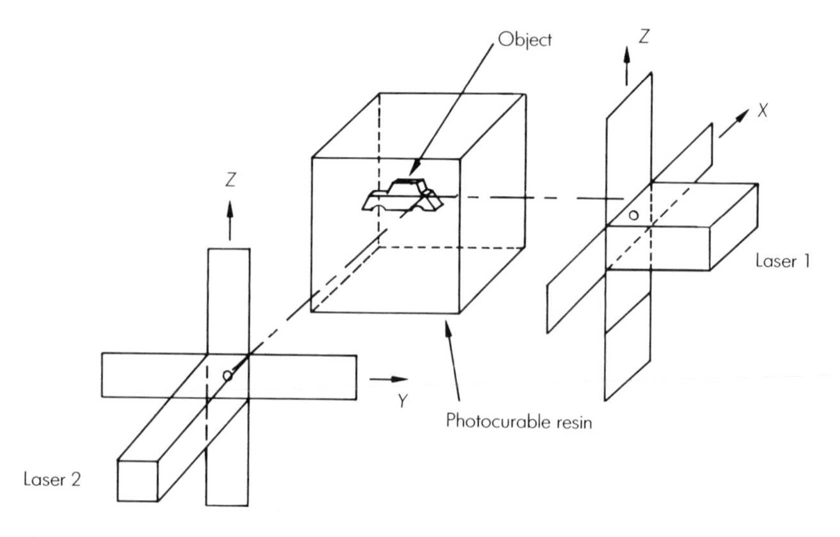

Figure 16-13. One possible realization of system proposed by Formigraphic Engine for fabricating part from solid material formed at crossing of two lasers. Lasers are traversed to scan part.

Babcock & Wilcox

Babcock & Wilcox is developing a manufacturing technique called "shape melting." The method uses arc welding to produce components directly from weld metal. The technique is similar to that currently used in industry to repair worn structures, building up the material by a welding process. The B & W process has been extended to produce near net shape forms using robotics to position the welder. This is another of the systems under development and detailed performance data is not available.

Several advantages accrue in using this technology for metal parts. These include higher strength and toughness, isotropic material properties that are uniform through the part, multimaterial parts, and tailored properties. As a result of the small amount of material that is molten at any instant, and the controlled thermal cycle achieved in this technique, uniform fine-grained microstructure is produced, providing increased strength and toughness.

Portions of this chapter were adapted from articles written by Adam L. Cohen for *Rapid Prototyping Report, The Newsletter of the Desktop Manufacturing Industry*, courtesy of CAD/CAM Publishing, San Diego, CA.

REFERENCES

1. Adapted from an article on DTM originally appearing in *Rapid Prototyping Report*, December 1991, p. 1, with permission of CAD/CAM Publishing, Inc.
2. Adapted from an article on MIT originally appearing in *Rapid Prototyping Report*, September 1991, p. 6, with permission of CAD/CAM Publishing, Inc.
3. Adapted from an article on Helisys originally appearing in *Rapid Prototyping Report*, June 1991, p. 1, with permission of CAD/CAM Publishing, Inc.
4. Adapted from an article on Cubital originally appearing in *Rapid Prototyping Report*, August 1991, p. 3, with permission of CAD/CAM Publishing, Inc.
5. Adapted from an article on Carnegie Mellon University originally appearing in *Rapid Prototyping Report*, October 1991, p. 2, with permission of CAD/CAM Publishing, Inc.
6. Adapted from an article on the U.S. Navy, David Taylor Research Center originally appearing in *Rapid Prototyping Report*, October 1991, p. 3, with permission of CAD/CAM Publishing, Inc.

Index

D

E

F

G

H

I

J

K

L

M

N

O

P

Q

R

W

Y

Z